Biochemistry
of Zinc

BIOCHEMISTRY OF THE ELEMENTS

Series Editor: Earl Frieden
Florida State University
Tallahassee, Florida

A Continuation Order Plan is available for this series. A continuation order will bring delivery of each new volume immediately upon publication. Volumes are billed only upon actual shipment. For further information please contact the publisher.

Biochemistry
of Zinc

Ananda S. Prasad

Department of Medicine
Division of Hematology and Oncology
Wayne State University School of Medicine
and
Harper Hospital
Detroit, Michigan
and
Veterans Administration Medical Center
Allen Park, Michigan

PLENUM PRESS • NEW YORK AND LONDON

Library of Congress Cataloging-in-Publication Data

Prasad, Ananda S. (Ananda Shiva)
    Biochemistry of zinc / Ananda S. Prasad.
        p.   cm. -- (Biochemistry of the elements ; v. 11)
    Includes bibliographical references and index.
    ISBN 0-306-44399-6
    1. Zinc--Physiological effect. 2. Zinc--Metabolism.   I. Title.
II. Series.
    [DNLM: 1. Zinc--deficiency. 2. Zinc--metabolism.   QU 130 B6144
1980 v. 11]
QP535.Z6P69 1993
612'.01524--dc20
DNLM/DLC
for Library of Congress   93-37102
                                                                  CIP

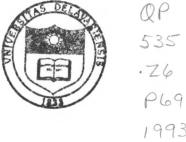

QP
535
.Z6
P69
1993

ISBN 0-306-44399-6

© 1993 Plenum Press, New York
A Division of Plenum Publishing Corporation
233 Spring Street, New York, N.Y. 10013

Printed in the United States of America

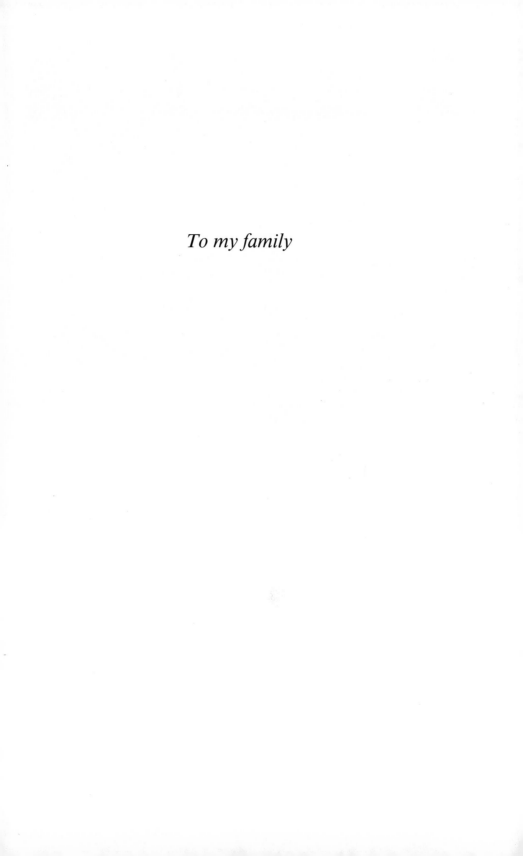

To my family

Preface

During the past three decades, remarkable advances have taken place in the field of zinc metabolism. Prior to 1963, its deficiency in humans was unknown. Today, it is recognized that a nutritional deficiency of zinc is common throughout the world, including the United States.

The growth retardation in children and adolescents commonly seen in developing countries is likely related to a nutritional deficiency of zinc. In the same areas variable immune deficiency disorders are common. Because of the important role of zinc in immunity, such disorders may also be a result of nutritional deficiency of zinc.

Diagnostic criteria for a mild or marginal deficiency of zinc, which appears to be common in the developed countries, have yet to be established, although considerable progress has been made recently in characterizing this state of zinc deficiency clinically, biochemically, and immunologically.

Besides nutritional deficiency of zinc, conditioned deficiency of this element caused by several disease states is probably common. This knowledge is likely to improve clinical management of various chronic illnesses. The discovery that a fatal genetic disorder, acrodermatitis enteropathica, is a genetically induced zinc deficiency syndrome has resulted in a complete cure of this condition, and indeed many lives have been saved by simple therapeutic oral administration of zinc. Mandatory use of supplemental zinc in fluids used for total parenteral nutrition has substantially reduced serious complications that were being induced by a severe deficiency of zinc prior to its parenteral administration. Recent recognition that therapeutic zinc is an effective means of decreasing copper burden in humans has now led to an effective nontoxic therapy for Wilson's disease. These are just a few examples of recent clinical advances in the field of zinc metabolism.

Very impressive advances have also taken place in our understanding of the basic aspects of zinc metabolism during the past three decades. Thirty years ago we knew of only three enzymes that require zinc for their activities; today we know of approximately 200 such enzymes. Recent increases in our knowledge of zinc-finger proteins and their roles in genetic expression of various growth factors and steroid receptors are truly exciting. This area is advancing rapidly, as evidenced by the number of papers being published. Zinc has been shown to control mRNA of metallothionein, a low-molecular-weight protein highly rich in cysteine. Rapid advances have taken place in the understanding of the structure and function of this protein. Its role in ameliorating heavy metal toxicity, in absorption of copper and zinc, and perhaps as a donor of zinc to apoenzymes has been documented only recently. Inasmuch as this protein is an excellent scavenger of hydroxyl ions, and zinc induces the synthesis of this protein, zinc may play an important role in free radical reactions.

Although the need for zinc for lymphocyte proliferation has been known since 1970, only recently has it been discovered that even a mild deficiency of zinc in humans may result in anergy, decreased production of IL-2 and IL-1, decreased natural killer cell activity, and decreased levels of active thymulin peptide. These observations may lead to correction of zinc-related immune disorders in the future and may have an impact on the clinical management of various patients.

The present volume does not discuss zinc toxicity except in its interaction with copper. In human studies, with the exception of copper deficiency, I have not been impressed with the toxic effects of zinc provided the level of zinc administration has been less than 50 mg/day orally.

I have included a chapter on technique, which deals with the assay of zinc in plasma and blood cells by atomic absorption spectrophotometry. My goal here is to encourage clinical laboratories to establish proper techniques for assessment of zinc status in humans.

An attempt has been made to cover important biochemical areas related to zinc and its functions. This field has grown so rapidly that it is impossible to do it justice in a monograph such as this. Nonetheless, this is a first attempt to bring together all the basic knowledge and relate it to the clinical effects of zinc deficiency in a readable fashion.

I sincerely hope that this book will provide stimulus for new research. The book should prove to be useful to students of nutrition and biochemistry, and to physicians who are likely to encounter many zinc-related problems in their practice.

I most sincerely thank Mary Yuhas, Mary Ann Gavura, Barbara Foulke, Sally Bates, and Tom Panczyszyn for their invaluable help in the preparation of this volume.

Ananda S. Prasad

Contents

Historical Aspects of Zinc ⟧

1.1 Zinc Deficiency in Microorganisms, Plants, and Animals

Raulin (1869) was the first to show that zinc was essential for the growth of *Aspergillus niger*. This was confirmed forty years later by Bertrand and Javillier (1911). In 1926, its essentiality for higher forms of plant life was established (Sommer and Lipman, 1926) (see Figs. 1-1 and 1-2).

Todd *et al.* (1934) established for the first time that zinc was needed for the growth of the rat. Earlier attempts by Bertrand and Benson (1922), McHargue (1926), and Hubbel and Mendel (1927) to demonstrate the essentiality of zinc in animals were unsuccessful, because the purified diets employed were deficient in other essential elements besides zinc.

Tucker and Salmon (1955) reported that zinc could cure and prevent the disease called parakeratosis in swine (see Figs. 1-3 and 1-4). O'Dell and his colleagues (1958) showed that zinc was essential for the growth of birds.

Zinc deficiency in suckling mice that were deprived of colostrum was reported by Nishimura (1953). The manifestations consisted of retarded growth and ossification, alopecia, thickening and hyperkeratinization of the epidermis, clubbed digits, deformed nails, and moderate congestion of certain visceral organs. Miller and Miller (1960) produced an experimental deficiency of zinc in calves. The main features were growth failure, testicular atrophy, and hyperkeratosis. Deficiency of zinc in the diet of breeding hens was shown to result in (1) lowered hatchability, (2) gross embryonic anomalies characterized by impaired skeletal development, and (3) varying degrees of weakness in chicks that hatched (Blamberg *et al.,* 1960).

By feeding dogs a diet low in zinc and high in calcium, a deficiency was produced whose clinical features included retardation of growth, emaciation, emesis, conjunctivitis, keratitis, general debility, and skin lesions on the abdomen and extremities (Robertson and Burns, 1963). In young Japanese quails, zinc deficiency was induced by feeding a low-zinc purified diet containing soy protein, which resulted in slow growth, abnormal feathering, la-

Figure 1-1. "Fern leaf" of Russet Burbank potatoes caused by a deficiency of zinc is shown on the right. [Reprinted with permission from Viets, F. G., 1966. Zinc deficiency in the soil–plant system, in *Zinc Metabolism* (A. S. Prasad, ed.), Thomas, Springfield, Ill. p. 90.]

bored respiration, incoordinate gait, and low content of zinc in the liver and tibias (Fox and Harrison, 1964).

1.2 Zinc Deficiency in Humans

Although the essentiality of zinc for animals has been recognized since 1934, its ubiquity made it seem unlikely that alterations in zinc metabolism could lead to significant problems in human nutrition or clinical medicine. This attitude has now changed.

In the fall of 1958, while I was visiting Iran, Dr. J. A. Halsted brought to my attention a 21-year-old male at Saadi Hospital, Shiraz, who looked like a 10-year-old boy. Besides severe growth retardation, his clinical features included hypogonadism, severe anemia, hepatosplenomegaly, rough and dry

Figure 1-2. Zinc-deficient onions on the right. [Reprinted with permission from Viets, F. G., 1966. Zinc deficiency in the soil–plant system, in *Zinc Metabolism* (A. S. Prasad, ed.), Thomas, Springfield, Ill., p. 90.]

skin, mental lethargy, and geophagia (Prasad *et al.,* 1961). The patient ate only bread made of unleavened wheat flour and the intake of animal protein was negligible. He ate one pound of clay per day. It became apparent to us later that the habit of geophagia (clay eating) was fairly prevalent in the villages around Shiraz. Our studies showed that the anemia was caused by a deficiency of iron but there was no evidence of blood loss. Hypopituitarism as an explanation of growth retardation was ruled out, inasmuch as ten additional similar cases were brought to my service for further studies within a very short period of time (see Fig. 1-5). We considered the following probable factors responsible for anemia in these cases: (1) The total amount of available iron in the diet was insufficient; (2) excessive sweating in a hot climate probably caused greater iron loss from the skin than would occur in a temperate climate; and (3) geophagia may have further decreased iron availability as was observed later by Minnich *et al.* (1968). The anemia was completely corrected by administration of oral zinc in every case.

Lemann (1910) had described similar clinical features in patients with hookworm infection but he did not relate these to a nutritional deficiency. Similar cases from Turkey were described by Reimann

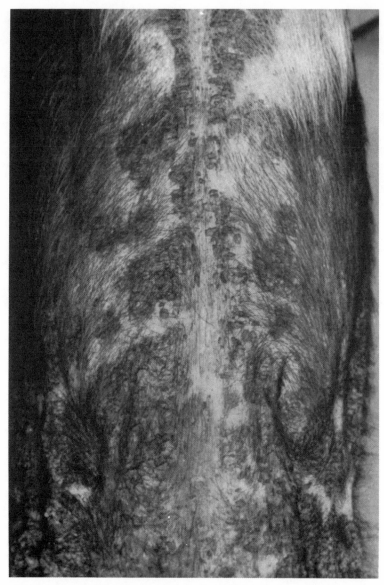

Figure 1-3. A view of the abdomen of a parakeratotic pig. [Reprinted with permission from Leucke, R. W., 1966. The role of zinc in animal nutrition, in *Zinc Metabolism* (A. S. Prasad, ed.), Thomas, Springfield, Ill., p. 202.]

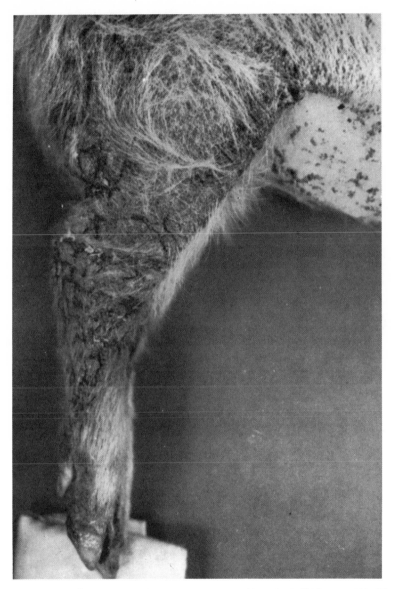

Figure 1-4. A view of the rear quarter of a parakeratotic pig. [Reprinted with permission from Leucke, R. W., 1966. The role of zinc in animal nutrition, in *Zinc Metabolism* (A. S. Prasad, ed.), Thomas, Springfield, Ill., p. 202.]

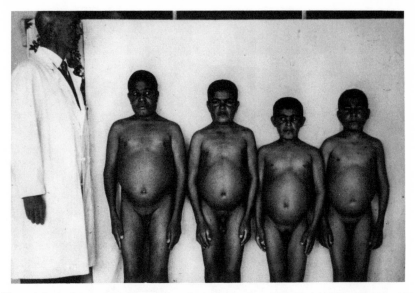

Figure 1-5. A picture of four dwarfs from Iran. From left to right: (1) Age 21, height 4 ft, 11½ in; (2) age 18, height 4 ft, 9 in; (3) age 18, height 4 ft, 7 in; (4) age 21, height 4 ft, 7 in. Staff physician at left is 6 ft in height. [Reprinted with permission from Prasad, A. S., 1966. Metabolism of zinc and its deficiency in human subjects, in *Zinc Metabolism* (A. S. Prasad, ed.), Thomas, Springfield, Ill., p. 250.]

(1955). Details of these cases were not provided and the author considered a genetic defect as a possible explanation for certain aspects of this clinical syndrome.

We provided a detailed description of this clinical syndrome in Iran in 1961 (Prasad *et al.,* 1961). Although we had no data to document zinc deficiency in those cases at that time, we speculated that a deficiency of zinc may have caused growth retardation, gonadal failure, skin changes, and mental lethargy (Prasad *et al.,* 1961).

1.2.1 Initial Studies in Iran

After therapy with orally administered ferrous sulfate (1 g daily) and a nutritious diet containing animal protein, the anemia was corrected, the hepatosplenomegaly improved, pubic hair grew, and the size of their genitalia increased (Prasad *et al.,* 1961). Liver function tests were unremarkable except for the serum alkaline phosphatase, which increased after treatment (Fig. 1-6). Retrospectively, this observation might be explained on two bases: (1) the

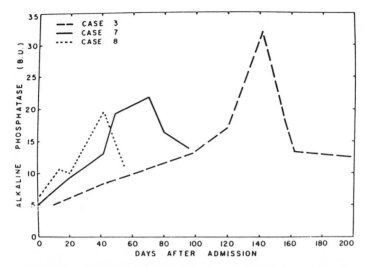

Figure 1-6. Changes in serum alkaline phosphatase associated with hospitalization and a well-balanced diet, in cases of Iranian dwarfs. [Reprinted with permission from Prasad, A. S., 1966. Metabolism of zinc and its deficiency in human subjects, in *Zinc Metabolism* (A. S. Prasad, ed.), Thomas, Springfield, Ill., p. 250.]

ordinary pharmaceutical preparation of iron might have contained appreciable quantities of zinc as a contaminant and (2) animal protein most likely supplied available zinc, thus inducing the activity of alkaline phosphatase, now known to be a zinc metalloenzyme.

Inasmuch as growth retardation and testicular atrophy are not observed as a result of iron deficiency in experimental animals, the possibility that zinc deficiency may have been present was considered. As noted earlier, zinc deficiency was known to produce retardation of growth and testicular atrophy in animals. Since heavy metals form insoluble complexes with phosphate, we speculated that some unknown dietary factors were complexing both iron and zinc, thus affecting adversely the availability of these elements for absorption. O'Dell and Savage (1960) reported that phytate (inositol hexaphosphate), which is present in cereal grains, markedly impaired the absorption of zinc. Changes in the activity of alkaline phosphatase after zinc supplementation to deficient animals were also similar to those observed by us in our subjects following administration of a diet containing adequate animal protein. Thus, in our Iranian subjects, dwarfism, testicular atrophy, retardation of skeletal maturation, and changes in serum alkaline phosphatase could have been explained on the basis of zinc deficiency.

1.2.2 Later Studies in Egypt

Later, in Egypt, I encountered patients who resembled the Iranian dwarfs. The clinical features were remarkably similar except for the following: The Iranian patients had more pronounced hepatosplenomegaly; they had a history of geophagia; and none had hookworm infections. In contrast, the Egyptian subjects had both schistosomiasis and hookworm infestations, and none had a history of geophagia.

We (H. H. Sandstead, A. Schulert, A. Miale, Z. Farid, and myself) carried out a detailed study of the Egyptian cases at the U.S. Naval Research Unit No. 3, Cairo (Prasad *et al.,* 1963a,b). The dietary history of the Egyptian cases was similar to that of the Iranians in that the consumption of animal protein was negligible. Their diet mainly consisted of bread and beans (*Vicia faba*). These subjects were demonstrated to have a deficiency of zinc (Fig. 1-7). This conclusion was based on the following: zinc levels in plasma, red cells, and hair decreased, and ^{65}Zn studies revealed that the plasma zinc turnover was greater; the 24-h exchangeable pool was smaller, and the excretion of ^{65}Zn in stool and urine was less in the subjects than in the controls (Prasad *et al.,* 1963a) (see Tables 1-1 and 1-2).

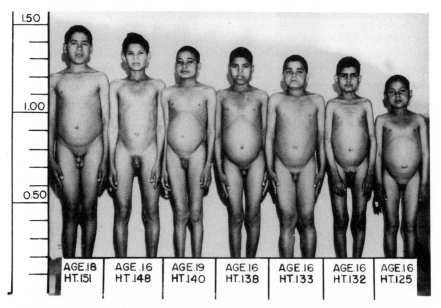

Figure 1-7. Seven of the dwarfs from delta villages near Cairo, Egypt. Height is shown in centimeters. [Reprinted with permission from Prasad, A. S., 1966. Metabolism of zinc and its deficiency in human subjects, in *Zinc Metabolism* (A. S. Prasad, ed.), Thomas, Springfield, Ill., p. 250.]

Table 1-1. Zinc Content of Plasma, Erythrocytes, and Hair[a]

	Plasma[b] (μg/100 ml)	RBC[b] (μg/ml)	Hair[b] (μg/g)
Normals	102 ± 13 (19)[c]	12.5 ± 1.2 (15)	99 ± 9 (10)
Dwarfs	67 ± 11 (17)	9.7 ± 1.1 (14)	65 ± 16 (10)

[a] Adapted with permission from Prasad, A. S., 1966. Metabolism of zinc and its deficiency in human subjects, in *Zinc Metabolism* (A. S. Prasad, ed.), Thomas, Springfield, Ill., p. 250.
[b] The differences between normals and dwarfs statistically significant ($p < 0.01$).
[c] Numbers in parentheses indicate the number of subjects included for each determination.

We carefully excluded chronic debilitating diseases and liver dysfunction in our subjects. Hypozincemia in humans, in the absence of advanced cirrhosis of the liver, had never been described before. Furthermore, in contrast to cirrhotics, who excrete abnormally high quantities of zinc in urine, our patients excreted less zinc in urine than did control subjects.

It was a common belief among the clinicians in Iran that the growth retardation and hypogonadism in these patients were the results of visceral leishmaniasis and geophagia. We failed to find evidence for visceral leishmaniasis in our subjects in Iran. The role of geophagia was not entirely clear; however, it was suspected that the excess amount of phosphate in the clay may have prevented absorption of both dietary iron and zinc. The predominantly wheat diet in the Middle East, now known to contain high quantities of phytate and fiber, may also have reduced the availability of zinc.

Table 1-2. Summary of Zinc—65 Studies[a]

	Turnover rate[b] (mg/kg/day)	24-h exchangeable pool[b] (mg/kg)	Urinary excretion[b] in % dose administered in 15 days	Excretion in stool[c] in % dose administered in 100 g of stool
Normals	1.00 ± 0.09 (9)[d]	7.0 ± 1.6 (8)	2.8 ± 0.56 (7)	0.66 ± 0.19 (7)
Dwarfs	1.50 ± 0.29 (10)	4.6 ± 1.2 (8)	1.6 ± 0.68 (7)	0.42 ± 0.13 (7)

[a] Adapted with permission from Prasad, A. S., 1966. Metabolism of zinc and its deficiency in human subjects, in *Zinc Metabolism* (A. S. Prasad, ed.), Thomas, Springfield, Ill., p. 250.
[b] The differences between normals and dwarfs statistically significant ($p < 0.01$).
[c] $p < 0.05$.
[d] Numbers in parentheses indicate the number of subjects included for each determination.

In Egypt, the cause of dwarfism was commonly considered to result from schistosomiasis. Chinese investigators had also implicated schistosomiasis as a causative factor for growth retardation (Eggleton, 1940). However, the existence of dwarfism and hypogonadism in areas such as Iran and Kharaga Oasis (Prasad *et al.,* 1963c), where schistosomiasis was not present, indicated that this parasitic infection was not responsible for these clinical findings. Furthermore, as noted earlier, iron deficiency in animals and human subjects does not cause growth retardation and hypogonadism. In view of the above findings and the similarity between the clinical features in our subjects and those seen in several animal species with zinc deficiency, it was a reasonable hypothesis to attribute growth retardation and hypogonadism in this syndrome to a deficiency of zinc.

It must be reemphasized that anemia in all cases was solely the result of iron deficiency, which was corrected completely by oral administration of iron salts. We were unable to account for the hepatosplenomegaly on the basis of liver disease. This left three possibilities: anemia, zinc deficiency, or a combination of these. Although our studies were inconclusive in this regard, in each case the size of the liver and spleen decreased significantly after zinc supplementation.

Our studies in the Middle East were limited to males (female subjects refused to participate in our studies). In later studies in Iran, Halsted *et al.* (1972) demonstrated that zinc deficiency in females manifesting growth retardation was probably prevalent.

Further studies in Egypt showed that the rate of growth was greater in patients who received supplemental zinc as compared with those who received iron instead, or those receiving only an adequate animal protein diet (Prasad, 1966; Sandstead *et al.,* 1967). The zinc-supplemented group gained approximately 5 inches in height on an annual basis (Fig. 1-8). Pubic hair appeared in all cases within 7–12 weeks after initiating zinc supplementation. Genitalia became normal and secondary sexual characteristics developed within 12–24 weeks in all patients receiving zinc. In contrast, no such changes were observed during a comparable length of time in the iron-supplemented group or in the group receiving an animal protein diet. Thus, we concluded that growth retardation and gonadal hypofunction in these subjects were caused by a deficiency of zinc. The anemia was the result of iron deficiency and responded to iron treatment.

1.2.3 Further Observations in Humans

It is now evident that nutritional as well as conditional deficiency of zinc may complicate many diseased states. MacMahon *et al.* (1968) demonstrated

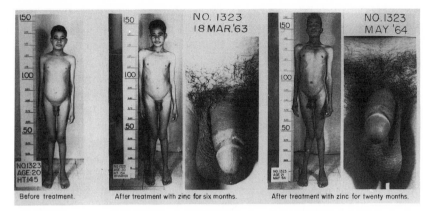

Figure 1-8. The general appearance of case 5 before and after zinc treatment. Notice the increase in height, growth of pubic hair, and increase in size of external genitalia following zinc therapy. The scale is in meters. (Reprinted with permission from Prasad, A. S., 1966. Metabolism of zinc and its deficiency in human subjects, in *Zinc Metabolism* (A. S. Prasad, ed.), Thomas, Springfield, Ill., p. 250.]

for the first time the occurrence of zinc deficiency in a patient with steatorrhea. Since then, several examples of zinc deficiency in patients with malabsorption have been reported (McClain *et al.*, 1988).

In the United States, Caggiano *et al.* (1969) were the first to report a case of zinc deficiency in a Puerto Rican subject with dwarfism, hypogonadism, hypogammaglobulinemia, giardiasis, strongyloidosis, and schistosomiasis. The patient responded to zinc therapy as far as growth and development were concerned.

In 1972, a number of children from middle-class families in Denver, Colorado, were found to exhibit evidence of symptomatic nutritional zinc deficiency (Hambidge *et al.*, 1972). Growth retardation, poor appetite, and impaired taste acuity were related to zinc deficiency in the children. Zinc supplementation corrected these features. Later, symptomatic zinc deficiency in U.S. infants was reported. Indeed, it is currently believed that risk of sub-optimal zinc nutrition may pose a problem for a substantial section of the U.S. population.

Halsted *et al.* (1972), in a study of zinc supplementation in Iran involving a group of 15 men and 2 women, confirmed the occurrence of zinc-responsive growth and gonadal dysfunctions in the Middle East. Their results were similar to those reported by Prasad and co-workers in Egypt (Prasad *et al.*, 1963a,b).

Zinc deficiency is prevalent in human populations throughout the world, although its incidence is unknown. Clinical pictures similar to those reported for zinc-deficient dwarfs have been observed in many countries. It is believed that zinc deficiency should be present in countries where primarily cereal proteins are consumed by the population. One would also expect to find a spectrum of zinc deficiency ranging from severe to marginally deficient cases in any given population.

In 1973, Barnes and Moynahan studied a 2-year-old girl with severe acrodermatitis enteropathica (a genetic disorder) who was being treated with diiodohydroxyquinoline and a lactose-deficient synthetic diet. The clinical response to this therapy was not satisfactory, and the physicians sought to identify contributing factors. It was noted that the concentration of zinc in the patient's serum was profoundly decreased; therefore, they administered oral zinc sulfate. The skin lesions and gastrointestinal symptoms cleared completely and the patient was discharged from the hospital. When zinc was inadvertently omitted from the child's regimen, she suffered a relapse that promptly responded to zinc administered orally. In their initial reports, the authors attributed zinc deficiency in this patient to the synthetic diet. It soon became clear that zinc might be fundamental to the pathogenesis of this rare inherited disorder and that the clinical improvement reflected improvement in the zinc status. This original observation was quickly confirmed in other cases throughout the world. The underlying pathogenesis of the zinc deficiency in these patients is, most likely, related to malabsorption of zinc, the mechanism of which is not well understood.

In 1974, a landmark decision to establish recommended dietary allowances for humans for zinc was made by the National Research Council, Food and Nutrition Board of the National Academy of Sciences.

Kay and Tasman-Jones (1975) reported the occurrence of severe zinc deficiency in subjects receiving total parenteral nutrition for prolonged periods without zinc. This observation is now well documented in the literature (Okada et al., 1976; Arakawa et al., 1976) and indeed in the United States zinc is being routinely included in total parenteral fluids for subjects who are likely to receive such therapy for extended periods.

An example of severe parakeratosis in humans related to deficiency of zinc was first reported by Klingberg et al. (1976) in a patient who received penicillamine therapy for Wilson's disease. Zinc supplementation completely reversed the clinical manifestations.

Recently, several studies have suggested that various clinical manifestations in patients with sickle-cell disease such as growth retardation, hypogonadism in the males, lack of prompt healing of chronic leg ulcers, abnormal dark adaptation, and abnormality in cell-mediated immunity are related to a deficiency of zinc (Prasad et al., 1975, 1981, 1988; Prasad and Cossack, 1984; Warth et al., 1981). The exact pathogenesis of zinc deficiency in sickle-cell disease is not well understood and further studies are needed. Hyperzincuria has been noted in such subjects, and this may be a contributing factor in the pathogenesis of zinc deficiency.

Although the role of zinc in humans has now been defined and its deficiency has been recognized in several clinical conditions, only recently has an experimental model been developed that allowed a study of specific effects of a mild zinc-deficient state in humans (Prasad *et al.*, 1978; Abbasi *et al.*, 1980; Rabbani *et al.*, 1987). The model also provided assessment of sensitive parameters that could be utilized clinically for diagnosing marginal zinc deficiency. A semipurified diet based on texturized soy protein was developed for consumption by human volunteers. This diet supplied necessary calories, protein, macro and micro elements, and vitamins according to recommended dietary allowances (National Academy of Science, Food and Nutrition Board) except for zinc, which varied as desired. A marginal deficiency of zinc in human subjects, as induced experimentally, is characterized by decreased zinc levels in granulocytes, lymphocytes, and platelets, testicular hypofunction, hyperammonemia, neurosensory changes, biochemical and immunological changes mainly affecting the thymus-dependent lymphocytes.

References

Abbasi, A. A., Prasad, A. S., Rabbani, P., and DuMouchelle, E., 1980. Experimental zinc deficiency in man: Effect on testicular function, *J. Lab. Clin. Med.* 96:544.

Arakawa, T., Tamura, T., and Igarashi, Y., 1976. Zinc deficiency in two infants during parenteral alimentation for diarrhea, *Am. J. Clin. Nutr.* 29:197.

Barnes, P. M., and Moynahan, E. J., 1973. Zinc deficiency in acrodermatitis enteropathica: Multiple dietary intolerance treated with synthetic zinc, *Proc. R. Soc. Med.* 66:327.

Bertrand, G., and Benson, R., 1922. Recherches sur l'importance du zinc dans l'alimentation des ainmaux, *C.R. Acad. Sci.* 175:289.

Bertrand, G., and Javillier, M., 1911. Influence du zinc et du manganese sur la composition minerale de l'aspergillus niger, *C.R. Acad. Sci.* 152:1337.

Blamberg, D. L., Blackwood, U. B., Supplee, W. C., and Combs, G. F., 1960. Effect of zinc deficiency in hens on hatchability and embryonic development, *Proc. Soc. Exp. Biol. Med.* 104:217.

Caggiano, V., Schnitzler, R., Strauss, W., Baker, R. K., Carter, A. C., Josephson, A. S., and Wallach, S., 1969. Zinc deficiency in a patient with retarded growth, hypogonadism, hypogammaglobulinemia, and chronic infection, *Am. J. Med. Sci.* 257:305.

Eggleton, W. E., 1940. The zinc and copper contents of the organs and tissues of Chinese subjects, *Biochem. J.* 34:991.

Fox, M. R. S., and Harrison, B. N., 1964. Use of Japanese quail for the study of zinc deficiency, *Proc. Soc. Exp. Biol. Med.* 116:256.

Halsted, J. A., Ronaghy, H. A., Abadi, P., Haghshenass, M., Amirhakimi, G. H., Barakat, R. M., and Reinhold, J. G., 1972. Zinc deficiency in man: The Shiraz experiment, *Am. J. Med.* 53:277.

Hambidge, K. M., Hambidge, C., Jacobs, M., and Baum, J. D., 1972. Low levels of zinc in hair, anorexia, poor growth, and hypogeusia in children, *Pediatr. Res.* 6:868.

Hofman, H. O., 1922. Word zinc first used in Europe, in *Metallurgy of Zinc and Cadmium,* McGraw–Hill, New York, p. 3.

Hubbel, R. B., and Mendel, L. B., 1927. Zinc and normal nutrition, *J. Biol. Chem.* 75:567.

Kay, R. G., and Tasman-Jones, C., 1975. Zinc deficiency and intravenous feeding, *Lancet* 2:605.

Klingberg, W. G., Prasad, A. S., and Oberleas, D., 1976. Zinc deficiency following penicillamine therapy, in *Trace Elements in Human Health and Disease* (A. S. Prasad, ed.), Academic Press, New York, p. 51.

Lemann, I. I., 1910. A study of the type of infantilism in hookworm disease, *Arch. Intern. Med.* 6:139.

McClain, C. J., Adams, L., and Shedlofsky, S., 1988. Zinc and the gastrointestinal system, in *Essential and Toxic Trace Elements in Human Health and Disease* (A. S. Prasad, ed.), Liss, New York, p. 55.

McHargue, J. S., 1926. Further evidence that small quantities of copper, manganese and zinc are factors in metabolism of animals, *Am. J. Physiol.* 77:245.

MacMahon, R. A., Parker, M. L., and McKinnon, M., 1968. Zinc treatment in malabsorption, *Med. J. Aust.* 2:210.

Miller, J. K., and Miller, W. J., 1960. Development of zinc deficiency in holstein calves fed a purified diet, *J. Dairy Sci.* 43:1854.

Minnich, V., Okevogla, A., Tarcon, Y., Arcasoy, A., Yorukoglu, O., Renda, F., and Demirag, B., 1968. The effect of clay on iron absorption as a possible cause for anemia of Turkish subjects with pica, *Am. J. Clin. Nutr.* 21:78.

Nishimura, H., 1953. Zinc deficiency in suckling mice deprived of colostrum, *J. Nutr.* 49:79.

O'Dell, B. L., and Savage, J. E., 1960. Effect of phytic acid on zinc availability, *Proc. Soc. Exp. Biol. Med.* 103:304.

O'Dell, B. L., Newberne, P. M., and Savage, J. E., 1958. Significance of dietary zinc for the growing chicken, *J. Nutr.* 65:503.

Okada, A., Takagi, Y., Itakura, T., Satani, M., Manabe, H., Iida, Y., Tanigaki, T., Iwasaki, M., and Kasaham, N., 1976. Skin lesions during intravenous hyperalimentation: Zinc deficiency, *Surgery* 80:629.

Prasad, A. S., 1966. Metabolism of zinc and its deficiency in human subjects, in *Zinc Metabolism* (A. S. Prasad, ed.), Thomas, Springfield, Ill. p. 250.

Prasad, A. S., and Cossack, Z. T., 1984. Zinc supplementation and growth in sickle cell disease, *Ann. Intern. Med.* 100:367.

Prasad, A. S., Halsted, J. A., and Nadimi, M., 1961. Syndrome of iron deficiency anemia, hepatosplenomegaly, hypogonadism, dwarfism, and geophagia, *Am. J. Med.* 31:532.

Prasad, A. S., Miale, A., Farid, Z., Schulert, A., and Sandstead, H. H., 1963a. Zinc metabolism in patients with the syndrome of iron deficiency anemia, hypogonadism, and dwarfism, *J. Lab. Clin. Med.* 61:537.

Prasad, A. S., Miale, A., Farid, Z., Sandstead, H. H., Schulert, A., and Darby, W. J., 1963b. Biochemical studies on dwarfism, hypogonadism and anemia, *AMA Arch. Intern. Med.* 111: 407.

Prasad, A. S., Schulert, A. R., Miale, A., Jr., Farid, Z., and Sandstead, H. H., 1963c. Zinc and iron deficiencies in male subjects with dwarfism and hypogonadism but without ancylostomiasis and schistosomiasis or severe anemia, *Am. J. Clin. Nutr.* 12:437.

Prasad, A. S., Schoomaker, E. B., Ortega, J., Brewer, G. J., Oberleas, D., and Oelshlegel, F. J., 1975. Zinc deficiency in sickle cell disease, *Clin. Chem.* 21, Am. Assn. Clinical Chemists, Washington, D.C., 582.

Prasad, A. S., Rabbani, P., Abbasi, A., Bowersox, F., and Fox, M. R. S., 1978. Experimental zinc deficiency in humans, *Ann. Intern. Med.* 89:483.

Prasad, A. S., Abbasi, A., Rabbani, P., and DuMouchelle, E., 1981. Effect of zinc supplementation on serum testosterone level in adult male sickle cell anemia subjects, *Am. J. Hematol.* 10: 119.

Prasad, A. S., Meftah, S., Abdallah, J., Kaplan, J., Brewer, G. J., Bach, J. F., and Dardenne, M., 1988. Serum thymulin in human zinc deficiency, *J. Clin. Invest.* 82:1202.

Rabbani, P. I., Prasad, A. S., Tsai, R., Harland, B. F., and Fox, M. R. S., 1987. Dietary model for production of experimental zinc deficiency in man, *Am. J. Clin. Nutr.* 45:1514.

Raulin, J., 1869. Etudes clinques sur la vegetation, *Ann. Sci. Nat. XI Bot.* 93.

Reimann, F., 1955. Wachstumsanomalien und Missbildungen bei Eisenmangelzustanden (Asiderosen), Proc. 5th Kongr. Eur. Esellschaft. Haematol. p. 546.

Robertson, B. T., and Burns, M. J., 1963. Zinc metabolism and the zinc deficiency syndrome in the dog, *Am. J. Vet. Res.* 24:997.

Sandstead, H. H., Prasad, A. S., Schulert, A. R., Farid, Z., Miale, A., Jr., Bassilly, S., and Darby, W. J., 1967. Human zinc deficiency, endocrone manifestations and response to treatment, *Am. J. Clin. Nutr.* 20:422.

Sommer, A. L., and Lipman, C. B., 1926. Evidence of indispensable nature of zinc and boron for higher green plants, *Plant Physiol.* 1:231.

Todd, W. R., Elvehjem, C. A., and Hart, E. B., 1934. Zinc in the nutrition of the rat, *Am. J. Physiol.* 107:146.

Tucker, H. F., and Salmon, W. D., 1955. Parakeratosis or zinc deficiency disease in pigs, *Proc. Soc. Exp. Biol. Med.* 88:613.

Warth, J. A., Prasad, A. S., Zwas, F., and Frank, R. N., 1981. Abnormal dark adaptation in sickle cell anemia, *J. Lab. Clin. Med.* 98:189.

Zinc and Enzymes

Zinc metalloenzymes catalyze approximately 50 important biochemical reactions. Many of these enzymes have been isolated from more than one species, resulting in identification of over 200 catalytically active zinc metalloproteins (Galdes and Vallee, 1983).

The first enzyme to be recognized as a zinc metalloenzyme was carbonic anhydrase, an enzyme essential for respiration in mammals (Keilin and Mann, 1940). At present zinc metalloenzymes have been recognized in all classes of enzymes: I, oxidoreductases (enzymes catalyzing oxidoreductions between two substrates); II, transferases (enzymes catalyzing transfer of a group other than hydrogen); III, hydrolases (enzymes catalyzing hydrolysis of esters, ether, peptide, glycosyl, acid anhydride, C–C, C–halide, or P–N bonds); IV, lyases (enzymes that catalyze removal of groups from substrates by mechanisms other than hydrolysis, leaving double bonds); V, isomerases (enzymes catalyzing interconversion of optical, geometric, or positional isomers); and VI, ligases (enzymes catalyzing the linking together of two components coupled to the breaking of a pyrophosphate bond in ATP or a similar compound) (see Table 2-1).

2.1 Functions of Zinc in Metalloenzymes

The functions of zinc in metalloenzymes are catalytic, structural, regulatory, and noncatalytic (Galdes and Vallee, 1983). Examples of enzymes in which zinc plays a catalytic role include carbonic anhydrase, carboxypeptidase, thermolysin, and aldolases. Zinc stabilizes the quaternary structure of oligomeric holoenzymes. It dimerizes *Bacillus subtilis* α-amylase without affecting its enzymatic activity and stabilizes the pentameric quaternary structure of aspartate-transcarbamylase. Zinc acts as an activator of bovine lens leucine aminopeptidase and inhibits the activities of porcine kidney leucine aminopeptidase and fructose-1,6-bisphosphatase.

Table 2-1. Zinc Metalloenzymes[a]

Name	Number	Source	Role[b]
Class I: oxidoreductases			
Alcohol dehydrogenase	9	Vertebrates, plants	A, D
Alcohol dehydrogenase	1	Yeast	A
D-Lactate dehydrogenase	1	Barnacle	?
D-Lactate cytochrome reductase	1	Yeast	?
Superoxide dismutase	12	Vertebrates, plants, fungi, bacteria	D
Class II: transferases			
Aspartate transcarbamylase	1	*E. coli*	B
Transcarboxylase	1	*Propionibacterium shermanii*	?
Phosphoglucomutase	1	Yeast	?
RNA polymerase	10	Wheat germ, bacteria, viruses	A
DNA polymerase	3	Sea urchin, *E. coli*, T_4 phage	A
Reverse transcriptase	3	Oncogenic viruses	A
Terminal dNT transferase	1	Calf thymus	A
Nuclear poly(A) polymerase	2	Rat liver, virus	A
Mercaptopyruvate sulfur transferase	1	*E. coli*	?
Class III: hydrolases			
Alkaline phosphatase	8	Mammals, bacteria	A, D
Fructose-1,6-bisphosphatase	2	Mammals	C
Phosphodiesterase (exonuclease)	1	Snake venom	A
Phospholipase C	1	*Bacillus cereus*	A
Nuclease P_1	1	*Penicillium cirtrinum*	?
α-Amylase	1	*B. subtilis*	B
α-D-Mannosidase	1	Jack bean	?
Aminopeptidase	10	Mammals, fungi, bacteria	A, C
Aminotripeptidase	1	Rabbit intestine	A
D-Carboxypeptidase	1	*Streptomyces albus*	A
Procarboxypeptidase A	2	Pancreas	A
Procarboxypeptidase B	1	Pancreas	A
Carboxypeptidase A	4	Vertebrates, crustaceans	A
Carboxypeptidase B	4	Mammals, crustaceans	A
Carboxypeptidase (other)	5	Mammals, crustaceans, bacteria	A
Dipeptidase	3	Mammals, bacteria	A
Angiotensin-converting enzyme	3	Mammals	A
Neutral protease	16	Vertebrates, fungi, bacteria	A
Collagenase	4	Mammals, bacteria	A
Elastase	1	*Pseudomonas aeruginosa*	?
Aminocyclase	1	Pig kidney	?
β-Lactamase II	1	*B. cereus*	A

Table 2-1. (*Continued*)

Name	Number	Source	Role[b]
Creatinase	1	*Pseudomonas putida*	?
Dihydropyrimidine aminohydrolase	1	Bovine liver	?
AMP deaminase	1	Rabbit muscle	?
Nucleotide pyrophosphatase	1	Yeast	A
Class IV: lyases			
Fructose-1,6-bisphosphate aldolase	4	Yeast, bacteria	A
L-Rhamnulose-1-phosphate aldolase	1	*E. coli*	A
Carbonic anhydrase	22	Animals, plants	A
δ-Aminolevulinic acid dehydratase	2	Mammalian liver, erythrocytes	A
Glyoxalase I	4	Mammals, yeast	A
Class V: isomerases			
Phosphomannose isomerase	1	Yeast	?
Class VI: ligases			
tRNA synthetase	3	*E. coli, Bacillus stearothermophilus*	A
Pyruvate carboxylase	2	Yeast, bacteria	?
Total	162		

[a] Adapted with permission from Galdes, A., and Vallee, B. L., 1983. Categories of zinc metalloenzymes, in *Metal Ions in Biological Systems* (H. Sigel, ed.), Dekker, New York, p. 1.
[b] A denotes a catalytic role, B a structural, C a regulatory, and D a noncatalytic role. A question mark indicates that available information is insufficient to make an assignment.

The precise mode of action of zinc in metalloenzymes is not well understood. Two types of mechanisms have been proposed to explain the manner in which the metal may affect catalysis. The first is the zinc carbonyl mechanism. According to this hypothesis, the substrate binds to zinc directly and displaces the metal-bound water molecule. The zinc is envisioned to act as a Lewis acid and polarizes the bound substrate, thereby facilitating the nucleophilic attack. This mechanism has been suggested for the role of zinc in aldolases and peptidases. The second is the zinc hydroxide mechanism in which zinc does not bind to substrate directly; rather, it mediates its function through the metal-bound water molecule. The zinc is believed to lower the pKa of the bound water molecule from -14 to -7. The resultant metal-bound hydroxide ion is then capable of attacking the substrate. This mechanism has been proposed for the role of zinc in carbonic anhydrase.

According to the integrated hypothesis, the substrate binds directly to the metal and does not displace the metal-bound water molecule, thus resulting in a pentacoordinate intermediate. In this model, the function of zinc would be both to polarize the substrate and to activate the water molecule which acts as a nucleophile. In addition, the metal through its flexible coordination geometry would act as a template and bring together the substrate and nucleophile. This suggestion is consistent with the entatic nature of the metal.

A fraction of the zinc molecules in certain metalloenzymes (equine and human alcohol dehydrogenase and *E. coli* alkaline phosphatase) is neither involved directly in catalysis nor essential for the maintenance of the tertiary structure of the enzymes. It may have some role in stabilization of the molecule but essentially its function remains obscure. This type of role for zinc is referred to as noncatalytic.

Zinc may be replaced by other metals in those enzymes where it plays either a catalytic or regulatory role but this replacement adversely affects the activity of the enzymes. If, on the other hand, the role of zinc is structural or noncatalytic, replacement by other metals produces only minor changes in enzymatic activity.

X-ray analysis of enzymes in which zinc plays a catalytic role shows that zinc is bound to three protein ligands and a water molecule. The presence of a water molecule bound to catalytic zinc signifies an open coordination site which is essential for the role of zinc in catalysis. Coordination of the geometry of catalytic zinc is distorted. Its properties fluctuate between tetracoordinate and pentacoordinate, suggesting this to be entatic. Entatic properties of zinc metalloenzymes may be important to catalysis, inasmuch as it lowers the energy barrier from the transition state and accelerates the conversion of substrate to products (Vallee and Williams, 1968). This view, however, remains controversial. In enzymes where zinc plays a structural or noncatalytic role, the zinc atom is fully coordinated by four protein ligands.

2.1.1 Catalytic Role of Zinc: Carbonic Anhydrase

Crystallographic studies of carbonic anhydrase show that CO_2 and H_2O are bound simultaneously to the catalytic zinc atom in this enzyme, thus supporting the integrated hypothesis (Galdes and Vallee, 1983):

$$H_2 + CO_2 \xrightarrow[\text{carbonic anhydrase}]{} H^+ + HCO_3^-$$

Carbonic anhydrase catalyzes a variety of hydration and hydrolysis reactions; however, not all are necessarily of biological significance. They involve the addition of hydroxide to a carbon–oxygen double bond or an analogue thereof. Available evidence suggests that zinc exerts its catalytic function through the bound water molecule (Fig. 2-1). It is believed that zinc greatly lowers the pKa of this water molecule and thereby affects its ionization. The

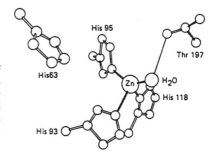

Figure 2-1. Schematic representation of the active site of human carbonic anhydrase (high activity), showing the metal-coordination geometry. [Reprinted with permission from Galdes, A., and Vallee, B. L., 1983. Categories of zinc metalloenzymes, in *Metal Ions in Biological Systems* (H. Sigel, ed.), Dekker, New York, p. 1.]

metal hydroxide ion so generated then attacks CO_2 which may be bound to the fifth site on the metal. If this hypothesis is correct, then zinc has multiple roles in this enzyme: (1) it is acting as a template to bring together the two reactants; (2) it is behaving as a Lewis acid, and it is producing the nucleophile (OH^-) and activating the electrophile (CO_2).

Erythrocytes of mammals have two isoenzymes of carbonic anhydrase: CAB (CA I), which has low activity, and CAC (CA II), which has high activity. The molecular weight of human CAC is 30,000 and it contains one atom of zinc per molecule. Human CAC has 259 amino acids. Human CAB has 260 amino acids and the two isoenzymes share 60% sequence homology. The three coordinating groups of zinc in this enzyme are histidines. All enzymes listed in Table 2-1 have one catalytic zinc with the exception of transcarbamylase and possibly superoxide dismutase.

2.1.2 Structural Role of Zinc: Aspartate Transcarbamylase

This enzyme is ubiquitous and catalyzes the condensation of carbamyl phosphate with L-aspartate to produce carbamyl aspartate. Carbamyl aspartate is the key precursor in the biosynthesis of pyrimidines (see Fig. 2-2).

The molecular weight of *E. coli* aspartate transcarbamylase is 310,000. Its activity is inhibited by several pyrimidine nucleotides such as cytosine

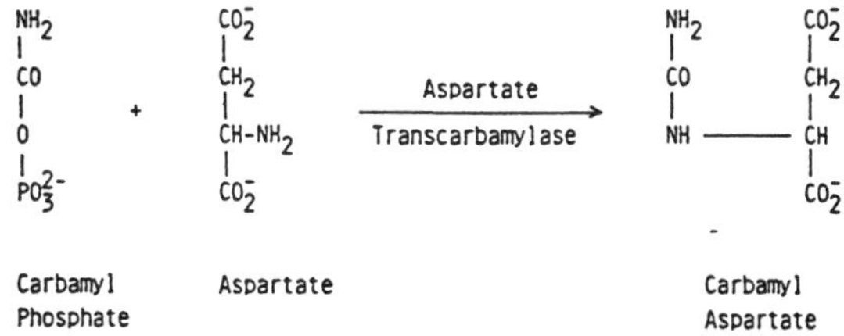

Figure 2-2. Reaction catalyzed by aspartate transcarbamylase. [Reprinted with permission from Galdes, A., and Vallee, B. L., 1983. Categories of zinc metalloenzymes, in *Metal Ions in Biological Systems* (H. Sigel, ed.), Dekker, New York, p. 1.]

triphosphate (CTP) and activity is enhanced by purine nucleotides such as adenosine triphosphate (ATP). The enzyme contains 6 g-atom of zinc per mol of the protein. One atom of zinc per chain (or two atoms per subunit) is bound to the R subunits and is not required for catalytic activity; rather, it appears to be essential for maintenance of the quaternary structure of the holoenzyme. The zinc binding site is located in the C-terminal region of the R chain. Zinc is bound by four cysteinyl residues in a tetrahedral coordination geometry.

2.1.3 Regulatory and Catalytic Roles of Zinc: Leucine Aminopeptidase

Leucine aminopeptidase catalyzes hydrolysis of N-terminal amino acid residues from protein, peptides, and amino acid amides, and generally requires a free α-amino group (or α-amino group) in the L configuration. Otherwise, their substrate specificity is broad and they can remove most amino acids from the N-terminus of amide linkages, particularly those with hydrophobic amino acids. These enzymes contain 2 g-atom of zinc per mol of subunit and their molecular weight is >200,000. Leucine aminopeptidases have been isolated from porcine intestinal mucosae and kidney, and bovine lens. In the bovine lens aminopeptidase, zinc does not play any structural role. The two metal atoms bound to leucine aminopeptidases function in two distinct ways: one has a catalytic function and the other regulates the activity induced by the zinc atom at the first site.

2.1.4 Noncatalytic and Catalytic Roles of Zinc: Alcohol Dehydrogenases

Alcohol dehydrogenases are NAD(H)-dependent enzymes that catalyze the interconversion of ethanol and their primary alcohols with the corresponding aldehydes (Fig. 2-3). Certain secondary alcohols and sterols also serve as substrates for these enzymes.

Equine liver alcohol dehydrogenase is dimeric and each of the two identical subunits has a molecular weight of 40,000 and is composed of a single polypeptide chain containing 374 amino acids (Fig. 2-3). There are three major isoenzymes which are of genetic origin. These are composed of two types of subunits E (ethanol active) and S (steroid active) as follows: EE, ES, and SS.

Each subunit of this enzyme contains two zinc atoms and binds one molecule of NAD(H). One zinc atom is essential for the catalytic activity and the function of the other zinc atom is unknown (noncatalytic). Inasmuch as complete removal of zinc does not affect the integrity of alcohol dehydrogenase

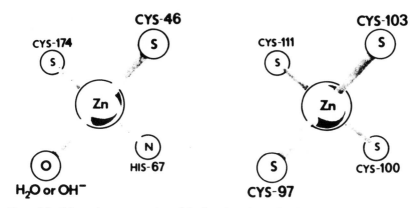

Figure 2-3. Schematic representation of the ligands and coordination geometry of the catalytic (left) and noncatalytic (right) zinc atoms of equine liver alcohol dehydrogenase. [Reprinted with permission from Galdes, A., and Vallee, B. L., 1983. Categories of zinc metalloenzymes, in *Metal Ions in Biological Systems* (H. Sigel, ed.), Dekker, New York, p. 1.]

structure, zinc is not required for the structure of these enzymes. However, treatment with guanidinium chloride or urea dissociates the dimers of the native enzyme into monomers with concomitant alterations of the tertiary structure and loss of activity.

2.2 Binding Sites of Zinc(II) in Enzymes

Zinc atoms of *E. coli* alkaline phosphatase serve a noncatalytic function. In exercising its regulatory function in bovine and porcine leucine amino-peptidase and fructose-1,6-bisphosphatase, zinc activates or inhibits but is not by itself essential for enzymatic activity.

Of the five zinc(II) enzymes whose active site structures have been established by x-ray, nuclear magnetic resonance (NMR), or metal substitution techniques, a combination of histidine imidazole, glutamate, or a separate COO^-, and cysteine (SH) are the enzyme-derived ligands which hold the metal in a distorted four or five coordinate geometry.

In carboxypeptidase (A and B) and thermolysin, the required Zn(II) is bound to the proteins by two histidine imidazoles and glutamate residue, with the water residing at a fourth ligand position (Fig. 2-4). Two mechanisms for the activity of carboxypeptidase A have been proposed: (1) The peptide binds to the enzyme displacing H_2O bound to zinc and the metal functions to polarize $C=O$, thereby rendering it more susceptible to attack by Glu-270 or a general base-delivered H_2O. (2) The second possibility is that Zn^+–

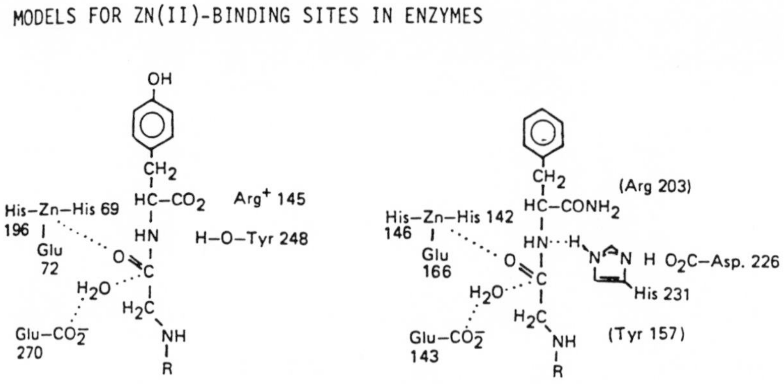

MODELS FOR ZN(II)-BINDING SITES IN ENZYMES

Figure 2-4. X-ray crystal structure of carboxypeptidase A and thermolysin, showing the binding modes of substrates or analogues into the active sites. [Reprinted with permission from Galdes A, and Vallee BL. 1983. Categories of zinc metalloenzymes. In: Sigel H, (ed.). Metal Ions in Biological Systems. New York and Basel, Marcel Dekker, Inc, p 1.]

OH attack is similar to that involved in carbonic anhydrase activity (Brown et al., 1983).

2.3 DNA and RNA Polymerases

DNA and RNA polymerases are nucleotidyl transferases that catalyze the replication and transcription of the cellular genome. The reaction of these enzymes requires the presence of extrinsic divalent metal ions, Mg(II) or Mn(II). In 1971, DNA and RNA polymerases from E. coli were the first two nucleotidyl transferases shown to be zinc metalloenzymes containing 1 and 2 g-atoms of tightly bound zinc per mol of the enzyme, respectively (Wu and Wu, 1983). The requirement of zinc for catalysis in DNA polymerase has been demonstrated in only two enzymes: DNA polymerase I from E. coli and reverse transcriptase for avian myeloblastosis virus (AMV).

Evidence suggesting that the intrinsic zinc ion of the polymerases interacts with the DNA template primer complex is as follows: (1) Nuclear quadrupolar relaxation studies of the role of bound zinc using $^{70}Br^-$ as a halide probe showed that the bound zinc ion had a large effect on the relaxation rate of Br^-. This effect was reduced by 70% on the addition of one molecule of polydeoxynucleotide primer per enzyme molecule but was not affected by the addition of substrate dTTP, suggesting that DNA primer but not the substrate competes out 70% of Br^- from the enzyme-bound zinc. (2) Kinetic analysis of the inhibition of DNA polymerase by 1,10-phenanthroline showed that the metal chelator competes with DNA but not with the substrate. A catalytic role of bound zinc in DNA polymerase has been suggested. According to this proposal, zinc coordinates with the 3'-OH primer terminus, thereby facilitating its deprotonation and preparing it for a nucleophilic attack on the α-phosphorus atom of the incoming nucleotide. It has been proposed that zinc ions play at least three possible roles in the function of the polymerases: (1) a catalytic role in the binding of substrate primer, or template; (2) a regulatory role in the specificity of gene replication and transcription; and (3) a structural role in the maintenance of the proper configuration of the polymerase.

The strongest evidence that the intrinsic zinc ions in *E. coli* RNA polymerase are required for enzymatic activity comes from the finding that the addition of zinc is necessary to reconstitute active enzyme from urea-denatured inactive apoenzyme. In addition, the participation of the intrinsic metal ions in substrate or template binding was evident from the characteristic spectral changes of Co–Co (Speckhard *et al.*, 1977), Co–Zn, and Ni–Zn enzymes induced by nucleotides or DNA template.

Comparative studies of *E. coli* Co–Co and Zn–Zn RNA polymerase and the observations that these two enzymes have different efficiencies in utilizing the A2 promoter versus the A1 + A3 promoters of T7 DNA suggest a regulatory role for the intrinsic metal ion.

The intrinsic zinc ions in DNA and RNA polymerases can affect the enzymatic activity indirectly by playing a structural role in maintaining proper enzyme conformation. The removal and restoration of the intrinsic metal ions has proven much more difficult for a multisubunit polymerase than a single-chain enzyme, suggesting that some of the metal ions in the multisubunit enzymes may affect the quaternary structure of the polymerase. The structural role of metal ions may be more important in eukaryotic RNA polymerase, which contain a greater number of subunits than do prokaryotic enzymes.

2.4 Role of Zinc in Protection of Essential Thiol Enzymes

Zinc forms a complex with proteins or apoenzymes by interaction with their electron-rich atoms such as oxygen, nitrogen, or sulfur. Thiol (SH) groups have a high affinity for metal ions of the II B series such as Zn^{2+}, Cd^{2+}, and Hg^{2+}.

Many biologically important proteins contain thiol groups which are essential for their functions (Finelli *et al.*, 1975). In some cases, zinc prevents the oxidation of thiols and thus protects and regulates their enzymatic activity. δ-Aminolevulinic acid dehydratase (δ-ALD) and dihydroorotase are two such examples of enzymes in which the role of zinc appears to be protection of thiol groups in these proteins.

δ-ALD (EC 4.2.1.24) catalyzes the condensation of two molecules of δ-aminolevulinate to form porphobilinogen, the second step in heme synthesis. The activity of this enzyme is decreased in lead toxicity. Zinc supplementation both *in vivo* and *in vitro* reverses the inhibitory effect of lead on this enzyme. The human red cell δ-ALD consists of eight subunits of 31 to 35 kDa. Each subunit contains one Zn^{2+} and treatment with oxygen or 5,5-dithiobis-(2-nitrobenzoic acid) (DTNB) causes loss of zinc in parallel with inactivation of enzymatic function. The integrity of both function and structure in the human and bovine enzyme is dependent on the protection of the thiol groups from oxidation (Tsukamoto *et al.*, 1980).

Bovine liver δ-ALD consists of eight 35-kDa subunits, each of which contains one Zn^{2+} and eight thiol groups (Tsukamoto *et al.*, 1980). Two cysteine and two histidine groups constitute the active site where the zinc resides. Apoenzyme (free of zinc and maintained under aerobic conditions) has eight free thiol groups per subunit and is fully active. Thus, it appears that zinc is not essential for catalytic activity but is essential for the maintenance of free thiols and the activity of this enzyme. When exposed to air, this apoenzyme loses two SH groups and all of its catalytic activity. Zinc cannot be introduced into the protein while in the oxidized state. Reduction with dithiothreitol restores activity and allows zinc to be complexed. Pb^{2+} replaces Zn^{2+} stoichiometrically and causes loss of activity of δ-ALD.

Treatment of δ-ALD apoenzyme with DTNB caused rapid loss of two SH groups leading to formation of an intramolecular disulfide bond (Gibbs *et al.*, 1985). This reaction was inhibited by the addition of 100 μM Zn^{2+} but not by Pb^{2+}. Exposure of apoenzyme to air resulted in the loss of activity and oxidation of the reactive thiols, but this reaction was blocked by Zn^{2+}, suggesting that zinc plays a critical role in stabilizing the reduced thiol configuration either by binding to thiol groups directly or by altering its reactivity by affecting a conformational change.

The activity of human erythrocyte δ-ALD is highly sensitive to Pb^{2+} at 70 to 100 μM concentration and it has been suggested that lead binds to one or more of the thiol groups. However, in the studies reported by Gibbs *et al.* (1985), Pb^{2+} had no effect on the reactivity of the thiol groups and ionic zinc (100 μM) reactivated the enzyme in the presence of 300 μM Pb^{2+}. It was therefore concluded that the two ions interact with the protein at different sites. Thus, relatively low concentrations of zinc appear to protect δ-ALD from oxidative damage as well as from lead inhibition, although it does not appear to be essential for its catalytic activity.

Zinc plays a regulatory role in another biologically important enzyme, dihydroorotase (DHO; EC 3.5.2.0) (Kelly *et al.*, 1986; Washabaugh and Collins, 1986). This enzyme catalyzes the formation of dihydroorotate from car-bamoylaspartate. This is the third step in biosynthesis of pyrimidines. In mammalian cells DHO activity resides in a 240-kDa protein that also has carbamoylphospate synthetase and aspartate transcarbamylase activities. Only the DHO domain contains zinc that is essential for catalytic activity. This 44-kDa domain contains seven thiols and one stable zinc. It is possible that some zinc ions may have been lost during isolation inasmuch as the zinc content of bacterial DHO varies with isolation procedure.

According to this model (as shown in Fig. 2-5), dimeric zinc-dependent enzyme (DHO) with three zinc atoms per unit (on the left) on reaction with substrate or a chelator yields an active enzyme whose monomers contain only one zinc at the active site. The addition of zinc ions reverses this effect. It appears that the two external sites are probably occupied by zinc *in vivo* when the substrate level is low and the enzyme is minimally active. The treatment of crude extracts or the 3-Zn DHO with a weak chelator such as Bio-Rex 70 or carbamyl aspartate increases the specific activity severalfold. Thus, ac-cording to this model, not only does zinc protect DHO oxidative loss of essential thiols when the substrate is limiting, but it also regulates its activity, decreasing the activity when the substrate concentration is low.

2.5 Cytochrome c Oxidase

In a recent study, cytochrome c oxidase from bovine heart muscle was shown to contain 1 zinc per 2 irons (Einarsdottir and Caughey, 1984). Assay for the metal contents of nine preparations showed that only Cu, Fe, and Zn were present in this enzyme. The average atomic ratios of Cu/Fe, Fe/Zn, and

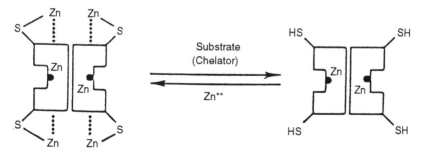

Figure 2-5. Dimeric zinc-dependent enzyme with three zinc atoms per subunit on the left which, on reaction with substrate or chelator, yields an enzyme whose monomers contain only one zinc at the active site, and enhanced activity. The addition of zinc ions reverses this effect. (Reprinted with permission from *Nutr. Rev.* 1986, 44:310.)

Cu/Zn were 1.3, 2.1, and 2.8, respectively. The zinc appeared to be tightly bound to this enzyme molecule.

Cytochrome c oxidase, the enzyme at the terminus of the respiratory chain, has a key bioenergetic role by catalyzing the following reaction:

$$O_2 + 4 \text{ cytochrome } c^{2+} + 4H^+ \rightarrow 2H_2O + 4 \text{ cytochrome } c^{3+}$$

Reduced cytochrome c molecules give up one electron each at the outer surface of the inner mitochondrial membrane. Present evidence indicates that these electrons are accepted first at one heme iron (heme a or Fe_c) and one copper (Cu_c) and are transferred later to the other heme (heme a_3 or Fe_L*) and the other copper (Cu_L). Fe_L and Cu_L, located near the inner surface of the membrane, are accessible to ligands and participate in the reduction of dioxygen to two water molecules. Thus, two iron and two copper centers appear to be involved in the overall function of cytochrome c oxidase. Analyses of cytochrome c oxidase preparations, however, have revealed a Cu/Fe ratio somewhat greater than 1 and the recent study shows the presence of zinc in the cytochrome c oxidase of bovine heart.

It has been suggested that zinc may have a catalytic as well as structural role at this dioxygen reduction site of cytochrome c oxidase. The positive charges of Zn^{2+} are considered important for the roles of zinc in many zinc metalloenzymes. Ligands with a potential proton for donation are rendered better proton donors upon binding to zinc. Proton as well as electron donations are requirements of the oxidase reaction. It has been suggested that a ligand bound to Zn^{2+} (e.g., imidazole or water) may promote catalysis of the reduction of dioxygen to two water molecules by donation of a proton to an oxygen atom. The Zn^{2+} not only could render the ligand a more effective proton donor but could also provide control over the stereochemistry of the proton donation to oxygen.

2.6 Enzyme Changes in Zinc Deficiency

Various studies by Vallee and associates in the late 1950s and early 1960s showed that zinc was essential for the activity of a number of enzymes (Vallee, 1959). It was therefore suggested that the level of zinc in cells may control the physiological processes through the formation and/or regulation of activity

---

* Fe_L, the iron of the heme A that binds ligands (the a_3 heme); Fe_C, the iron of the heme A that does not bind ligands (the a heme); Cu_L, the copper most closely associated with Fe_L; Cu_C, the copper that is not closely associated with Fe.

of zinc-dependent enzymes. However, until 1966, evidence supporting this concept in experimental animals did not exist in the literature. Since then our studies showed for the first time that in zinc-deficient rats (Prasad *et al.*, 1967, 1971) and young pigs (Prasad *et al.*, 1969) the activities of various zinc-dependent enzymes and zinc content in sensitive tissues were reduced relative to their pair-fed controls. These early studies utilized histochemical techniques for the assay of enzyme activities.

In the rats, the zinc content of testis, bone, esophagus, kidney, and muscle was decreased, whereas the iron content was increased in the testis of zinc-deficient rats compared with restrictedly fed control rats (Prasad *et al.*, 1967). Histochemical enzyme determinations revealed reduced activities of certain enzymes in the testis, bone, esophagus, and kidney. In the testis, lactic dehydrogenase (LDH), malic dehydrogenase (MDH), alcohol dehydrogenase (ADH), and NADH diaphorase; in the bone, LDH, MDH, ADH, and alkaline phosphatase; in the esophagus, MDH, ADH, and NADH diaphorase; and in the kidney, MDH and alkaline phosphatase were decreased in the zinc-deficient rats compared with restrictedly fed controls. Succinic dehydrogenase (SDH), an iron-dependent enzyme, revealed no significant changes under the conditions of experiments in various groups of rats that were investigated.

In a "repleted" group of rats (zinc sufficient), content of zinc in the testis and bone increased significantly relative to the deficient group. In the testis, bone, esophagus, and kidney, the activities of various enzymes increased after repletion with zinc. Inasmuch as the major manifestations of zinc deficiency in these rats included growth retardation, testicular atrophy, and esophageal parakeratosis, our results suggested that the content of zinc in the above tissues most likely controlled the physiological process through the formation of zinc-dependent enzymes.

We also studied the effects of zinc deficiency in young pigs (Prasad *et al.*, 1969). Growth retardation was evident by the end of the first week and moderately severe parakeratosis was observed by the end of the second week. The weight of most organs decreased in proportion to the body weight. Thymus showed a greater decrease, whereas adrenal, pituitary, and kidney were heavier in zinc-deficient pigs. Serum zinc and alkaline phosphatase activity were decreased significantly in the zinc-deficient pigs. Zinc content of the bone, pancreas, and liver and activities of various zinc-dependent enzymes in bone, pancreas, testis, skin, esophagus, and pituitary were decreased in zinc-deficient pigs compared with the restrictedly fed controls. SDH showed no differences in its activities between the two groups. Iron content in pancreas, copper in bone and liver, calcium in intestine, and manganese in liver and pancreas were increased in the zinc-deficient group, suggesting physiological competition between zinc and other elements for similar binding sites in the tissues.

Our results suggested that in the young pigs certain tissues were more susceptible to effects of zinc deficiency than others. Bone and pancreas appeared to be very sensitive to zinc depletion. These results in the young pigs were also supportive of the hypothesis that the content of zinc in various tissues controls the physiological processes through the formation or regulation, or both, of activities of zinc-dependent enzymes.

In another study carried out in young pigs (Prasad *et al.*, 1971), RNA and DNA content and activities of several zinc-dependent enzymes were assayed in tissue homogenates under optimal conditions and expressed in terms of per milligram DNA. The following enzymes were assayed: ADH, LDH, and aldolase (ALD) in liver; ADH, ALD, and alkaline phosphatase (AP) in kidney; carboxypeptidase (CPD) and LDH in pancreas; and ADH, AP, LDH, and ALD in bone. SDH and isocitric dehydrogenase, a manganese-dependent enzyme, were also assayed in the kidney as controls.

Zinc content and RNA/DNA ratios were significantly reduced in the liver, kidney, pancreas, and bone, but DNA per milligram of net weight remained unaffected in zinc-deficient animals compared with their pair-fed controls (Table 2-2). Only zinc-dependent enzymes revealed significantly reduced activities in zinc-deficient tissues; exceptions were LDH in the liver and pancreas, and ALD in bone. Addition of small amounts of ethylene diaminetetraacetate (EDTA) to tissue homogenates significantly reduced activities of ADH, AP, and CPD, but not the activities of LDH and ALD, thus revealing the differences in affinity for zinc of the various enzymes.

The likelihood of detecting any biochemical changes is greatest in tissues that are sensitive to zinc depletion. In the present studies, liver, kidney, bone, and pancreas of pigs revealed a marked decrease in the zinc and RNA content and concomitantly, the activity of various zinc-dependent enzymes was demonstrated to be decreased as a result of zinc deficiency. Unfortunately, many investigators in the past studied only the liver of zinc-deficient rats for biochemical changes, an organ which does not show any decrease in zinc content and, therefore, does not appear to be a sensitive tissue in this regard. Thus, whereas one would consider the pig liver to be a zinc-sensitive tissue, the rat liver is not sensitive to zinc depletion. Perhaps a study of the testis, bone, pancreas, thymus, and kidney of rats would be more likely to yield positive results inasmuch as these tissues show a decrease in zinc content as a result of zinc deficiency.

One should not expect that all zinc-dependent enzymes would be affected similarly in all tissues of zinc-deficient animals. The differences in susceptibility of the enzymes may be caused by differences in the affinity for zinc of the various zinc-dependent enzymes and by the turnover rate of the proteins and the cells of tissues involved.

Table 2-2. Zinc, DNA, RNA, and Protein Content and Activities of Various Enzymes in Tissues of Pair-Fed Controls and Zinc-Deficient Animals[a,b]

Tissue[d]	Zinc (µg/g dry wt)	DNA (µg/mg wet wt)	RNA (µg/µg DNA)	Protein (mg/mg DNA)	ADH[c] Initial	ADH After addition of Zn[e]	ADH After addition of EDTA[f]	LDH	ALD	SDH	ICDH	AP (U/mg DNA)[c]	CPD Initial	CPD After addition of Zn[e]	CPD After addition of EDTA[f]
Liver															
A	150.8 ± 12	4.2 ± 0.42	3.5 ± 0.34	40 ± 2.9	6.4 ± 0.69	89 ± 2	79 ± 4	46 ± 4	5.7 ± 0.7	—	—	—	—	—	—
B	96.1 ± 8	5.1 ± 0.3	2.5 ± 0.18	35 ± 1.9	3.6 ± 0.39	95 ± 3	77 ± 1	50 ± 4	3.0 ± 0.3	—	—	—	—	—	—
p[g]	<0.005	NS	<0.05	NS	<0.005			NS	<0.005						
Kidney															
A	97.8 ± 2.6	3.7 ± 0.16	1.4 ± 0.06	23.4 ± 0.8	0.162 ± 0.009	91 ± 4	NA	—	2.54 ± 0.11	1.6 ± 0.17	6.9 ± 0.72	154.8 ± 16.4	—	—	—
B	87.0 ± 2.8	3.9 ± 0.20	1.1 ± 0.05	21.2 ± 1.9	0.123 ± 0.008	96 ± 8	NA	—	1.86 ± 0.09	1.7 ± 0.19	7.2 ± 0.19	80.2 ± 9.5	—	—	—
p[g]	<0.025	NS	<0.005	<0.05	<0.025				<0.001	NS	NS	<0.005			
Bone															
A	98.5 ± 2.08	0.53 ± 0.07	3.04 ± 0.26	57.5 ± 5	2.30 ± 0.35	60 ± 10	NA	81 ± 10	21 ± 3.4	—	—	2640 ± 420	—	—	—
B	48.1 ± 2.74	0.83 ± 0.13	1.81 ± 0.22	36.8 ± 5	0.64 ± 0.07	50 ± 24	NA	44 ± 7	12 ± 2.5	—	—	480 ± 120	—	—	—
p[g]	<0.001	<0.1 > 0.05	<0.005	<0.025	<0.001			<0.025	<0.1 > 0.05			<0.001			
Pancreas															
A	133.3 ± 6	4.7 ± 0.28	5.0 ± 0.15	25 ± 1.8	—	—	—	14 ± 1.6	—	—	—	—	2.8 ± 0.1	82 ± 7	40 ± 16
B	96.8 ± 4	4.0 ± 0.30	4.1 ± 0.1	22 ± 2.2	—	—	—	14 ± 1.0	—	—	—	—	1.3 ± 0.2	70 ± 12	35 ± 15
p[g]	<0.001	<0.2 > 0.1	<0.05	NS				NS					<0.001		

Note: ADH[c] columns "After addition of Zn[e]" and "After addition of EDTA[f]" express Change from initial (%). Enzyme activities are expressed as ΔOD/min/mg DNA[c]. CPD[c] columns "After addition of Zn[e]" and "After addition of EDTA[f]" express Change from initial (%).

[a] Adapted with permission from Prasad, A. S., 1978. Zinc, in *Trace Elements and Iron in Human Metabolism*, Plenum Press, New York, p. 251.
[b] Values are means ± S.E. NS, not significant; NA, no activity.
[c] ADH, alcohol dehydrogenase; ALD, aldolase; SDH, succinic dehydrogenase; CPD, carboxypeptidase; LDH, lactic dehydrogenase; ICDH, isocitric dehydrogenase; AP, alkaline phosphatase.
[d] A, pair-fed controls; B, zinc-deficient animals.
[e] Final concentration of Zn: Liver, 8×10^{-5} M; kidney, 3.3×10^{-5} M; bone, 1.7×10^{-5} M; pancreas, 1.8×10^{-5} M.
[f] Final concentration of EDTA: Liver, 6.7×10^{-3} M; kidney, 3.3×10^{-3} M; bone, 6.7×10^{-3} M; pancreas, 7.3×10^{-3} M.
[g] Comparison of means.

In a study in rats, the zinc content of the liver, kidney, testis, pancreas, bone, and thymus were shown to be significantly decreased in the deficient animals compared with their pair-fed controls (Prasad and Oberleas, 1971). ADH in the liver, kidney, testis, and bone, AP in the kidney, testis, pancreas, bone, and thymus, and CPD in the pancreas were significantly reduced in the deficient rats compared with their pair-fed controls (Table 2-3). No such changes were observed with respect to isocitric dehydrogenase (a manganese-dependent enzyme) and SDH (an iron-dependent enzyme) in those tissues. All of the above-mentioned studies thus established that zinc-dependent enzymes changed in zinc-deficient tissues of the experimental animals.

2.7 Enzymes Involved in Nucleic Acids

Lieberman *et al.* (1963) reported that in *in vitro* experiments with kidney cells from rats, zinc ions were necessary for DNA synthesis. Removal of zinc from the growth medium resulted in inhibition of both DNA synthesis and DNA polymerase and thymidine kinase activity. Terhune and Sandstead (1972) studied the effect of zinc deficiency on the activity of DNA-dependent RNA polymerase in the nuclei of liver cells of suckling rats born of zinc-deficient mothers. The activity of this enzyme steadily decreased in the young from the tenth day of life, demonstrating that zinc was necessary for the activity of nuclear DNA-dependent RNA polymerase in mammalian liver. In a similar study on the brains of prenatal zinc-deficient rats, reduced RNA polymerase activity was observed in addition to smaller brain size and diminished DNA synthesis.

In another study of rats, activities of ribonuclease (RNase) and deoxyribonuclease (DNase) were investigated in the testis, kidney, bone, and thymus of zinc-deficient, continuously pair-fed and ad libitum-fed control rats (Prasad and Oberleas, 1973). Nucleases were measured at both acid and alkaline pHs in all of the tissues, except for alkaline RNase and DNase in the bone and alkaline DNase in the thymus, where the activities were too low to be measured. Whereas DNase activities showed no difference between the zinc-deficient and pair control rats, the activities of RNase were increased in the zinc-deficient tissues. *In vitro,* zinc is known to regulate the activity of yeast RNase (Ohtake *et al.,* 1963). Thus, our results suggest that the increased RNase activity in zinc-deficient tissues may be a partial explanation for a decreased RNA/DNA ratio and decreased protein synthesis in zinc-deficient tissues which have been observed previously (Prasad and Oberleas, 1973).

Fernandez-Madrid *et al.* (1973) and Somers and Underwood (1969) suggested that zinc may inhibit RNase, whereby RNA degradation would be lowered. An enhanced activity of this enzyme has been observed in zinc-deficient tissues of experimental animals (Prasad and Oberleas, 1973). In one study, the testis of zinc-deficient rats contained less zinc, RNA, DNA, and protein, and at the same time exhibited an elevated RNase activity. Although the RNase activity was increased in zinc-deficient tissue such as testis, kidney, bone, and thymus of zinc-deficient rats, the activity of DNase was not affected. It was suggested that the increased RNase activity in zinc deficiency may account for lowered RNA/DNA ratio, lowered protein synthesis, and growth depression so commonly observed in many species including humans as a result of zinc deficiency.

Elevated plasma RNase activity has been observed in zinc-deficient sickle-cell anemia subjects, in acrodermatitis enteropathica, and in zinc-deficient patients with chronic renal diseases, and in volunteers in whom a mild deficiency of zinc was produced experimentally by dietary means (Prasad, 1982).

2.7.1 Deoxythymidine Kinase

Effects of zinc deficiency on DNA synthesis and activity of thymidine kinase (TK) in rapidly regenerating collagen connective tissue in rats were investigated (Prasad and Oberleas, 1974). Three groups of rats, namely (1) zinc deficient, (2) restricted-fed zinc supplemented, and (3) ad libitum-fed zinc supplemented, were investigated. Total duration of the zinc-deficient state in experimental animals in three experiments was 6, 13, and 17 days, respectively. One polyvinyl sponge was implanted subcutaneously either 6 or 10 days prior to sacrifice in order to stimulate production of collagen connective tissue. Incorporation of [2-^{14}C]thymidine as a measure of DNA synthesis in the collagen connective tissue was determined *in vitro* in 6-day experiments in zinc-deficient and restricted-fed control animals. TK activity in 6-day experiments was as follows (mean $\pm$ S.E.): deficient, 1.04 ± 0.14; restricted-fed, 3.57 ± 0.36; and ad libitum-fed, 3.37 ± 0.36 (Table 2-4). The difference in activity between tissues from deficient animals and either control was statistically significant ($p < 0.025$). Similar results were obtained in the 13-day experiment. In the 17-day experiment TK activity in the deficient tissues was not measurable. [2-^{14}C]-Thymidine incorporation into DNA was significantly reduced in the 6-day experiments in the deficient tissues compared with the restricted-fed control animals ($p < 0.001$) (Tables 2-5 and 2-6).

This was the first demonstration of an adverse effect of zinc deficiency on the activity of TK in animals. Because this enzyme is essential for DNA synthesis and cell division, our data suggested that this early metabolic defect as a result of zinc deficiency may play an important role in causing growth retardation in zinc-deficient rats.

Table 2-3. Zinc and Specific Activities of Various Enzymes in Tissues of Controls and Zinc-Deficient Rats[a,b]

Tissue[c]	Zinc (μg/g dry wt)	Activity of enzymes (ΔOD/min per mg protein)[d]					Alkaline phosphatase (sigma U/mg protein)	CPD (μmol β-naphthol liberated/mg protein)[c]
		ADH	Aldolase	LDH	ICDH	SDH		
Liver								
A	91.7 ± 4.7	0.021 ± 0.0022	0.107 ± 0.0064		0.43 ± 0.02			
B	107.3 ± 3.1	0.022 ± 0.0013	0.134 ± 0.0098		0.38 ± 0.03	0.151 ± 0.012		
C	87.6 ± 3.5	0.013 ± 0.0008	0.124 ± 0.0100		0.38 ± 0.02	0.152 ± 0.017		
A vs B[e]	$p < 0.025$	NS	$p < 0.05$		NS			
A vs. C	NS	$p < 0.001$	NS		NS			
B vs. C	$p < 0.001$	$p < 0.001$	NS		NS	NS		
Kidney								
A	93 ± 2.2	0.0031 ± 0.00025	0.20 ± 0.01	2.9 ± 0.27	0.57 ± 0.03		3.3 ± 0.22	
B	87 ± 0.8	0.0031 ± 0.00012	0.22 ± 0.01	2.67 ± 0.07	0.48 ± 0.03	0.188 ± 0.0075	4.5 ± 0.25	
C	80 ± 0.9	0.0014 ± 0.00013	0.21 ± 0.01	2.53 ± 0.04	0.49 ± 0.04	0.187 ± 0.0056	2.3 ± 0.17	
A vs. B	NS	NS	NS	NS	NS		$p < 0.005$	
A vs. C	$p < 0.001$	$p < 0.001$	NS	NS	NS		$p < 0.005$	
B vs. C	$p < 0.001$	$p < 0.001$	NS	NS	NS	NS	$p < 0.001$	
Testis								
A	191.0 ± 3.1	0.022 ± 0.0018	0.161 ± 0.0034	1.20 ± 0.08	0.063 ± 0.0067		2.75 ± 0.11	
B	197.1 ± 2.9	0.028 ± 0.0030	0.161 ± 0.0047	1.22 ± 0.09	0.046 ± 0.0052	0.128 ± 0.0046	2.80 ± 0.10	
C	114.7 ± 4.9	0.012 ± 0.0014	0.164 ± 0.0112		0.090 ± 0.0076	0.127 ± 0.0054	1.70 ± 0.10	
A vs. B	NS	NS	NS		$p < 0.025$		NS	
A vs. C	$p < 0.001$	$p < 0.001$	NS		$p < 0.025$		$p < 0.001$	
B vs. C	$p < 0.001$	$p < 0.001$	NS	NS	$p < 0.001$	NS	$p < 0.001$	

Pancreas						
A	79.5 ± 3.2	0.0020 ± 0.0001		0.18 ± 0.02	1.8 ± 0.18	64.8 ± 11.2
B	88.9 ± 3.7	0.0026 ± 0.0001		0.16 ± 0.02	1.7 ± 0.11	61.19 ± 10.29
C	74.4 ± 3.2	0.0009 ± 0.0001		0.14 ± 0.01	1.4 ± 0.05	10.69 ± 6.41
A vs. B	NS	$p < 0.001$		NS	NS	NS
A vs. C	NS	$p < 0.001$		NS	$p < 0.5$	$p < 0.001$
B vs. C	$p < 0.01$	$p < 0.001$		NS	$p < 0.025$	$p < 0.001$
Bone						
A	180 ± 7.7	0.128 ± 0.000	0.82 ± 0.04		3.46 ± 0.23	
B	211 ± 5.4	0.068 ± 0.006	0.76 ± 0.04		4.69 ± 0.16	
C	139 ± 9.6	0.078 ± 0.009			2.69 ± 0.14	
A vs. B	$p < 0.01$	$p < 0.001$			$p < 0.001$	
A vs. C	$p < 0.01$	$p < 0.001$			$p < 0.025$	
B vs. C	$p < 0.001$	NS	NS		$p < 0.001$	
Thymus						
A	83.7 ± 1.9	0.079 ± 0.008	4.41 ± 0.34		6.2 ± 0.54	
B	98.8 ± 3.5	0.080 ± 0.006	6.78 ± 0.52		5.7 ± 0.62	
C	81.1 ± 2.2	0.077 ± 0.005			3.5 ± 0.25	
A vs. B	$p < 0.005$	NS	NS		NS	
A vs. C	NS	NS			$p < 0.001$	
B vs. C	$p < 0.005$	NS			$p < 0.001$	

[a] Adapted with permission from Prasad, A. S., 1978. Zinc, in *Trace Elements and Iron in Human Metabolism*. Plenum Press, New York, p. 251.

[b] Values are means ± S.E. NS, not significant ($p > 0.05$).

[c] A, ad libitum-fed controls; B, pair-fed controls; C, zinc-deficient rats.

[d] ADH, alcohol dehydrogenase; LDH, lactic dehydrogenase; ICDH, isocitric dehydrogenase; SDH, succinic dehydrogenase; CPD, carboxypeptidase.

[e] Comparison of means.

Table 2-4. Thymidine Kinase Activity in Regenerating Tissue
(nmol TMP Formed/h per mg Protein)[a]

	Deficient	Restricted-fed	Ad libitum
6 day[b]	1.04 ± 0.14[c] (12)	3.57 ± 0.36 (13)	3.37 ± 0.36 (12)
13 day[d]	0.58 ± 0.02 (6)	2.40 ± 0.59 (5)	1.65 ± 0.12 (5)
17 day	None (5)	2.70 ± 0.6 (5)	2.68 ± 0.7 (5)

[a] Adapted with permission from Prasad, A. S., and Oberleas, D., 1974. Thymidine kinase activity and incorporation of thymidine into DNA in zinc-deficient tissue, *J. Lab. Clin. Med.* 83:634.
[b] Six day: deficient versus either control, $p < 0.001$.
[c] Mean ± S.E.
[d] Thirteen day: deficient versus either control, $p = 0.025$.

The activity of TK was also assayed in implanted sponge connective tissue in three groups of human subjects: (1) five normal controls (having normal levels of plasma and red cell zinc), (2) four patients with sickle-cell anemia who had low zinc in red cells and hair, and (3) two volunteers (under strict dietary control), after 6 months of zinc depletion (intake 2.7 mg/day) and repeated after 3 months of zinc repletion (intake 30 mg/day). Total proteins, total collagen, RNA/DNA, and TK activity were measured by established techniques (Prasad *et al.*, 1979). In sickle-cell anemia subjects, TK activity was not detected, and RNA/DNA, total collagen, and total protein contents were decreased compared with normal controls. In human volunteers TK activity was not detected during the zinc restriction phase. After supplementation with zinc, TK activity became 70% of normal control levels, and RNA/DNA, total collagen, and total protein contents of sponge connective tissue increased. Thus, these studies confirmed the importance of zinc for maintenance of TK activity and for the synthesis of protein in human subjects as well.

A markedly reduced activity of TK in 12-day-old embryos of zinc-deficient rats was also reported (Dreosti *et al.*, 1986). Interestingly, the reduced

Table 2-5. [¹⁴C]Thymidine Incorporation into DNA
(6-Day Experiment) (dpm/mg DNA)[a]

	Deficient	Restricted-fed	t[b]
Expt I	18.0 ± 5.7 × 10³[c] (4)	136.2 ± 15.2 × 10³ (4)	7.24+
Expt II	14.4 ± 2.9 × 10³ (6)	54.7 ± 7.5 × 10³ (5)	4.99+

[a] Adapted with permission from Prasad, A. S., and Oberleas, D., 1974. Thymidine kinase activity and incorporation of thymidine into DNA in zinc-deficient tissue, *J. Lab. Clin. Med.* 83:634.
[b] $p < 0.001$.
[c] Mean ± S.E.

Table 2-6. Gain in Body Weight of Rats, Sponge Connective Tissue (SCT) Weight, and Concentration of DNA, RNA, Protein, and Zinc in Sponge Connective Tissue in 6-Day Experiments[a]

	Total gain in body wt (g)	SCT dry wt (mg)	DNA (μg)/ mg tissue	RNA (μg)/ mg tissue	Protein (mg)/ mg tissue	Zn (μg)/ mg tissue
Zn-deficient	18 ± 1.6[b] (24)[a]	117.8 ± 10.6 (16)	6.2 ± 0.39 (12)	11.3 ± 0.39 (12)	0.54 ± 0.03 (12)	0.10 ± 0.008 (12)
Restricted-fed control animals	14 ± 1.2 (24)	153.3 ± 16.2 (15)	7.54 ± 0.37 (12)	13.4 ± 1.0 (12)	0.56 ± 0.02 (12)	0.14 ± 0.007 (12)
Ad libitum-fed control animals	53 ± 2.0 (24)	176.5 ± 17.7 (15)	6.6 ± 0.26 (12)	14.6 ± 1.5 (12)	0.52 ± 0.03 (12)	0.12 ± 0.19 (12)
p value						
Zn-deficient vs. restricted-fed animals	0.05	NS	0.025	NS	NS	0.005
Zn-deficient vs. ad libitum-fed animals	0.001	0.01	NS	NS	NS	NS
Restricted-fed vs. ad libitum-fed animals	0.001	NS	NS	NS	NS	NS

[a] Adapted with permission from Prasad, A. S., and Oberleas, D., 1974. Thymidine kinase activity and incorporation of thymidine into DNA in zinc-deficient tissue, J. Lab. Clin. Med. 83:634.
[b] Mean ± S.E.
[c] Number in parentheses is the number of rats.

activity of TK became evident even when pregnant females were given the zinc-deficient diet only for days 9 to 12 of pregnancy. Thus, it appears that the reduced activity of this enzyme is one of the earliest and most sensitive biochemical effects of zinc deficiency and could ultimately prove to be responsible for the rapid decline in growth rate and other manifestations of zinc deficiency.

2.7.2 Urea Cycle Enzymes

In order to characterize clinical and biochemical changes in humans as a result of zinc deficiency, we have established an experimental human model, and by dietary means we have produced a mild state of zinc deficiency in human subjects (Prasad *et al.,* 1978). During the course of this study, an elevation of plasma ammonia levels was unexpectedly observed during the zinc-restriction phase. In order to confirm this, we carried out related studies in rats (Rabbani and Prasad, 1978).

The effects of zinc deficiency on the activity of hepatic ornithine carbamoyl transferase (OCT) and plasma ammonia were studied in rats. One group received (ad libitum) a zinc-deficient diet containing 2 ppm zinc and the other group received a diet containing 110 ppm zinc (group pair-fed control) equal to the amount consumed by zinc-deficient rats during the previous 24 h. Rats were killed at weekly intervals. Blood urea nitrogen (BUN), plasma ammonia, and hepatic OCT activity were determined. By the end of the first week on the zinc-deficient diet, the plasma ammonia levels were significantly higher than those of the controls, and remained elevated throughout the study period. BUN increased initially for 2 weeks in the deficient rats, but by the end of the fourth week the levels were lower than the controls. The hepatic OCT activity in deficient animals was significantly lowered compared with the controls by the third week (Figs. 2-6–2-8). It was concluded that an increase in plasma ammonia may occur as a result of zinc deficiency.

In another study, male rats were randomized into zinc-deficient and pair-fed control groups (Cossack and Prasad, 1987). Dietary zinc levels were 1.5 ppm for the zinc-deficient group and 100 ppm for the control group. Twelve rats from each group were sacrificed after 4 weeks of dietary treatment. Zinc was measured in plasma, tibia, liver, muscle, and brain. We assayed the activity of glutamate dehydrogenase (GDH), aspartate aminotransferase (AST), and carbamoyl phosphate synthetase-1 (CPS) in the liver. Glutamine synthetase (Glns) was assayed in the liver, muscle, and brain. Plasma NH_4^+ and BUN were also assayed in experimental groups. Our results showed a significant increase in the activities of GDH and CPS in the zinc-deficient group (Table

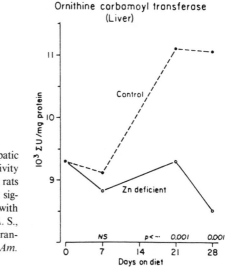

Figure 2-6. Effect of zinc deficiency on hepatic ornithine carbamoyl transferase (OCT) activity of male rats. Liver OCT activity in deficient rats gradually decreases, while in control rats it significantly increases with time. (Reprinted with permission from Rabbani, P., and Prasad, A. S., 1978. Plasma ammonia and liver ornithine transcarbamoylase activity in zinc deficient rats, *Am. J. Physiol.* 235:E203.)

2-7). Activities of AST in liver and Glns in liver, muscle, and brain showed no significant alterations in zinc-deficient animals.

The urea cycle is a major biochemical pathway where ammonia is produced and subsequently removed from the body by the synthesis of urea, and thus the assay of the activities of the urea cycle-related enzymes may provide an insight into the mechanism for hyperammonemia in zinc deficiency.

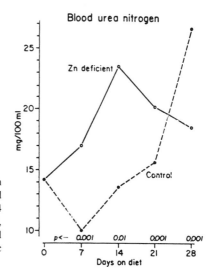

Figure 2-7. Progressive effect of zinc deficiency on BUN level in male rats. A significantly decreased level occurs in deficient animals at the end of 4 weeks. (Reprinted with permission from Rabbani, P., and Prasad, A. S., 1978. Plasma ammonia and liver ornithine transcarbamoylase activity in zinc deficient rats, *Am. J. Physiol.* 235:E203.)

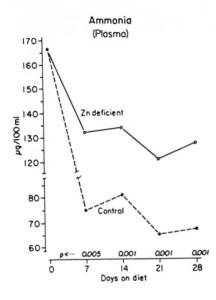

Figure 2-8. Progressive effect of zinc deficiency on plasma ammonia concentration in male rats. Plasma ammonia remained significantly higher in deficient animals throughout the experimental period. (Reprinted with permission from Rabbani, P., and Prasad, A. S., 1978. Plasma ammonia and liver ornithine transcarbamoylase activity in zinc deficient rats, *Am. J. Physiol.* 235: E203.)

GDH catalyzes the conversion of glutamate to α-ketoglutaric acid and ammonia. The present results confirm a previous report by Rahmatullah *et al.* (1980). These authors reported a significant increase in the activity of liver GDH in rats fed 5.6 ppm of zinc, for 15 weeks, as compared with their pair-fed controls fed 100 ppm of zinc. In contrast, Kfoury *et al.* (1968) observed no change in the activity of liver GDH in rats fed 2–4 ppm of zinc compared with rats fed 20–30 ppm. Burch *et al.* (1975) also found that zinc deficiency in the pig did not result in a change in the activity of hepatic GDH. This discrepancy may relate to the difference in the levels of zinc being fed and the length of the experimental period. For instance, in the study of Kfoury *et al.* (1968), pair-fed rats were given 20–30 ppm of zinc for 45 to 180 days while in our study and that of Rahmatullah *et al.* (1980), the pair-fed control group was given 100 ppm of zinc for 4 and 15 weeks, respectively.

Coleman and Foster (1970) reported an absence of zinc in bovine GDH. They indicated that the low, nonstoichiometric content of zinc and the lack of proportionality between zinc content and the activity of GDH suggest that zinc is not an integral part of this enzyme. Furthermore, they found that GDH was inhibited by extrinsic zinc ($K_i = 3 \times 10^{-7}$ M) and under similar conditions binds 1 g-atom of zinc per peptide chain. Thus, the possibility that a dietary level of zinc of 100 ppm may have an inhibitory effect on GDH cannot be ruled out.

We observed no change in the activity of AST caused by zinc deficiency, which is in accord with a previous report (Rahmatullah *et al.*, 1980). This

Table 2-7. Plasma NH_3, BUN, and Activities of GDH, AST, CPS, and GlnS
in Rat Tissues (Mean ± S.D.)[a]

	Zinc deficient	Pair-fed control	p value
Plasma ammonia (μg/100 ml)	120.65 ± 12.99	75.60 ± 15.20	<0.0005
Blood urea nitrogen (mg/100 ml)	22.50 ± 1.20	30.75 ± 3.1	<0.025
Liver			
GDH (units/mg protein)[b]	9.02 ± 1.13	5.56 ± 1.10	<0.005
AST (IU/mg protein)[c]	3.85 ± 0.71	3.29 ± 0.37	N.S.
CPS (nmol/min per protein)[d]	69.54 ± 13.57	23.31 ± 4.63	<0.001
GlnS (units/mg protein)[e]	6.55 ± 1.60	5.90 ± 0.89	N.S.
Brain GlnS (units/mg protein)[e]	9.17 ± 1.76	9.27 ± 2.25	N.S.
Muscle GlnS (units/mg protein)[e]	7.69 ± 1.28	7.80 ± 1.21	N.S.

[a] Adapted with permission from Cossack, Z. T., and Prasad, A. S., 1987. Hyperammonemia in zinc deficiency: Activities of urea cycle related enzymes, *Nutr. Res.* 7:1161.
[b] Glutamate dehydrogenase: One unit will reduce 1.0 μmol of α-ketoglutarate to L-glutamate per min at pH 7.3 at 25°C, in the presence of NH_4^+.
[c] Aspartate aminotransferase: One IU (international unit) of an enzyme will transform 1 μmol of substrate per min.
[d] Carbamoyl phosphate synthetase: nmol of hydroxyurea produced per min per mg protein.
[e] Glutamine synthetase: μmol of product produced per mg protein per min.

enzyme catalyzes the production of aspartate and ammonia from oxaloacetate and glutamic acid.

In the brain, the major mechanism for removal of ammonia is glutamine synthesis from glutamic acid, an ammonia detoxification process which is catalyzed by GlnS (White *et al.*, 1973). Formation of glutamine in the brain must be preceded by synthesis of glutamate in the brain itself because the supply of blood glutamate is inadequate to account for the increased amount of glutamine formed in the brain in the presence of high levels of blood ammonia. GlnS has been found in a number of organs and tissues, and, inasmuch as the activity of this enzyme is high in the liver, brain, and muscle (Iqbal and Ottaway, 1970; Lund, 1970; Meister, 1974), we considered it necessary to assay its activity in these tissues. Our data showed no significant effect of zinc deficiency on the activity of this enzyme in the liver, brain, or muscle.

Interestingly, we observed a threefold increase in the activity of liver CPS in a zinc-deficient as compared with a pair-fed group of rats. Carbamoyl phosphate synthesis is considered to be a key reaction regulating the urea cycle (White *et al.*, 1973). While the equilibrium constant of GDH favors the formation of glutamate rather than that of ammonia, the removal of ammonia of the CPS reaction serves to favor glutamate catabolism and ammonia is produced at a rate which is beyond that of utilization by the CPS reaction. Cheung and Raijman (1980) have demonstrated the stimulatory effect of am-

DEGRADATION OF PURINES

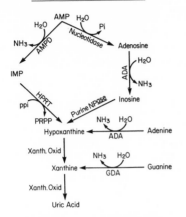

AMP = Adenosine Monophosphate
IMP = Inosine Monophosphate
ADA = Adenosine Deaminase

HPRT = Hypoxanthine Phosphoribosyl Transferase
PRPP = Phosphoribosylpyrophosphate
AMPD = Adenosine Monophosphate Deaminase
GDA = Guanine Deaminase

Figure 2-9. Purine catabolic pathway. (Reprinted with permission from Prasad, A. S., and Rabbani, P., 1981. Nucleoside phosphorylase in zinc deficiency, *Trans. Assoc. Am. Physicians* 94:314.)

monia on CPS activity. They indicated that CPS in the mitochondria is approximately half saturated with acetyl glutamate, and that the most important involvement of acetyl glutamate in the urea synthesis is in bringing about the inactivation of CPS when the supply of ammonia is diminished. Thus, the increased concentration of ammonia, as observed in zinc deficiency, is expected to activate CPS. Furthermore, the decreased activity of OCT, as reported by Rabbani and Prasad (1978), results in the accumulation of ornithine, which has been reported (Raijman and Jones, 1976) to stimulate the activity of CPS (15–20% increase in the activity of CPS).

Thus, our results showed a significant alteration of urea cycle enzymes in zinc-deficient animals. Further studies are needed in order to fully characterize the observed changes in plasma NH_4^+ and BUN as a result of zinc deficiency and also to investigate the role of zinc in short-term and long-term regulation of the urea cycle enzymes.

2.7.3 Purine Catabolic Enzymes

Zinc may play a role in purine metabolism through enzymatic regulation of purine catabolism. Giblett *et al.* (1972) reported a deficiency of the enzyme adenosine deaminase (ADA) in two children with severely impaired B-lymphocyte function. ADA catalyzes the irreversible deamination of adenosine to inosine and deoxyadenosine to deoxyinosine (Fig. 2-9). It is believed that

ADA deficiency leads to increased levels of deoxyadenosine which is toxic to B lymphocytes, although the mechanism of this cytotoxicity is not well understood. Giblett *et al.* (1975) reported on a 5-year-old girl with severely defective T-cell immunity, normal B-cell function, and nucleoside phosphorylase deficiency. Nucleoside phosphorylase catalyzes the conversion of inosine and deoxyinosine to hypoxanthine and guanosine and deoxyguanosine to xanthine. In this case it is believed that accumulation of deoxyguanosine is cytotoxic to T lymphocytes. The above observations have been confirmed by other investigators (Hirschhorn *et al.,* 1979; Meuwissen and Pollara, 1978; Stoop *et al.,* 1977; Cohen *et al.,* 1978; Mitchell and Kelly, 1980).

Several recent observations suggested to us that zinc may play a regulatory role in purine metabolism and T-cell dysfunction may be observed as a result of a deficiency of zinc. Fraker *et al.* (1977, 1978) demonstrated that zinc deficiency in young adult A/J mice primarily impaired T-helper cell function. In the A-46 lethal variant of Fresian cattle, dermatitis, diarrhea, failure of thymic development, and severe immunodeficiency occur, which are cured by zinc administration (Brummerstedt *et al.,* 1971). Acrodermatitis enteropathica, a childhood disease of high morbidity and mortality that is genetically transmitted as an autosomal recessive trait and is associated with characteristic skin disease, bowel and central nervous system malfunctions, and immunodeficiency, is also completely correctable by administration of zinc (Endre *et al.,* 1975; Oleske *et al.,* 1979; Pekarek *et al.,* 1979).

Several investigators, such as Good and Fernandes (1979), Chandra (1980), Fraker *et al.* (1977, 1978), and Frost *et al.* (1977, 1981), have shown that not only T-helper cells, but also T-suppressor and T natural killer cells may be zinc dependent. Because zinc appears to have a specific role in T-cell functions, it occurred to us that one possible mechanism by which zinc may play such a role is to regulate the activity of the enzyme nucleoside phosphorylase.

Brody *et al.* (1977) have demonstrated an increased activity of muscle AMP deaminase, an enzyme of the purine catabolic pathway, and suggested that an increase in plasma ammonia level in zinc deficiency may be the result of an increased deamination of adenosine. Yoshino *et al.* (1978) have reported that AMP deaminase is inhibited by zinc *in vitro;* thus, an increase in the activity of this enzyme may be expected to occur in zinc deficiency. The possibility that zinc deficiency may adversely affect the activity of nucleoside phosphorylase and thus affect T-cell functions and cause a secondary increase in the activity of AMP deaminase, was considered, and therefore we assayed the activities of nucleoside phosphorylase and ADA in zinc-deficient experimental animals (Prasad and Rabbani, 1981).

White male rats of the Holtzman strain were individually housed in stainless steel cages. The animals were fed a standard diet for 5 days, after which they received a semipurified diet containing 2 ppm of zinc. The rats were randomly allotted by weight to two experimental groups.

The zinc-deficient group received (ad libitum) the basal diet with additional 0.2% phytic acid. The pair-fed rats were given an amount of the control diet equal to the average intake of deficient rats during the previous 24-h period.

The pair-fed control rats gained more weight than the deficient rats. The zinc level of the tibias also showed a marked decrease in the deficient group relative to the pair-fed controls.

A significant decrease in the activity of nucleoside phosphorylase was noted as early as the end of 3 weeks in the zinc-deficient animals relative to the pair-fed controls. A significantly lower level of nucleoside phosphorylase activity in the deficient group relative to the pair-fed controls persisted throughout the entire experiment (5 weeks).

The activity of AMP deaminase was significantly higher in the deficient group relative to the pair-fed and ad libitum-fed controls at the end of 6 weeks.

Our results thus demonstrated that the activity of nucleoside phosphorylase in red cells and muscle was adversely affected as a result of zinc deficiency in rats. Because T-cell function is known to be adversely affected in a genetic disorder associated with nucleoside phosphorylase deficiency in human subjects, our results are supportive of the hypothesis that zinc may partially affect T-cell function by its regulatory role on the enzyme nucleoside phosphorylase.

In our experiments, the activity of nucleoside phosphorylase in red cells of the pair-fed control rats declined with age, and in the muscle its activity was less than the ad libitum-fed controls at the end of 5 weeks, suggesting that an increased caloric and food intake in ad libitum-fed controls may have been responsible for the higher activity of this enzyme in this group of animals. Thus, a decreasing activity of nucleoside phosphorylase in the red cells of pair-fed controls with increasing age may have resulted from either a restriction of food intake and/or an effect of age of the animals itself caused by some other mechanism. The difference in the activity of this enzyme between the pair-fed controls and the zinc-deficient rats, however, was the result of zinc intake alone, thus establishing the role of zinc in the regulation of the activity of nucleoside phosphorylase.

We were able to observe a statistically significant increase in the activity of AMP deaminase in the muscle of zinc-deficient rats only at the end of 6 weeks. Because the changes in the enzyme occurred later than those observed for nucleoside phosphorylase, it is possible that the effect was secondary to the primary effect of zinc deficiency on nucleoside phosphorylase.

An increased food intake in ad libitum-fed control rats appeared to increase the muscle AMP deaminase activity relative to the pair-fed controls.

In spite of this, however, zinc-deficient rats whose food intake was much less than the ad libitum-fed controls had higher muscle AMP deaminase activity than both control groups. Since ammonia is released by deamination in this step of the metabolic reaction, it is likely that one probable explanation for hyperammonemia (increased plasma ammonia level) in later stages of zinc deficiency may be the increased activity of AMP deaminase.

We have also observed that the activity of nucleoside phosphorylase in erythrocytes and lymphocytes of zinc-deficient sickle-cell anemia (SCA) and non-SCA subjects is decreased and, with zinc supplementation, the activity is increased to normal levels (Ballester and Prasad, 1983). Thus, zinc may play a role in cell-mediated immune function through its effect on purine catabolic pathway enzymes.

In conclusion, our studies show that nucleoside phosphorylase activity is adversely affected in zinc deficiency, and it is suggested that a regulatory role of zinc on nucleoside phosphorylase activity may at least partially account for its effect on T-lymphocyte functions.

Normal human immune functions decline with age (Boss et al., 1980). The decline is most pronounced with T-cell-dependent functions and correlates with age-related atrophy of the thymus (Kay, 1979; Boyd, 1932). A decline in B-cell functions also occurs but this is later in life and is less marked than the decline in T-cell functions. The activity of lymphocyte ecto 5'-nucleotidase (5'-NT) decreases with advancing age (Fig. 2-10). T-lymphocyte 5'-NT activity begins to fall after the age of 40 and subjects in the 41 to 50, 51 to 60, 61 to 75, and 76 to 85 age ranges have only 57, 52, 38, and 19%, respectively, of the T-lymphocyte 5'-NT activity of subjects under the age of 40 (Boss et al., 1980). B-lymphocyte 5'-NT activity remains stable until age 60 and subjects over 60 have 42% of the B-lymphocyte 5'-NT activity of subjects under 60. These data suggest that lymphocyte 5'-NT activity may be a biochemical marker of immune system function. There is no evidence, however, at the present time that the activity of lymphocyte 5'-NT in any way determines the functions of the immune system. In another study, Pilz et al. (1982) have shown that plasma membrane 5'-NT can exist as an inactive apoenzyme and zinc plays a unique role in the expression of plasma membrane 5'-NT activity.

Because elderly subjects have decreased intake of zinc and also the absorption of zinc is decreased with age (Prasad, 1988), a mild deficiency of zinc in the elderly may account for changes in 5'-NT in the lymphocytes as discussed above.

2.8 Alkaline Phosphatase

In the serum of rats, Kirchgessner et al. (1976) found the activity of alkaline phosphatase to decline 25% after 2 days and 50% after 4 days

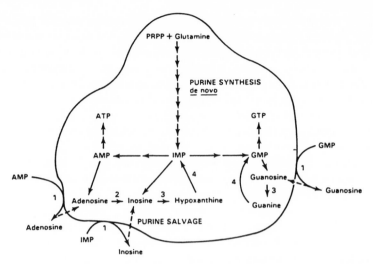

Figure 2-10. Cellular purine metabolism. (1) Ecto 5′-nucleotidase; (2) adenosine deaminase; (3) purine nucleoside phosphorylase; (4) hypoxanthine guanine phosphoribosyltransferase. (Reprinted with permission from Thompson, L. F., 1985. Inosine 5′ monophosphate vs inosine and hypoxanthine as substrates for purine salvage in human lymphoid cells, *Proc. Soc. Exp. Biol. Med.* 179:432.)

of dietary zinc restriction. The decrease in the activity of this enzyme preceded the decreased growth and reduced food intake which one notices commonly as a result of zinc deficiency in rats. The enzyme activity was restored very promptly after zinc supplementation to the deficient animals.

A decreased activity of alkaline phosphatase in the serum and in the granulocytes has also been reported in human subjects with zinc deficiency (Prasad *et al.,* 1978). Historically speaking, the first demonstration that zinc supplementation had a marked enhancing effect on the activity of serum alkaline phosphatase in zinc-deficient human subjects was published in the early 1960s from the Middle East (Prasad *et al.,* 1961, 1963). In an experimental human model in which a mild deficiency of zinc was induced by dietary means, the activity of alkaline phosphatase in serum declined and was subsequently corrected by zinc supplementation (Prasad *et al.,* 1978). Low serum zinc and decreased activity of serum alkaline phosphatase have been reported in patients with acrodermatitis enteropathica and in subjects who received total parenteral nutrition without zinc and who subsequently developed zinc deficiency. After zinc therapy the activity of serum alkaline phosphatase was corrected in both cases.

2.8.1 Carbonic Anhydrase

In zinc-deficient subjects with sickle-cell anemia, the activity of carbonic anhydrase in red blood cells correlated closely with the zinc content (Prasad *et al.*, 1975). Experimental zinc-deficient rats showed a reduction in the activity of carbonic anhydrase expressed as per unit of erythrocytes. This was observed soon after initiating zinc depletion and persisted throughout the stage of extreme zinc deficiency (Roth and Kirchgessner, 1974). Also, in the intestine and stomach of the zinc-deficient rats, the activity of carbonic anhydrase was reduced by 33 and 47%, respectively (Iqbal, 1971).

In another study, no differences in renal carbonic anhydrase activity were found between zinc-deficient rats and pair-fed controls (Ross *et al.*, 1982). In patients with chronic hemodialysis, one group of investigators have reported increased levels of zinc and carbonic anhydrase I and II relative to healthy controls (Goriki *et al.*, 1982). The isoenzymes were determined using the single radial immunodiffusion method. Thus, the changes in carbonic anhydrase activity as related to zinc status either in experimental animals or in humans are not well settled, and clearly more studies are needed.

2.8.2 Carboxypeptidase

Pancreatic carboxypeptidase A and B are important zinc metalloenzymes in protein digestion. The activity of carboxypeptidase A in the pancreas has been shown to decrease remarkably in zinc-deficient rats and pigs (Hsu *et al.*, 1966; Prasad and Oberleas, 1971). The activity of this enzyme decreased about 25% in the pancreas of rats after just 2 days of zinc restriction (Kirchgessner *et al.*, 1976). In repletion studies the activity of carboxypeptidase A returned to nearly normal levels within 3 days after zinc supplementation. Also, in the case of pancreatic carboxypeptidase B, a reduction of its activity by approximately 50% was found in zinc-deficient rats compared with pair-fed and ad libitum-fed controls.

2.8.3 Dehydrogenases

In experimental zinc deficiency in human subjects, the serum lactate dehydrogenase activity decreased as a result of zinc restriction and increased following zinc supplementation (Prasad *et al.*, 1975). A decreased activity of alcohol dehydrogenase was observed in various zinc-deficient tissues of experimental animals (Prasad and Oberleas, 1971).

2.8.4 Effect of Zinc Supplementation on Retinal Reductase in Zinc-Deficient Rats

Retinal reductase (RR) activity was assayed in the retina of zinc-restricted and zinc-repleted rats (Cossack et al., 1982). Male rats were randomized into three experimental groups: zinc restricted (ZD), pair-fed controls (PF), and zinc restricted-repleted (ZDR). Rats from each group were sacrificed after 0, 2, and 4 weeks of zinc restriction and 2, 4, 6, and 8 weeks of repletion. Results indicated a significant difference in the activity of RR (OD/min per mg protein, mean ± S.D.) after 2 weeks of zinc restriction (ZD, 0.052 ± 0.017 versus PF, 0.111 ± 0.021, $p = 0.025$). The activity of RR returned to normal after 6 weeks of repletion (0.096 ± 0.039). A significant positive correlation was shown between tibia zinc concentration (x) and the activity of RR (y): ($r = 0.88$, $p < 0.025$). Although the effect of zinc deficiency in RR was seen as early as 2 weeks, it takes much longer to reverse this effect with zinc supplementation. This observation may explain the slow reversibility of dark adaptation following zinc supplementation to zinc-deficient human subjects (Cossack et al., 1982).

2.8.5 Aspartate Transcarbamylase

The activity of aspartate transcarbamylase (EC 2.1.3.2) was significantly reduced ($p < 0.05$) in regenerating liver of the zinc-deficient rats in comparison with the ad libitum-fed and pair-fed control animals (Duncan, 1984). The enzyme activity was not affected by deficiency of zinc in nonregenerating liver of different groups of animals. In vitro addition of zinc and a number of other divalent metal ions to enzyme extracts of regenerating liver from zinc-deficient and control rats had no significant effect on the activity of this enzyme. Dialysis of enzyme extracts against EDTA also had no effect on enzyme activity. These studies indicate that aspartate transcarbamylase from regenerating rat liver is a zinc-activated enzyme and that zinc associated with the enzyme is probably firmly bound.

Aspartate transcarbamylase is one of the regulatory enzymes involved in the DNA synthetic pathway and has been shown to be a zinc metalloenzyme in E. coli. This enzyme is essential for the de novo synthesis of pyrimidine nucleotides. The above findings indicate that aspartate transcarbamylase from regenerating liver requires adequate zinc nutrition for optimal activity. The exact role of zinc in this enzyme, however, remains unknown.

2.8.6 Angiotensin-Converting Enzyme

The enzyme angiotensin I-converting enzyme or kinase II (EC 3.4.14.1) is a glycoprotein similar to pancreatic carboxypeptidases and is found in plasma and bound to the endothelium of lung, kidney, and other organs. Ionic zinc forms a tetravalent coordination complex within the enzyme. Amino acid residues of the enzyme occupy three of the positions and the remaining site binds to the carbonyl oxygen of a peptide bond of the substrates, thereby polarizing it (Lockett *et al.,* 1983).

Angiotensin I-converting enzyme catalyzes the transformation of the decapeptide angiotensin I of the vasoactive octapeptide angiotensin II and the dipeptide residue, histidylleucine. As a kinase II, the enzyme catalyzes the hydrolysis of the dipeptide phenylalanine-arginine. Since the enzyme cleaves the peptidyl-dipeptide bonds in these reactions, it is also referred to as a peptidyl-dipeptidehydrolase or peptidyl-dipeptidase.

Antihypertensive diuretics may decrease the activity of the angiotensin-converting enzyme by increasing urinary zinc excretion. Captopril normally inhibits the angiotensin-converting enzyme, but also affects electrolyte excretion; thus, its use as an antihypertensive agent is clinically important.

References

Ballester, O. F., and Prasad, A. S., 1983. Anergy, zinc deficiency, and decreased nucleoside phosphorylase activity in patients with sickle cell anemia, *Ann. Intern. Med.* 98:180.

Boss, G. R., Thompson, L. F., Spiegelberg, H. L., Pichler, W. J., and Seegmiller, J. E., 1980. Age dependency of lymphocyte ecto 5′ nucleotidase activity, *J. Immunol.* 125:679.

Boyd, E., 1932. The weight of the thymus gland in health and disease, *Am. J. Dis. Child.* 43:1162.

Brody, M. S., Steinberg, J. R., Svinger, B. A., and Luecke, R. E., 1977. Increased purine nucleotide cycle activity associated with dietary zinc deficiency, *Biochem. Biophys. Res. Commun.* 78:144.

Brown, R. S., Huguet, J., and Curtiss, N. S., 1983. Models for Zn (II)-binding sites in enzymes, in *Metal Ions in Biological Systems* (H. Sigel, ed.), Dekker, New York, p. 55.

Brummerstedt, E., Flagstad, T., Basse, A., and Andersen, E., 1971. The effect of zinc on calves with hereditary thymus hypoplasia (lethal trait A-46), *Acta Pathol. Microbiol. Scand. Sect. A* 79:686.

Burch, R., Williams, R., Hahn, H., Jetton, M., and Sullivan, J., 1975. Serum and tissue enzyme activity and trace element content in response to zinc deficiency in the pig, *Clin. Chem.* 21: Am. Assn. Clin. Chemists, Washington, D.C., 568.

Chandra, R. K., 1980. Single nutrient-deficiency and cell-mediated immune response. I. Zinc, *Am. J. Clin. Nutr.* 33:736.

Cheung, C. W., and Raijman, L., 1980. The regulation of carbamoyl phosphate synthetase (ammonia) in rat liver mitochondria, *J. Biol. Chem.* 255:5051.

Cohen, A., Gudas, L. J., Ammann, A. J., Staal, G. E. J., and Martin, D. W., Jr., 1978. Deoxyguanosine triphosphate as a possible toxic metabolite in the immune deficiency associated with purine nucleoside phosphorylase deficiency, *J. Clin. Invest.* 61:1405.

Coleman, R. F., and Foster, D. S., 1970. The absence of zinc in bovine liver glutamate dehydrogenase, *J. Biol. Chem.* 245:6190.

Cossack, Z. T., and Prasad, A. S., 1987. Hyperammonemia in zinc deficiency: Activities of urea cycle related enzymes, *Nutr. Res.* 7:1161.

Cossack, Z. T., Prasad, A. S., and Koniuch, D., 1982. Effect of zinc supplementation on retinal reductase in zinc deficient rats, *Nutr. Rep. Int.* 26:841.

Dreosti, I. E., Buckley, R. A., and Record, I. R., 1986. The teratogenic effect of zinc deficiency and accompanying feeding patterns in mice, *Nutr. Res.* 6:159.

Duncan, J. R., 1984. Aspartate transcarbamylase from regenerating rat liver—A zinc activated enzyme, *Nutr. Res.* 4:93.

Einarsdottir, O., and Caughey, W. S., 1984. Zinc is a constituent of bovine heart cytochrome *c* oxidase preparations, *Biochem. Biophys. Res. Commun.* 124:836.

Endre, L., Katona, Z., and Gycitkovits, K., 1975. Zinc deficiency and cellular immune deficiency in acrodermatitis enteropathica, *Lancet* 1:1196.

Fernandez-Madrid, F., Prasad, A. S., and Oberleas, D., 1973. Effect of zinc deficiency on nucleic acids, collagen, and non-collagenous protein of the connective tissue, *J. Lab. Clin. Med.* 82: 951.

Finelli, V. N., Klauder, D. S., Karaffz, M. A., and Petering, H. G., 1975. Interaction of zinc and lead on δ-aminolevulinate dehydrase, *Biochem. Biophys. Res. Commun.* 65:303.

Fraker, P. J., Haas, S. M., and Leucke, R. W., 1977. Effect of zinc deficiency on the immune response of the young adult A/J mouse, *J. Nutr.* 107:1889.

Fraker, P. J., DePasquale-Jardieu, P., Zwickl, C. M., and Luecke, R. W., 1978. Regeneration of T-cell helper function in zinc deficient adult mice, *Proc. Natl. Acad. Sci. USA* 75:5660.

Frost, P., Rabbani, P., Smith, J., and Prasad, A. S., 1977. The effect of zinc deficiency on the immune response, in *Zinc Metabolism: Current Aspects in Health and Disease* (G. J. Brewer and A. S. Prasad, eds.), Liss, New York, p. 143.

Frost, P., Rabbani, P., Smith, J., and Prasad, A. S., 1981. Cell mediated cytotoxicity and tumor growth in zinc deficient mice, *Proc. Soc. Exp. Biol. Med.* 167:333.

Galdes, A., and Vallee, B. L., 1983. Categories of zinc metalloenzymes, in *Metal Ions in Biological Systems* (H. Sigel, ed.), Dekker, New York, p. 1.

Gibbs, P. N. B., Gore, M. G., and Jorand, P. M., 1985. Investigation of the effect of metal ions on the reactivity of thiol groups in human 5-aminolevulinate dehydratate, *Biochem. J.* 255: 573.

Giblett, E. R., Anderson, J. E., Cohen, F., Pollara, B., and Meuwissen, H. J., 1972. Adenosine deaminase deficiency in two patients with severely impaired cellular immunity, *Lancet* 2: 1067.

Giblett, E. R., Ammann, A. J., Wara, D. W., Sandman, R., and Diamond, L. K., 1975. Nucleoside phosphorylase deficiency in a child with severely defective T-cell immunity and normal B-cell immunity, *Lancet* 1:1010.

Good, R. A., and Fernandes, G., 1979. Nutrition, immunity, and cancer—A review, *Clin. Bull.* 9:3.

Goriki, K., Wada, K., Hata, J., Kobayashi, M., Hirabayashi, A., Shigemoto, K., Hamaguchi, N., Yorioka, N., and Yamakido, M., 1982. The relationship between carbonic anhydrases and zinc concentration of erythrocytes in patients under chronic hemodialysis, *Hiroshima J. Med. Sci.* 2:31.

Hirschorn, R., Vawter, G. F., Kirkpatrick, J. A., and Rosen, F. S., 1979. Adenosine deaminase deficiency: Frequency and comparative pathology in autosomally recessive severe combined immunodeficiency, *Clin. Immunol. Immunopathol.* 14:107.

Hsu, J. M., Anilane, J. K., and Scanlan, D. E., 1966. Pancreatic carboxypeptidase: Activities in zinc deficient rats, *Science* 153:882.

Iqbal, M., 1971. Activity of alkaline phosphatase and carbonic anhydrase in male and female zinc deficient rats, *Enzyme* 12:33.

Iqbal, M., and Ottaway, J. H., 1970. Glutamine synthetase in muscle and kidney, *Biochem. J.* 119:145.

Kay, M. M. B., 1979. An overview of immune aging, *Mech. Ageing Dev.* 9:39.

Keilin, D., and Mann, T., 1940. Carbonic anhydrase. Purification and nature of the enzyme, *Biochem. J.* 34:1163.

Kelly, R. E., Mally, M. J., and Evans, D. R., 1986. The dihydrooxatase domain of the multifunction protein GAD, *J. Biol. Chem.* 261:6073.

Kfoury, G., Reinhold, J., and Simonian, S. J., 1968. Enzyme activities in tissues of zinc-deficient rats, *J. Nutr.* 95:102.

Kirchgessner, M., Roth, H. P., and Weigand, E., 1976. Biochemical changes in zinc deficiency, in *Trace Elements in Human Health and Disease,* Volume I (A. S. Prasad, ed.), Academic Press, New York, p. 189.

Lieberman, I., Abrams, R., Hunt, N., and Ove, P., 1963. Levels of enzyme activity and deoxyribonucleic acid synthesis in mammalian cells cultured from the animal, *J. Biol. Chem.* **238:** 3955.

Lockett, C. J., Reyes, A. J., Leary, W. P., Alcocer, L., and Olhaberry, J. V., 1983. Zinc, angiotensin I-converting enzyme and hypertension, *S. Afr. Med. J.* 64:1022.

Lund, P. A., 1970. A radiochemical assay for glutamine synthetase and activity of the enzyme in rat tissues, *Biochem. J.* 118:35.

Meister, A., 1974. Glutamine synthesis, in *The Enzymes* Boyer P. D. ed.), Academic Press, New York, p. 443.

Meuwissen, H. J., and Pollara, B., 1978. Combined immunodeficiency and inborn errors of purine metabolism, *Blut* 37:173.

Mitchell, B. S., and Kelly, W. N., 1980. Purinogenic immunodeficiency disease: Clinical features and molecular mechanisms, *Ann. Intern. Med.* 92:826.

Ohtake, Y., Uchida, K., and Sakai, T., 1963. Purification and properties of ribonuclease for yeast, *J. Biochem.* 54:322.

Oleske, J. M., Westphal, M. L., Shore, S., Gorden, D., Bogden, J. D., and Nahmias, A., 1979. Zinc therapy of depressed cellular immunity in acrodermatitis enteropathica, *Am. J. Dis. Child.* 133:915.

Pekarek, R. S., Sandstead, H. H., Jacob, R. A., and Barcome, D. F., 1979. Abnormal cellular immune responses during acquired zinc deficiency, *Am. J. Clin. Nutr.* 32:1466.

Pilz, R. B., Willis, R. C., and Seegmiller, J. E., 1982. Regulation of human lymphoblast plasma membrane 5′ nucleotidase by zinc, *J. Biol. Chem.* 257:13544.

Prasad, A. S., 1982. Clinical and biochemical spectrum of zinc deficiency in human subjects, in *Clinical, Biochemical and Nutritional Aspects of Trace Elements* (A. S. Prasad, ed.), Liss, New York, p. 4.

Prasad, A. S., 1988. Clinical spectrum and diagnostic aspects of human zinc deficiency, in *Essential and Toxic Trace Elements in Human Health and Disease* (A. S. Prasad, ed.), Liss, New York, p. 3.

Prasad, A. S., and Oberleas, D., 1971. Changes in activities of zinc-dependent enzymes in zinc deficient tissues of rats, *J. Appl. Physiol.* 31(6):842.

Prasad, A. S., and Oberleas, D., 1973. Ribonuclease and deoxyribonuclease activities in zinc-deficient tissues, *J. Lab. Clin. Med.* 82:461.

Prasad, A. S., and Oberleas, D., 1974. Thymidine kinase activity and incorporation of thymidine into DNA in zinc-deficient tissue, *J. Lab. Clin. Med.* 83:634.

Prasad, A. S., and Rabbani, P., 1981. Nucleoside phosphorylase in zinc deficiency, *Trans. Assoc. Am. Physicians* 94:314.

Prasad, A. S., Halsted, J. A., and Nadimi, M., 1961. Syndrome of iron deficiency anemia, hepatosplenomegaly, hypogonadism, dwarfism, and geophagia, *Am. J. Med.* 31:532.

Prasad, A. S., Miale, A., Farid, Z., Schulert, A., and Sandstead, H. H., 1963. Zinc metabolism in patients with the syndrome of iron deficiency anemia, hypogonadism and dwarfism, *J. Lab. Clin. Med.* 61:537.

Prasad, A. S., Oberleas, D., Wolf, P., and Horwitz, J. P., 1967. Studies on zinc deficiency: Changes in trace elements and enzyme activities in tissues of zinc deficient rats, *J. Clin. Invest.* 46: 549.

Prasad, A. S., Oberleas, D., Wolf, P., Horwitz, J. P., Miller, E. R., and Luecke, R. W., 1969. Changes in trace elements and enzyme activities in tissues of zinc deficient pigs, *Am. J. Clin. Nutr.* 22:628.

Prasad, A. S., Oberleas, D., Miller, E. R., and Luecke, R. W., 1971. Biochemical effects of zinc deficiency: Changes in activities of zinc-dependent enzymes and ribonucleic acid and deoxyribonucleic acid content of tissues, *J. Lab. Clin. Med.* 77:144.

Prasad, A. S., Schoomaker, E. B., Ortega, J., Brewer, G. J., Oberleas, D., and Oelschlegel, F. J., 1975. Zinc deficiency and sickle cell disease, *Clin. Chem.* Washington, D.C. 21: 582.

Prasad, A. S., Rabbani, P., Abbasi, A., Bowersox, E., and Fox, M. R. S., 1978. Experimental zinc deficiency in humans, *Ann. Intern. Med.* 89:483.

Prasad, A. S., Fernandez-Madrid, F., and Ryan, J. F., 1979. Deoxythymidine kinase activity of human implanted sponge connective tissue in zinc deficiency, *Am. J. Physiol.* 236: E272.

Rabbani, P., and Prasad, A. S., 1978. Plasma ammonia and liver ornithine transcarbamoylase activity in zinc deficient rats, *Am. J. Physiol.* 235:E203.

Rahmatullah, M., Louise, Y., Fong, L., and Boyde, T., 1980. Zinc-deficiency and activities of urea cycle-related enzymes in rats, *Experientia* 36:1281.

Raijman, L., and Jones, M. E., 1976. Purification, composition and some properties of rat liver carbamoyl phosphate synthetase (ammonia), *Arch. Biochem. Biophys.* 175:270.

Ross, P. K., Noordewier, B., Hook, J. B., and Bond, J. T., 1982. Zinc deficiency and the kidney, *Miner. Electrolyte Metab.* 7:257.

Roth, H. P., and Kirchgessner, M., 1974. Zur Aktivital der Blut-Carboanhydrase bei Zn-Mangel waschsender Ratten, *Z. Tierphysiol. Tierernaehr. Futtermittelkd.* 32:296.

Somers, M., and Underwood, E. J., 1969. Ribonuclease activity of nucleic acid and protein metabolism in the testes of zinc deficient rats, *Aust. J. Biol. Sci.* 22:1277.

Speckhard, D. C., Wu, F. Y. H., and Wu, C. W., 1977. Role of the intrinsic metal in RNA polymerase from Escherichia coli. In vivo substitution of tightly bound zinc with cobalt, *Biochemistry* 16:5228.

Stoop, J. W., Zegers, B. J. M., Hendrickx, G. F. M., Van Heukelom, L. H. S., Staal, G. E. J., de Bree, P. K., Wadman, S. K., and Ballieux, R. E., 1977. Purine nucleoside phosphorylase deficiency associated with selective cellular immunodeficiency, *N. Engl. J. Med.* 296: 651.

Terhune, M. W., and Sandstead, H. H., 1972. Decreased RNA polymerase activity in mammalian zinc deficiency, *Science* 177:68.

Tsukamoto, I., Yoshinaga, T., and Sanos, S., 1980. Zinc and cysteine residues in the active site of bovine liver delta-aminolevulinic acid dehydratase, *Int. J. Biochem.* 12:751.

Vallee, B. L., 1959. Biochemistry, physiology and pathology of zinc, *Physiol. Rev.* 39:443.

Vallee, B. L., and Williams, R. J. P., 1968. Metalloenzymes: The entatic nature of their active sites, *Proc. Natl. Acad. Sci. USA* 59:498.

Washabaugh, M. W., and Collins, K. D., 1986. Dihydrooratase from Escherichia coli. Sulphydroxyl group–metal interactions, *J. Biol. Chem.* 261:5920.

White, A., Handler, P., and Smith, E., 1973. Metabolism of ammonia, in *Principles of Biochemistry*, 5th ed., McGraw–Hill, New York, p. 645.

Wu, F. Y., and Wu, C., 1983. The role of zinc in DNA and RNA polymerase, in *Metal Ions in Biological Systems* (H. Sigel, ed.), Dekker, New York, p. 157.

Yoshino, M., Murakami, K., and Tusushima, K., 1978. Inhibition of AMP deaminase by zinc ions, *Biochem. Pharmacol.* 27:2651.

Zinc and Gene Expression ③

Recent studies show that zinc has a very important role in gene expression (Chesters, 1982; Falchuk, 1988; Vallee, 1983; Miller *et al.,* 1985). A role of zinc in growth and development and the teratological abnormalities of zinc deficiency in fetal development have been known for many years. However, its role in cell differentiation and gene expression has been appreciated only recently.

Earlier studies in *Euglena gracilis* showed that zinc deficiency affected growth, morphology, cell cycle, and mitosis and it was postulated that zinc had a role in gene expression, possibly through its effects on zinc-dependent enzymes (Falchuk *et al.,* 1975a,b, 1976, 1978, 1986; Falchuk, 1988; Stankiewicz *et al.,* 1983). It was shown that in zinc-deficient media, the DNA content of zinc-deficient cells was twice normal. Growth was arrested, and cells did not divide but remained viable.

Analysis of the DNA content of intact cells by laser-induced cytofluorometry permits dynamic studies of the cell cycle in the synchronously dividing eukaryote *E. gracilis* (Falchuk *et al.,* 1975b). The DNA content of zinc-deficient cells was characteristic of a population of cells blocked in S/G_2 phase with a small fraction in G_1 (see Fig. 3-1). Cells synchronized in G_1, and placed in zinc-deficient media did not progress into S phase. These studies show that zinc is essential for the biochemical events of the premitotic state which include initiation of DNA synthesis, and progression from G_2 to mitosis.

In a recent study, T lymphocytes from normal human controls and zinc-deficient sickle-cell disease (SCD) patients were isolated from peripheral blood and cultured for 72 h following addition of phytohemagglutinin (PHA). The ratio of the fraction of cells in DNA synthesis (S phase) over the fraction in G_2 phase (S/G_2) was significantly higher in SCD patients than controls (mean $\pm$ S.D.) (4.01 ± 0.78 versus 2.78 ± 0.76, $p < 0.02$) (see Fig. 3-2). Following *in vivo* zinc supplementation to two SCD patients, the S/G_2 ratio was normalized. This study showed that the distribution of T lymphocytes

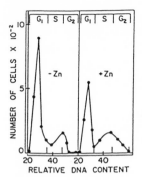

Figure 3-1. Comparison of the histograms of DNA content of *E. gracilis* incubated in (−Zn) medium prior to and following addition of zinc. Following cessation of cell division, the majority of the (−Zn) cells are in G_1 with a small fraction in S. On addition of zinc, the number of cells blocked in G_1 decreases and a histogram typical of dividing log-phase cells results. (Reprinted with permission from Falchuk, K. H., Krishnan, A., and Vallee, B. L., 1975. DNA distribution in the cell cycle of Euglena gracilis, Cytofluorimetry of zinc deficient cells, *Biochemistry* 14:3449.)

in the cell cycle was altered in zinc-deficient SCD patients and that this effect was zinc-dependent (Abdallah *et al.*, 1988).

In *E. gracilis* as a result of zinc deficiency, decreased rate of incorporation of [^{3}H]uridine into RNA and [^{3}H]leucine into proteins was observed, and nucleotides, peptides, and amino acids accumulated in the cells. The composition of mRNA was altered, and the stability of the ribosomes which rRNA generated was markedly decreased in zinc-deficient cells. Zinc-deficient cells generated considerable amounts of arginine- and asparagine-rich polypeptides relative to the zinc-sufficient control cells. The zinc-deficient cells contained only one unusual type of RNA polymerase ("X") as opposed to RNA polymerases I, II, and III of zinc-sufficient cells.

Although the DNA content of zinc-deficient cells (*E. gracilis*) doubled, the total RNA of these cells was unchanged. The amount of rRNA was slightly lower, but mRNA was almost twice that of the zinc-sufficient cells. The purine and pyrimidine content of rRNA from both types of cells was identical. The guanine content of tRNA decreased from 34% to 24%, and cytosine increased

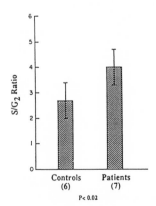

Figure 3-2. S/G_2 cell ratio in T lymphocytes following PHA stimulation in SCD and control subjects. (Reprinted with permission from Abdallah, J. M., Kukuraga, M., Nakeff, A., and Prasad, A. S., 1988. Cell cycle distribution defect in PHA-stimulated T lymphocytes of sickle cell disease patients, *Am. J. Hematol.* **28**:279.)

only slightly from 27% to 28% in the zinc-deficient cells. No change in adenine and uracil content was observed as a result of zinc deficiency. The mRNA of zinc-deficient cells contained several minor unidentified additional bases not found in normal cells besides uracil, guanine, cytosine, and adenine. These data showed that there were changes in the composition of mRNA; however, the functional capacity as expressed by translation was not affected in zinc-deficient cells.

The normal zinc-sufficient cells produced a number of polypeptides of molecular weight > 10,000 in contrast to zinc-deficient cells which did not synthesize higher-molecular-weight polypeptides. Also, the zinc-deficient cells synthesized arginine-rich polypeptides characteristically some of which inhibited the activity of *E. gracilis* RNA polymerase II.

In view of the fact that the usual three RNA polymerases (I, II, and III) were not synthesized, and instead an unusual "X" form of RNA polymerase was formed in zinc-deficient cells, one may consider that, most probably, zinc has a selective effect on gene expression. Both RNA polymerase, and histones which bind to DNA and facilitate its interaction with RNA polymerases are known to play important roles in gene expression.

Zinc is known to be present in the nucleus and is involved in diverse aspects of the metabolism of chromatin proteins, RNA polymerases, and genomic DNA (Vallee, 1983). It stabilizes RNA and is essential for the catalytic activity of RNA polymerases. It is also essential for the function of at least two chromatin proteins: transcription of TFIIIA and replication of g32P (Falchuk, 1988).

In all eukaryotes chromatin is a complex structure which is comprised of DNA, RNA transcripts, DNA and RNA polymerases, gene activators and repressors, histones, and nonhistone proteins. The basic unit of DNA is the nucleosome which is comprised of DNA segments of approximately 150–250 base pairs (bp) associated with five major histones (Kornberg, 1977). According to Falchuk (1988), the organization of the chromatin of zinc-deficient *E. gracilis* cells differs significantly from that of zinc-sufficient cells. The chromatin in zinc-deficient cells is arranged more compactly which reduces the digesting capability of nucleases (Fig. 3-3 and Table 3-1).

Figure 3-3. Comparison of acid-soluble material resulting from micrococcal nuclease (50 or 500 units) digestions of (+Zn) chromatins (400 μg DNA). The digestion of (+Zn) chromatin with 500 units of nuclease was nearly complete within 10 min, and hence, the material is solubilized totally. A_{260} supernatants of these digests. [Reprinted with permission from Falchuk, K. H., 1988. Zinc deficiency and the E. gracilis chromatin, in *Essential and Toxic Trace Elements in Human Health and Disease* (A. S. Prasad, ed.), Liss, New York, p. 75.]

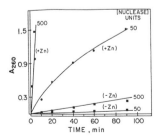

Table 3-1. Size of DNA Fragments of (+Zn)
and (−Zn) Chromatin Digested with
Micrococcal Nuclease[a]

Nuclease (units)	+Zn (bp)	−Zn (bp)
50	59 ± 14	
	115 ± 9	
	213 ± 12	
	287 ± 10	
	468 ± 24	
	536 ± 33	>2000
	656 ± 23	
	831 ± 2	
	956 ± 50	
500		53 ± 8
		182 ± 17
		260 ± 13
		454 ± 16
	<50	559 ± 27
		690 ± 19
		814 ± 1
		1001 ± 46

[a] Adapted with permission from Falchuk, K. H., 1988. Zinc deficiency
and the E. gracilis chromatin, in *Essential and Toxic Trace Elements
in Human Health and Disease* (A. S. Prasad, ed.), Liss, New York,
p. 75.

The ordered structure of chromatin is presumed to be maintained by the
binding of histones H2A, H2B, H3, and H4 to DNA which results in a repeating globular complex (a nucleosome), and by H1 binding to their intervening DNA segments between these complexes (Fig. 3-4). Extensive phos-
phorylation, acetylation, or methylation of histones, and/or their interaction
with other nuclear proteins, is presumably important for conversion of the
ordered, inactive chromatin structure into one that replicates and transcribes
actively. The histones from zinc-deficient cells migrated differently in an electrical field than did zinc-sufficient cells and a fraction migrating in the most
cathodic region of the zinc-deficient histones (which was absent in the zinc-
sufficient cells) comprised the arginine-rich peptide fraction. These changes
might alter the binding of histones to DNA resulting in either activation or
repression of specific genes.

DNA binding proteins which induce the ordered structure of chromatin
and impart its resistance to nuclease digestion are characteristically acid soluble
(Table 3-1). The solubilized basic proteins from zinc-sufficient cells separate
on polyacrylamide gel electrophoresis into bands ranging from 3000 to greater

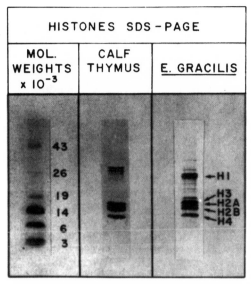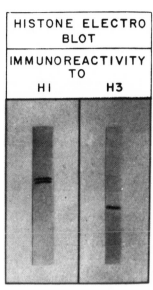

Figure 3-4. Electrophoretic and immunologic properties of (+Zn) *E. gracilis* histones. Their molecular weights are 20,500, 15,300, 14,200, 13,200, and 11,000, respectively, similar to those of calf thymus histones, and have been assigned as H1, H3, H2A, H2B, and H4, respectively. [Reprinted with permission from Falchuk, K. H., 1988. Zinc deficiency and the E. gracilis chromatin, in *Essential and Toxic Trace Elements in Human Health and Disease* (A. S. Prasad, ed.), Liss, New York, p. 75.]

than 43,000 D. The electrophoretic patterns of proteins corresponding to log- and stationary-phase zinc-sufficient cells are nearly identical. The chromatin protein patterns of zinc-deficient cells extracted from log phase differ from the patterns of zinc-sufficient cells in that many bands are absent in the deficient cells and in the extract of cells from the stationary phase. Only one fraction migrating at the most cathodic region of the gel with an apparent molecular mass of 3000 Da is seen (Falchuk, 1988; Czupryn *et al.*, 1987). This chromatin protein band is not present in zinc-sufficient cells. It has been suggested that this basic polypeptide in zinc-deficient cells may be partly responsible for resistance to nuclease digestion. It appears that the inhibitory effect of the 3-kDa polypeptide on the template properties of DNA and the decreased number of RNA polymerase classes in zinc-deficient cells may participate in the repression of overall transcription of zinc-deficient cells.

3.1 "Zinc Fingers": A Novel Protein Motif for Gene Expression

An essential part of gene expression and regulation is the binding of a regulatory protein to the recognition sequence of the appropriate gene. Many such proteins have in their structures a domain (or motif) which binds to DNA. Miller *et al.* (1985) reported that the *Xenopus* transcription factor IIIA (TFIIIA) contains small sequence units repeated in tandem, and they proposed that each unit is folded about a zinc atom to form separate structural domains (Fig. 3-5). Similar units have now been found in the amino acid sequence of other transcription factors and, more generally, in nucleic acid binding proteins. Thus, a second and apparently more commonly used structure motif for DNA recognition has emerged, conveniently called the "zinc finger."

3.2 Repetitive Zn-Binding Domains in TFIIIA

Xenopus TFIIIA is required for correct initiation of transcription of *Xenopus* 5 S RNA genes by RNA polymerase III (Klug and Rhodes, 1987). This 40-kDa protein binds to a 50-bp region located within the coding sequence of 5 S RNA genes (the internal control region). In the ovaries of immature frogs, large quantities of TFIIIA are stored as a 7 S complex with its own gene product 5 S RNA. TFIIIA binds to both DNA and RNA, thus presenting an intriguing structural problem as to how a small protein interacts with a long tract of DNA.

Preparations of TFIIIA 7 S complex, purified to homogeneity, contained 7–11 atoms of zinc per mol (Fig. 3-5). This result was consistent with the fact that this protein contained a large number of histidine and cysteine residues, the most common ligands for zinc in enzymes and other proteins. This observation also explained the finding that zinc is necessary for transcription of 5 S RNA genes. Later it was suggested that the 30-kDa domain of TFIIIA may contain a periodic arrangement of small, compact domains of 3 kDa. If each of the domains contained one zinc atom, then the observed zinc content would be accounted for.

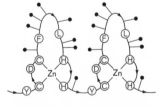

Figure 3-5. Schematic folding model for a linear arrangement of repeated domains. Each is centered on a tetrahedral arrangement of Zn ligands. Circled residues are the conserved amino acids which include the Cys and His Zn ligands, the negatively charged Asp-11, and the three hydrophobic groups that may form a structural core. Solid circles mark the most probable DNA-binding side chains. (Reprinted with permission from Miller, J., McLachlan, A. D., and Klug, A., 1985. Repetitive zinc-binding domains in the protein transcription factor IIIA from xenopus oocytes, *EMBO J.* 4:1609.)

It has been proposed that most of the TFIIIA protein has a repeating structure in which each of the nine 30-amino-acid units folds around a zinc atom to form a small independent structural domain. Figure 3-5 shows a schematic representation of the proposed folding of a TFIIIA domain in which most of the 30 amino acid residues are in the loop formed around the central zinc atom, and a few amino acids provide the linkers between the consecutive fingers. The zinc atoms form the basis of the folding by being tetrahedrally coordinated to the two invariant pairs of cysteines and histidines. Each repeat also contains besides this unique conserved pattern of Cys-Cys . . . His-His, several other conserved amino acids, namely Tyr-6 (or Phe-6), Phe-17, and Leu-23, all of which are hydrophobic. The largest number of basic and polar residues are present in the region between the second cysteine and the first histidine which is known to be important for binding to DNA.

Table 3-2 is a partial list of proteins that contain amino acid units bearing a high degree of sequence similarity to the repeating unit of TFIIIA, which might therefore be expected to form zinc binding domains. Recently, other subclasses of putative zinc binding domains have also started to emerge. Instead of the classical arrangement of Cys-Cys . . . His-His of TFIIIA with a constant spacing of 12 or 13 amino acids between the inner ligands, a sequence unit having the same spacing but containing Cys-Cys . . . Cys-Cys is found. Such sequence units are seen in the first half of the two putative zinc finger domains of several of the hormone receptor proteins, yeast regulatory protein GAL4, and PPRI. A unique feature of these Cys-Cys . . . Cys-Cys domains is that they do not contain conserved hydrophobic amino acids, but instead contain acidic residues at invariant positions. It is possible that these amino acids form salt bridges with some of the basic residues to substitute for the hydrophobic cluster found in the TFIIIA-type domains. A third subclass of nucleic acid binding proteins shows combinations of His and Cys as seen in the *gag* gene of retroviruses and T4 phase gene 32 protein. The number of residues between the inner ligands is small (4 to 5 amino acids), but could coordinate zinc to form "stubbier" fingers (Table 3-2). It is possible that these shorter-sequence units are not interacting directly with nucleic acids but are being utilized for stabilization.

3.3 Zinc: A Necessary Cofactor for Domain Binding to DNA

Gene deletion mutant studies of the hormone receptor proteins from human, chicken, and the yeast transcriptional activators GAL4, ADRI, and SW15 show that the region containing the zinc-binding domains is essential for binding to DNA (Green and Chambon, 1987; Giguere *et al.,* 1986; Keegan *et al.,* 1986; Blumberg *et al.,* 1987). Frankel *et al.* (1987) have expressed a

Table 3-2. A Partial List of Proteins (or DNA Sequences) That Contain Sequences Homologous to the Zn Finger Motif of TFIIIA: Grouping Is by Classes According to the Type of Proved, or Potential Ligands to the Zn Atom[a]

	Name of sequence or protein	Type	Evidence for Zn- or DNA-binding
Xenopus	TFIIIA	CC–HH	Zn. DNA/RNA
	Xfin	CC–HH (one CC–HC)	
Drosophila	Serendipity	CC–HH	
	Krüppel	CC–HH	
	Hunchback	CC–HH	
Mouse	mk1, mk2	CC–HH	
Mammalian cells (SV40)	Sp1	CC–HH	DNA
Yeast	ADR1	CC–HH	DNA
	SW15	CC–HH (one CC–HC)	Zn. DNA
	GAL4	CC–CC	Zn, DNA
	PPR1	CC–CC	
	ARGRII	CC–CC	
Human	Estrogen receptor	CC–CC	Zn, DNA
	Glucorticoid receptor	CC–CC	DNA
	C-erb A (thyroid hormone receptor)	CC–CC	
Rat	Glucocorticoid receptor	CC–CC	DNA
Chicken	Estrogen receptor	CC–CC	
	Progesterone receptor	CC–CC	
E. coli	UvrA	CC–CC	Zn, DNA
Retroviruses	Nucleic acid binding proteins	CC–HC	RNA
E. coli	Gene 32 protein	CH–CC	Zn, ss DNA, ss RNA

[a] Reprinted with permission from Klug, A., and Rhodes, D., 1987. "Zinc fingers": A novel protein motif for nucleic acid recognition, *Trends Biochem. Sci.* 12:461.

single finger of TFIIIA in *E. coli* and shown that zinc is necessary for folding of the synthetic amino acid chain but have not demonstrated specific binding to DNA. Klug and Rhodes (1987) have been studying transcription factor SW15 from yeast which contains three tandem units, each of 30 amino acids, homologous to the TFIIIA repeat, a region implicated in DNA binding. This unit has been expressed in *E. coli*, purified in an unfolded state, refolded in the presence of zinc, and shown to bind specifically to a region about 15 bp of DNA. The DNA binding domain of the protein alone is not sufficient for the complete biological function and other parts of the protein, the so-called activation domain, are needed for making essential contact with other components of the transcriptional machinery.

The role of zinc in the DNA-binding finger appears to be structural. A property peculiar to zinc, and not copper and iron, is the absence of redox chemistry. As such, it cannot be reduced in the reducing atmosphere inside the cell. Thus, zinc might be used in situations where the presence of redox reaction would lead to damaging radicals that would hydrolyze RNA or DNA and possibly even the protein chain. In experiments in which various metals including Co or Cd were tested for their ability to restore the partially dissociated 7 S complex, only zinc was able to do so. Detailed understanding of the tertiary structure of a zinc finger is not available and must await crystallographic or NMR studies.

The zinc finger module can be used singly without reference to DNA symmetry and can be repeated in tandem to recognize DNA (or RNA) sequences of different lengths. Each domain is based on a similar framework and each interacts with a small number of base pairs. Modulations in the amino acid sequence of each domain and synthetic variations in the DNA sequence enable spatial registration of the interaction to be precise. The strength of the interaction and the number of zinc fingers can be varied, as can the spacing between them, thus achieving a high level of specificity in recognition. This modular design offers a large number of possible combinations for the specific recognition of DNA. This is probably the reason why it is found widespread throughout so many different types of organisms.

3.4 *GAL* Genes of *Saccharomyces cerevisiae*

Studies of gene regulation in fungi have contributed greatly to our understanding of mechanisms of eukaryotic gene expression. Among the best understood genetic regulatory circuits in yeast cells is the one acting on *GAL* genes, which encode the enzymes of galactose utilization (Johnston, 1987a,b).

Galactose is utilized by *S. cerevisiae* by its conversion to glucose-6-phosphate, catalyzed by the enzymes of the Leloir pathway (Johnston, 1987a,b; Johnston and Dover, 1987). These enzymes are encoded by GAL 1 (kinase), GAL 7 (transferase), GAL 10 (epimerase), and GAL 5 (mutase) (see Figs. 3-6–3-10). Galactose in the form of melibiose is also available to yeast cells after cleavage of this disaccharide catalyzed by D-galactosidase, encoded by the *MEL1* gene. Galactose enters yeast cells through a specific permease, encoded by *GAL2* (Johnston, 1987b). Expression of the genes encoding these enzymes is regulated except for *GAL5*, which is unregulated. The regulated genes are induced by growth on galactose and repressed during growth on glucose.

The *GAL1, GAL7,* and *GAL10* genes are clustered on chromosome II but separately transcribed from individual promoters. *GAL2* and *MEL1* lie on chromosome XII. The *GAL4* gene is present on chromosome XVI and

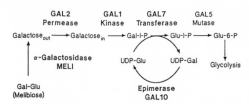

Figure 3-6. Pathway of galactose utilization. The enzymes are galactokinase (encoded by *GAL1*), galactose-1-phosphate uridyl transferase (encoded by *GAL7*), uridine diphosphoglucose 4-epimerase (encoded by *GAL10*), phosphoglucomutase (encoded by *GAL5*), and α-galactosidase (encoded by *MEL1*). Shown in boldface are the genes whose expression is regulated by galactose. (Reprinted with permission from Johnston, M., 1987. A model fungal gene regulatory mechanism: The GAL genes of Saccharomyces cerevisiae, *Microbiol. Rev.* 51:458.)

encodes GAL4 protein, which activates transcription of the above five genes by binding to sites located upstream of each gene. The GAL80 protein binds directly to the GAL4 protein preventing it from activating transcription. The inducer prevents GAL80 protein from inhibiting GAL4, presumably by binding to GAL80 protein. During growth on galactose, the inducer prevents GAL80 protein from inhibiting GAL4 protein function, thus allowing expression of *GAL1, GAL10, GAL7, GAL2,* and *MEL1* genes. In the absence of galactose, GAL80 protein binds to GAL4 protein, thus preventing expression of the above fingers. Growth on glucose (even in the presence of galactose) prevents *GAL* gene expression through a regulatory circuit termed catabolic repression about which not much is known.

GAL4 protein is required for the transcription of *GAL1, -2, -7,* and *-10. GAL4* mutants fail to transcribe these genes. GAL4 protein is also known to increase the basal level of *MEL1* transcription. GAL4 protein binds to DNA upstream of these genes.

The events that occur following binding to DNA and leading to transcription activation are unknown. The GAL4 protein-binding site appears to be a sequence module whose role is to bring GAL4 protein near a promoter.

The *GAL4* gene has been sequenced and encodes a large protein of 881 amino acids (99,350 Da). Only one or two molecules of protein are present per cell in *S. cerevisiae.*

The DNA-binding domain resides in the N-terminal 74 amino acids of GAL4 protein. Mutations that affect the DNA-binding activity of GAL4 protein are, however, limited to amino acids 10 to 51, as shown in Fig. 3-8. This region of the protein is homologous to regions in several other eukaryotic DNA-binding and transcriptional regulatory proteins and is thought to form a structure called the "cysteine-zinc DNA binding finger" (see Fig. 3-8). The two pairs of cysteine residues are proposed to chelate a zinc ion, with the intervening amino acids helping to form a "finger" that is thought to contact DNA. The cysteine-zinc finger of GAL4 protein is especially homologous to

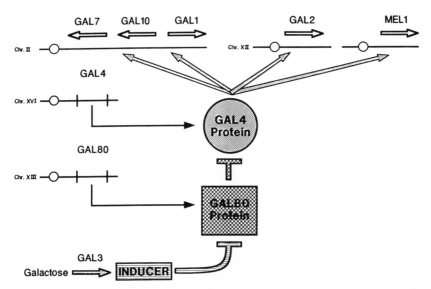

Figure 3-7. Components of the *GAL* gene regulatory circuit. Bold lines with arrows denote stimulation of activity; those with bars denote inhibition of activity. (Reprinted with permission from Johnston, M., 1987. A model fungal gene regulatory mechanism: The GAL genes of Saccharomyces cerevisiae, *Microbiol. Rev.* 51:458.)

similar regions in several other transcriptional regulating proteins for *S. cerevisiae* and one from the filamentous fungus *Neurospora crassa*. The consensus sequence given at the bottom of Fig. 3-8 shows that these proteins have identical amino acids at 8 of 23 positions; 6 other positions have amino acids with very similar properties in at least 5 of the 7 proteins. The regions between the pairs of cysteines are rich in amino acids known to interact with DNA.

Direct evidence that the cysteine-zinc finger is involved in DNA binding comes from the identification of *GAL4* mutations that alter this structure and abolish the DNA-binding activity of GAL4 (Johnston, 1987a,b; Johnston and Dover, 1987). Inasmuch as the cysteine-zinc fingers of the seven fungal proteins (Fig. 3-8) are highly homologous, despite the fact that they bind to different DNA sequences, this structure may be a general DNA-binding domain that does not provide sequence specificity. The sequence specificity of GAL4 protein DNA binding might reside in the region adjacent to the cysteine-zinc finger which shows little apparent homology among the five proteins listed in Fig. 3-8. The DNA binding domain of GAL4 protein thus may consist of two regions: one (the cysteine-zinc finger) that binds nonspecifically to DNA, and another (the specific domain) that provides for the sequence specificity of binding.

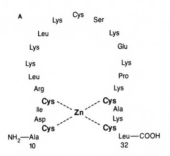

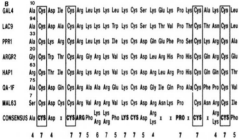

Figure 3-8. Proposed structure of the cysteine-zinc DNA binding finger of GAL4 protein (A) and homologies among the cysteine-zinc DNA binding fingers of several fungal transcriptional regulatory proteins (B). (A) The sequence of GAL4 protein from amino acids 10 to 32 is shown. The two pairs of cysteine residues shown in boldface are thought to chelate a zinc ion. (B) The number above the first amino acid of each protein is its residue number in that protein. Shown at the bottom is the consensus sequence (Phohydrophobic amino acid; X = any amino acid). Shown below the consensus sequence is the number of proteins that each consensus amino acid appears in. The amino acids in boldface capital letters are present in all seven proteins. (Reprinted with permission from Johnston, M., 1987. A model fungal gene regulatory mechanism: The GAL genes of Saccharomyces cerevisiae, *Microbiol. Rev.* 51: 458.)

There is good evidence that the cysteine-zinc finger of GAL4 protein indeed contains a zinc ion that is required for DNA-binding activity. One of the mutations that alter this structure (Pro-26 → Leu) results in a GAL4 protein that is unable to bind DNA because of its reduced affinity for zinc ions (Johnston, 1987a,b; Johnston and Dover, 1987). The GAL phenotype of mutants carrying this *GAL* mutation is corrected by high concentration of zinc in the medium. Also, the ability of the altered GAL4 protein to bind to DNA *in vitro* is restored in the presence of zinc. It seems likely that residue No. 26 in the mutant causes a bend in the protein and alters its ability to bind to DNA. Zinc probably plays an essential structural role in the DNA-binding domain of GAL4 protein. The portion of GAL4 protein responsible for activation of transcription lies in the C-terminal 90% of the protein, and this domain functions independently of the DNA-binding domain. Thus, it appears that the role of the DNA-binding domain is to appropriately position the transcription activation domain near promoters. It seems likely that GAL4 protein activates gene expression through contacts between its transcription activation domain and other proteins more directly responsible for transcription such as RNA polymerase. In summary, the available evidence suggests that GAL4 protein activates transcription after it is positioned near the appropriate promoter elements by the DNA-binding domain by interacting between two distinct acidic transcription activation domains and other proteins of the transcriptional apparatus.

How the GAL4 protein is able to activate transcription over variable distances is a question central to the mechanism of GAL4 protein function and is relevant to the mechanism of action of enhancer elements of higher eukaryotes. It has been suggested that this flexibility is a property of DNA which bends to bring the GAL4 protein-binding site (or enhancer) close to the sites upon which the protein acts. Another possibility (not necessarily exclusive of the first model) is that the flexibility resides in the GAL4 (or enhancer-binding) protein, which might consist of transcription activation domains linked to the DNA-binding domain by a flexible arm. In this model (see Fig. 3-9), the only function of most of the protein would be to allow the transcription activation domains to extend to their site of action on DNA (presumably the TATA box).

Since the GAL4 protein is synthesized in the cytoplasm but acts in the nucleus, it must be able to pass through the nuclear envelope. It seems likely that a specific transport system accomplishes this task, because the large size of GAL4 protein (99 kDa) compared with the nuclear pore (9 nm) makes it difficult to imagine that it is a simple process of diffusion (see Fig. 3-10). It appears that GAL4 protein sequences responsible for its nuclear localization lie in the N-terminal region that also contains the DNA-binding domain.

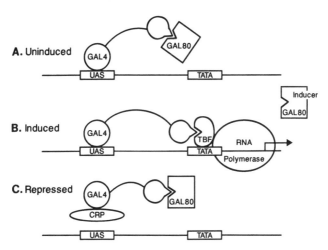

Figure 3-9. Model for mechanism of action of GAL4 and GAL80 proteins. See text for explanation. TBF, TATA box binding factor; CRP, catabolite-repressing protein; UAS, upstream activation sequence. (Reprinted with permission from Johnston, M., 1987. A model fungal gene regulatory mechanism: The GAL genes of Saccharomyces cerevisiae, *Microbiol. Rev.* 51:458.)

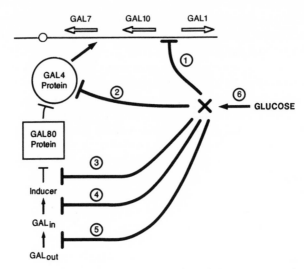

Figure 3-10. Possible mechanism of catabolite repression of *GAL* gene expression. See text for explanation. (Reprinted with permission from Johnston, M., 1987. A model fungal gene regulatory mechanism: The GAL genes of Saccharomyces cerevisiae, *Microbiol. Rev.* **51**:458.)

GAL4 protein must be able to diffuse into the nucleus at some level, because hybrid GAL4 proteins that lack the nuclear localization domain and are therefore not specifically concentrated in the nucleus, are still able to activate gene expression. It appears that the domains responsible for DNA binding and nuclear localization appear to be functionally separate.

The mechanism of *GAL* gene regulation is relevant to understanding how genes are expressed in higher eukaryotes. The steroid hormones induce gene expression by binding to receptor proteins in the cytoplasm, which causes these proteins to enter the nucleus, bind to DNA, and activate transcription of certain genes. This family of proteins shares many properties with GAL4 protein.

The glucocorticoid hormone receptor binds to short sequence elements— glucocorticoid response elements (GRE)—that lie upstream of genes that it regulates. Like GAL4 protein, GRE work over variable distances as modules that allow any gene that carries this sequence (in the proper location) to be induced by the hormone. The receptor protein probably uses a cysteine-zinc finger to bind to the GRE sequence. The glucocorticoid receptor also contains a domain responsible for activation of transcription that presumably interacts with a protein(s) of the transcriptional apparatus. Unlike GAL4 protein, however, this domain does not appear to be separable from the DNA-binding domain.

The most striking similarity of the mechanism of action of steroid hormone receptors with the *GAL* gene regulating circuit is the manner by which the inducer activates gene expression. The hormone binds to a domain in the receptor protein whose function is remarkably analogous to that of GAL80 protein. In the absence of hormone, this domain of the protein inhibits transcription activation, possibly by binding to the sequences of the receptor that interact with proteins of the transcriptional apparatus. Like GAL80 protein, this region of the receptor is presumably prevented from inhibiting transcription activation when inducer binds to it. Furthermore, deletion of the hormone-binding domain has the same effect on receptor function as deletion of GAL80 has on GAL4 protein: It causes constitutive activation of transcription. It is possible that higher eukaryotes regulate gene expression by using mechanisms similar to those responsible for regulation of the yeast *GAL* genes.

A recent article has reviewed the evidence (1) that the finger-loop domains have been highly conserved during evolution, (2) that they furnish one of the fundamental mechanisms for regulating gene expression, and (3) that zinc is required for binding of finger-loops to DNA and for their biological functions (Sunderman and Barber, 1988). Certain DNA-binding proteins that regulate gene expression contain one or several copies of short polypeptide sequences (approximately 30 residues long), consisting of combinations of four Cys-His residues at defined spacing so that zinc ion is complexed in tetrahedral coordination with the respective SH and/or imidazole-nitrogen atoms. Zinc plays a structural role and stabilizes folding of the domain into a "finger-loop" which is capable of site-specific binding to double-stranded DNA.

3.5 Hormone Receptors

Recent studies have shown that zinc coordinates with cysteine and histidine residues in certain peptides and confers a tertiary structure which has affinity for unique stretches of DNA in promoter gene regions (e.g., zinc-finger proteins, zinc-thiolate clusters). Binding of zinc-finger proteins to promoter area enhances DNA transcription and increases synthesis of the protein product for that mRNA sequence.

Hollenberg *et al.* (1985) determined the amino acid sequence of human glucocorticoid receptor (AGR) and Giguere *et al.* (1987) established important structural features of the proteins. The receptor comprises three major segments: (1) an immunogenic domain in the amino-terminal half of the molecule, (2) the DNA-binding domain (containing two zinc-finger loops) near the middle, and (3) the glucocorticoid-binding domain near the carboxy-terminus (see Fig. 3-11). Recent studies have shown that the dual finger-loop

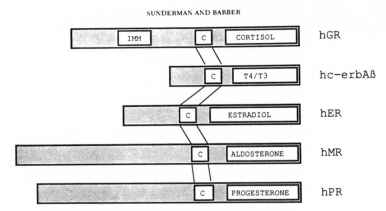

SUNDERMAN AND BARBER

Figure 3-11. Schematic diagram of the DNA-binding domains (labeled "C") of five steroid hormone receptors. Abbreviations: hGR, human glucocorticoid receptor (with the immunogenic domain labeled "IMM"); hTR2, human thyroid hormone receptor (which is identical to human erbA onc-protein); hER, human estrogen receptor; hMR, human mineralocorticoid receptor; hPR, human progesterone receptor. The DNA-binding domains of these receptors contain the dual finger-loop sequences listed in Table 3-3. (Reprinted with permission from Sunderman, F. W., Jr., and Barber, A. M., 1988. Finger-loops, oncogenes, and metals, *Ann. Clin. Lab. Sci.* 18:267.)

domains of the glucocorticoid receptor bear homology to domains in the human mineralocorticoid receptor (Arriza *et al.,* 1987), human and avian estrogen receptors (Greene *et al.,* 1986; Krust *et al.,* 1986; Kumar *et al.,* 1986), human and rat androgen receptors (Chang *et al.,* 1988; Lubahn *et al.,* 1988), human and rat thyroid hormone receptors (Thompson *et al.,* 1987; Weinberger *et al.,* 1986), rat and avian vitamin D_3 receptors (Burmester *et al.,* 1988; McDonnell *et al.,* 1987), and human retinoic acid receptor (Giguere *et al.,* 1987; Petkovich *et al.,* 1987). The erbA onc-protein has been shown to represent a high-affinity receptor for triiodothyronine and thyroxine (Sabbath *et al.,* 1987; Weinberger *et al.,* 1985, 1986, 1987) containing the dual finger-loops that are conserved throughout the family of hormone receptors. Figure 3-11 and Table 3-3 illustrate some of the above points.

A genetic lesion in the pathway of vitamin D action has been suspected to occur in patients with vitamin D-resistant rickets. This rare disorder is characterized by hypocalcemia, secondary hyperparathyroidism, and early onset of rickets, in spite of an increase of 1,25-dihydroxyvitamin D_3 [1,25-$(OH)_2D_3$] in the plasma (Liberman *et al.,* 1983). In many cases, total alopecia has been reported (Hirst *et al.,* 1985; Hochberg *et al.,* 1984). Progress in unraveling the molecular pathogenesis of this disorder has been slow because it is difficult to obtain receptor containing "target" tissue such as intestine and bone from human subjects. Fibroblasts from human skin have been shown

Table 3-3. Finger-loops in Thyroid and Steroid Hormone Receptors, Vitamin-D_3 Receptor, and Retinoid Acid Receptor[a]

Receptor[b] and residues	Sequence	Ref.
hTR2 102	C VV C GDKATGYHYRCIT C EG C	Wharton et al. (1985)
hGRα 421	C LV C SDEASGCHYGVLT C GS C	Hollenberg et al. (1985)
hMR 603	C LV C GDEASGCHYGVVT C GS C	Arriza et al. (1987)
hER 185	C AV C NDYASGYHYGVWS C EG C	Krust et al. (1986), Kumar et al. (1986)
hPR 567	C LI C GDEASGCHYGVLT C GS C	Misrabi et al. (1987)
hAR	C LI C GDEASGCHYGALT C GS C	Chang et al. (1988), Lubahn et al. (1988)
cVDR 37	C GV C GDRATGFHFNAMT C EG C	McDonnell et al. (1987)
hRAR 58	C FV C QDKSSGYHYGVSA C EG C	Giguere et al. (1987), Petkovich et al. (1987)
hTR2 140	C KYEGK C VIDKVTRNQ C QE C RFKK C	
hGRα 458	C AGRND C IIDKIRRKN C PA C RYRK C	
hMR 639	C AGRND C IIDKIRRKN C PA C RLQK C	
hER 221	C PATNQ C TIDKNRRKS C QA C RLRK C	
hPR 603	C AGRND C IVDKIRRKN C PL C RLRK C	
hAR	C ASRND C TIDKFRRKN C PS C RLRK C	
cVDR 81	C PFNGD C KITKDNRRH C QA C RLKR C	
hRAR 94	C HRDKN C IINKVTRNR C QY C RLQK C	

[a] Reprinted with permission from Sunderman, F. W., Jr., and Barber, A. M., 1988. Finger-loops, oncogenes and metals, *Ann. Clin. Lab. Sci.* 18:267.

[b] hTR2, human thyroid hormone receptor; hGRα, human glucocorticoid receptor; hMR, human mineralocorticoid receptor; hER, human estrogen receptor; hPR, human progesterone receptor; hAR, human androgen receptor; cVDR, chicken vitamin D_3 receptor; hRAR, human retinoic acid receptor.

to contain the 1,25-(OH)$_2$D$_3$ receptor and to exhibit biological response (Feldman *et al.*, 1980). A decreased or absent DNA binding of the 1,25-(OH)$_2$D$_3$ receptor and an inability to induce the enzyme 25-hydroxyvitamin D-24-hydroxylase have been observed in skin fibroblast culture of patients and families (Hirst *et al.*, 1985; Chen *et al.*, 1984; Castells *et al.*, 1986; Gamblin *et al.*, 1985). Recently, two families with affected children homozygous for this autosomal recessive disorder were studied for abnormalities in the intracellular vitamin D receptor and its gene (Hughes *et al.*, 1988; Baker *et al.*, 1988). In each family, a different single nucleotide mutation was found in the DNA-binding domain of the "zinc-finger" protein. In one family, the mutations involved the tip of the first zinc-finger protein (Gly → Asp), and in the other, the tip of the second zinc-finger protein (Arg → Gly) (see Fig. 3-12).

The steroid hormones, thyroid hormones, vitamin D$_3$, and retinoic acid enter cells by facilitated diffusion and combine with the respective receptors, either before or after entering the nucleus. It appears that the complexation of a hormone by its specific receptor initiates a conformational change that unmasks the finger-loops, so that they bind to high-affinity sites on chromatin, activating transcriptional enhancers that are termed response elements (Johnston and Dover, 1987; Weinberger *et al.*, 1987). Kumar *et al.* (1986) have shown that mutations in the first finger-loop of human estrogen receptor disrupt its DNA binding. Green and Chambon (1987) reported that mutational substitution of two Cys by two His residues in the first finger-loop domain of human estrogen receptor blocks transcriptional activation. These studies sug-

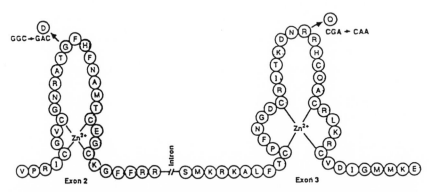

Figure 3-12. Amino acid sequence and hypothetical structure of the VDR (vitamin D receptor) DNA-binding domain. The deduced amino acids from the VDR cDNA are shown as two potential zinc-finger arrays each encoded by separate gene exons. (Reprinted with permission from Hughes, M. R., Malloy, P. J., Kieback, D. G., Kesterson, R. A., Wesley-Pike, J., Feldman, D., and O'Malley, B. W., 1988. Point mutations in the human vitamin D receptor gene associated with hypocalcemic rickets, *Science* 242:1702.)

gest that the dual finger-loop region may determine the steroid hormone receptor's specificity for target genes.

Colvard and Wilson (1984) observed that Zn^{2+} (and Ni^{2+} to a lesser extent) potentiates *in vitro* binding of androgen receptor to isolated nuclei from rat prostate whereas other metals such as Ca^{2+}, Mg^{2+}, Mn^{2+}, Co^{2+}, Cu^{2+}, and Cd^{2+} had no effects. Sabbath *et al.* (1987) showed that DNA binding of activated bovine estrogen receptor was blocked by 1,10-phenanthroline (a zinc chelator). These studies show that metal ions play an important role in DNA binding of the hormone–receptor complex.

Sunderman and Barber (1988) searched amino acid sequences of 38 transforming proteins and identified possible finger-loop domains in the myc, fms, fps, raf-I, rfp, src, syn, yes, erb A, int-1, and TGF-alpha gene products. The search also identified a possible finger-loop domain in human insulin receptor, which may provide a mechanistic explanation as to how insulin after binding to its cell surface receptor, is translocated to hepatocyte nuclei and becomes bound to chromatin. Sunderman and Barber (1988) have proposed that Zn^{2+} coordination sites in finger-loop domains may be potential targets of metal toxicity and substitution of Ni^{2+}, Co^{2+}, or Cd^{2+} for Zn^{2+} in finger-loops of transforming proteins is a hypothetical mechanism for metal carcinogenesis.

Recently, Frankel *et al.* (1988) have shown that Tat, the trans-activating protein from human immunodeficiency virus (HIV), forms a metal-linked dimer with metal ions bridging cysteine-rich regions from each monomer. This unique arrangement is different from the "zinc-finger" domain observed in other eukaryotic regulatory proteins. It appears that the metal binding has its primary effects in the cysteine-rich region and relatively little effect on the folding of other regions. The cysteine-rich region is strikingly similar to the metallothioneins.

HIV encodes several regulatory proteins that are absent in simpler retroviruses. The Tat protein (one of the regulatory proteins) trans-activates genes that are expressed from the HIV long terminal repeat (LTR) and Tat is essential for replication of virus *in vitro* (Frankel *et al.,* 1988). The Tat protein contains 86 amino acids. Its sequence includes a highly basic region (two lysines and six arginines within nine residues) that might participate in nucleic acid binding. Tat also contains a cysteine-rich region (7 cysteines within 16 residues), which appears to be highly conserved.

Inasmuch as Tat is essential for viral replication, it has been suggested that a modification of Tat protein may provide a therapeutic tool for HIV infection (Frankel *et al.,* 1988). Dietary intake of metals or use of specific metal chelating agents may prove to be useful in this regard. Drugs designed to prevent Tat dimerization could also be utilized as an antiviral agent.

References

Abdallah, J. M., Kukuruga, M., Nakeff, A., and Prasad, A. S., 1988. Cell cycle distribution defect in PHA-stimulated T lymphocytes of sickle cell disease patients, *Am. J. Hematol.* 28:279.

Arriza, J. L., Weinberger, C., Cerelli, G., Glaser, T. M., Handelin, B. L., Houseman, D. E., and Evans, R. M., 1987. Cloning of human mineralocorticoid receptor complementary DNA: Structural and functional kinship with the glucocorticoid receptor, *Science* 237:268.

Baker, A. R., McDonnell, D. P., Hughes, M., Crisp, T. M., Mangelsdorf, D. J., Hanssler, M. R., Pike, J. W., Shine, J., and O'Malley, B. W., 1988. Cloning and expression of full-length cDNA encoding human vitamin D receptor, *Proc. Natl. Acad. Sci. USA* 85:3294.

Blumberg, H., Eisen, A., Sledziewski, A., Bader, D., and Young, E. T., 1987. Two zinc fingers of a yeast regulatory protein shown to be essential for its functions, *Nature* 328:443.

Burmester, J. K., Maeda, N., and DeLuca, H. F., 1988. Isolation and expression of rat 1,25-dihydroxyvitamin D$_3$ receptor of cDNA, *Proc. Natl. Acad. Sci. USA* 85:1005.

Castells, S., Greig, F., Fusi, M. A., Finberg, L., Yasumura, S., Liberman, U. A., Eil, C., and Marx, S. J., 1986. Severely deficient binding of 1,25-dihydroxyvitamin D to its receptors in a patient responsive to high doses of this hormone, *J. Clin. Endocrinol. Metab.* 63:252.

Chang, C., Kokontis, J., and Liao, S., 1988. Molecular cloning of human and rat complementary DNA encoding androgen receptors, *Science* 240:324.

Chen, T. L., Hirst, M. A., Cone, C. M., Hochberg, Z., Tietze, H. V., and Feldman, D., 1984. 1,25-hydroxyvitamin resistance, rickets, and alopecia: Analysis of receptors and bioresponse in cultured fibroblasts from patients and parents, *J. Clin. Endocrinol. Metab.* 59:383.

Chesters, J. K., 1982. Metabolism and biochemistry of zinc, in *Clinical, Biochemical, and Nutritional Aspects of Trace Elements* (A. S. Prasad, ed.), Liss, New York, p. 221.

Colvard, D. S., and Wilson, E. M., 1984. Zinc potentiation of androgen receptor binding to nuclei in vitro, *Biochemistry* 23:3471.

Czupryn, M., Falchuk, K. H., and Vallee, B., 1987. Zinc deficiency and metabolism of histones and non-histone proteins in Euglena gracilis, *Biochemistry* 26:8263.

Falchuk, K. H., 1988. Zinc deficiency and the E. gracilis chromatin, in *Essential and Toxic Trace Elements in Human Health and Disease,* (A. S. Prasad, ed.), Liss, New York, p. 75.

Falchuk, K. H., Fawcett, D., and Vallee, B. L., 1975a. Role of zinc in cell division of E gracilis, *J. Cell Sci.* 17:57.

Falchuk, K. H., Krishan, A., and Vallee, B. L., 1975b. DNA distribution in the cell cycle of Euglena gracilis. Cytofluorimetry of zinc deficient cells, *Biochemistry* 14:3449.

Falchuk, K. H., Mazus, B., Ulpino, L., and Vallee, B. L., 1976. Euglena gracilis DNA dependent RNA polymerase II: A zinc metalloenzyme, *Biochemistry* 15:4468.

Falchuk, K. H., Hardy, C., Ulpino, L., and Vallee, B. L., 1978. RNA metabolism, manganese, and RNA polymerases of zinc supplemented and zinc deficient Euglena gracilis, *Proc. Natl. Acad. Sci. USA* 75:4175.

Falchuk, K. H., Gordon, P. R., Stankiewizc, A., Hilt, K. L., and Vallee, B. L., 1986. Euglena gracilis chromatin: Comparison of effects of zinc, iron, magnesium, or manganese deficiency and cold shock, *Biochemistry* 25:5388.

Feldman, D., Chen, T., Hirst, M., Colston, K., Karaski, M., and Cone, C., 1980. Demonstration of 1,25-dihydroxyvitamin D3 receptors in human skin biopsies, *J. Clin. Endocrinol. Metab.* 51:1463.

Frankel, A. D., Berg, J. M., and Pabo, C. O., 1987. Metal-dependent folding of a single zinc finger from transcription factor IIIA, *Proc. Natl. Acad. Sci. USA* 84:4841.

Frankel, A. D., Bredt, D. S., and Pabo, C. O., 1988. Tat protein from human immunodeficiency virus forms a metal-linked dimer, *Science* 240:70.

Gamblin, G. T., Liberman, U. A., Eil, C., Downs, R. W., Jr., DeGrange, D. A., and Marx, S. J., 1985. Vitamin D-dependent rickets type II. Defective induction of 25-hydroxyvitamin D3-24-hydroxylase by 1,25-dihydroxyvitamin D3 in cultured skin fibroblasts, *J. Clin. Invest.* 75: 954.

Giguere, V., Hollenberg, S. M., Rosenfeld, M. G., and Evans, R. M., 1986. Functional domains of the human glucocorticoid receptor, *Cell* 48:645.

Giguere, V., Ong, E. S., Segui, P., and Evans, R. M., 1987. Identification of a receptor of the morphogen retinoic acid, *Nature* 330:624.

Green, S., and Chambon, P., 1987. Oestradiol induction of a glucocorticoid responsive gene by a chimaeric receptor, *Nature* 325:75.

Greene, G. L., Gilna, P., Waterfield, M., Baker, A., Hort, Y., and Shine, J., 1986. Sequence and expression of human estrogen receptor complementary DNA, *Science* 231:1150.

Hirst, M. A., Hochman, H. E., and Feldman, D., 1985. Vitamin D resistance and alopecia: A kindred with normal 1,25 dihydroxyvitamin D binding, but decreased receptor affinity for deoxyribonucleic acid, *J. Clin. Endocrinol. Metab.* 60:490.

Hochberg, Z., Benderlin, A., Levy, J., Vardi, P., Weisman, Y., Chen, T., and Feldman, D., 1984. 1,25-Dihydroxyvitamin D resistance, rickets, and alopecia, *Am. J. Med.* 77:805.

Hollenberg, S. M., Weinberger, C., Ong, E. S., Cerelli, G., Oro, A., Labo, R., Thompson, E. B., Rosenfield, M. G., and Evans, R. M., 1985. Primary structure and expression of a functional human glucocorticoid receptor cDNA, *Nature* 318:635.

Hughes, M. R., Malloy, P. J., Kieback, D. G., Kesterson, R. A., Wesley-Pike, J., Feldman, D., and O'Malley, B. W., 1988. Point mutations in the human vitamin D receptor gene associated with hypocalcemic rickets, *Science* 242:1702.

Johnston, M., 1987a. Genetic evidence that zinc is an essential co-factor in the DNA binding domain of GAL 4 protein, *Nature* 328:353.

Johnston, M., 1987b. A model fungal gene regulatory mechanism: The GAL genes of Saccharomyces cerevisiae, *Microbiol. Rev.* 51:458.

Johnston, M., and Dover, J., 1987. Mutations that indicate a yeast transcriptional regulatory protein cluster in an evolutionarily conserved DNA binding domain, *Proc. Natl. Acad. Sci. USA* 84:2401.

Keegan, L., Gill, G., and Ptashne, M., 1986. Separation of DNA binding from the transcription-activating function of a eukaryotic regulatory protein, *Science* 231:699.

Klug, A., and Rhodes, D., 1987. "Zinc fingers": A novel protein motif for nucleic acid recognition, *Trends Biochem. Sci.* 12:461.

Kornberg, R. D., 1977. Structure of chromatin, *Annu. Rev. Biochem.* 46:931.

Krust, A., Green, S., Argos, P., Kumar, V., Walter, P., Bornert, J. M., and Chambon, P., 1986. The chicken oestrogen receptor sequence: Homology with v-erbA and the human oestrogen and glucocorticoid receptors, *EMBO J.* 5:891.

Kumar, V., Green, S., Staub, A., and Chambon, P., 1986. Localization of the oestradiol-binding and putative DNA-binding domains of the human oestrogen receptor, *EMBO J.* 5:2231.

Liberman, U. A., Eil, C., and Marx, S. J., 1983. Resistance to 1,25-dihydroxyvitamin D. Association with heterogenous defects in cultured skin fibroblasts, *J. Clin. Invest.* 71:192.

Lubahn, D. B., Joseph, D. R., Sullivan, P. M., Willard, H. F., French, F. S., and Wilson, E. M., 1988. Cloning of human androgen receptor complementary DNA and localization of the X chromosome, *Science* 240:327.

McDonnell, D. P., Mangelsdorf, D. J., Pike, J. W., Haussler, M. R., and O'Malley, B. W., 1987. Molecular cloning of complementary DNA encoding the avian receptor for vitamin D, *Science* 235:1214.

Miller, J., McLachlan, A. D., and Klug, A., 1985. Repetitive zinc-binding domains in the protein transcription factor IIIA from xenopus oocytes, *EMBO J.* 4:1609.

Misrahi, M., Atger, M., D'Auriol, L., Loosfelt, H., Meriel, C., Fridlansky, F., Guiochon-Mantel, A., Galibert, F., and Milgrom, E., 1987. Complete amino acid sequence of the human progesterone receptor deduced from cloned cDNA, *Biochem. Biophys. Res. Commun.* 143:740.

Petkovich, M., Brand, N. J., Krust, A., and Chambon, P., 1987. A human retinoic acid receptor which belongs to the family of nuclear receptors, *Nature* 330:444.

Sabbath, M., Redeuilh, G., Secco, C., and Baulieu, E. E., 1987. The binding activity of estrogen receptor to DNA and heat shock protein (M,90,000) is dependent on receptor-bound metal, *J. Biol. Chem.* 262:8631.

Stankiewicz, A. J., Falchuk, K. H., and Vallee, B. L., 1983. Composition and structure of zinc deficient Euglena gracilis chromatin, *Biochemistry* 22:5150.

Sunderman, F. W., Jr., and Barber, A. M., 1988. Finger-loops, oncogenes, and metals, *Ann. Clin. Lab. Sci.* 18:267.

Thompson, C. C., Weinberg, C., Lebo, R., and Evans, R. M., 1987. Identification of a novel thyroid hormone receptor expressed in the mammalian central nervous system, *Science* 237: 1610.

Vallee, B. L., 1983. A role of zinc in gene expression, *J. Inherited Metab. Dis.* 6(Suppl. 1):31.

Weinberger, C., Hollenberg, S. M., Rosenfeld, M. G., and Evans, R. M., 1985. Domain structure of human glucocorticoid receptor and its relationship to the v-erb-A oncogene product, *Nature* 318:670.

Weinberger, C., Thompson, C. C., Ong, E. S., Lebo, R., Gruol, D. J., and Evans, R. M., 1986. The c-erb-A gene encodes a thyroid hormone receptor, *Nature* 324:641.

Weinberger, C., Giguere, V., Hollenberg, S. M., Thompson, C., Arriza, J., and Evans, R. M., 1987. Human steroid receptors and erb-A gene products form a superfamily of enhancer-binding proteins, *Clin. Physiol. Biochem.* 5:179.

Wharton, K. A., Johansen, K. M., Xu, T., and Artavanis-Tsakonas, S., 1985. Nucleotide sequence from the neurogenic locus notch implies a gene product that shares homology with proteins containing EGF-like repeats, *Cell* 43:567.

Biochemistry of Metallothionein

4.1 Introduction

Historically the term *metallothionein* (MT) has been used to designate the Cd-, Zn-, and Cu-containing sulfur-rich protein from equine renal cortex (Kagi and Vallee, 1960). The major characteristics of this protein were defined by Kagi *et al.* (1974) and Kojima *et al.* (1976).

Margoshes and Vallee (1957) discovered MT in equine kidney cortex, a cadmium-binding protein responsible for natural accumulation of cadmium in the kidneys. Cadmium is one of several metals such as zinc and copper which bind to this protein.

Piscator (1964) reported that the MT content of the liver in rabbits exposed to Cd was remarkably increased and he postulated that Cd induced the biosynthesis of MT and that it played a role in metal detoxification. Pulido *et al.* (1966) showed that MT was present in human kidney and that this protein contained Cd, Zn, and Cu, and under some circumstances Hg.

MTs have an extremely high metal and sulfur content. The mammalian forms of MT have a molecular weight of 6000–7000, contain approximately 60 amino acid residues of which 20 are cysteine (Cys), binding a total of 7 equivalents of bivalent metal ions. Aromatic amino acids are absent and all Cys occur in the reduced form. The metal ions are coordinated through mercaptide bonds, giving rise to metal thiolate complexes and metal thiolate clusters.

Characteristic features of MT include:

1. Low molecular weight (6000–7000)
2. High content of heavy metal
3. Optical properties characteristic of metal thiolate (mercaptides), i.e., metal bound only by clusters of thiolate bonds
4. An amino acid composition containing 23–33% Cys with no disulfide bonds, aromatic amino acids, or histidines
5. Amino acid sequence with conserved distribution of Cys residues

Zn-MT refers to MT containing zinc only and Cd-MT refers to MT containing Cd only. When MT contains more than one metal such as 5 g-atoms of Cd and 2 g-atoms of Zn, the construction Cd_5Zn_2-MT should be used. "Thionein" can be used to denote metal-free protein.

MTs occur throughout the animal kingdom. They are also found in higher plants, eukaryotic microorganisms, and some prokaryotes (Kagi and Nordberg, 1979). In animals this protein is most abundant in liver, kidney, pancreas, and intestines. Many factors are known to affect the levels of MT in tissues. These include age, stage of development, dietary regimen, metal administration, stress, hormones, and other factors which are not well defined.

Although MT is a cytoplasmic protein, its presence has been noted in lysosomes, and in the nucleus during development (Kagi and Nordberg, 1979).

4.2 Classification

MTs have been subdivided into three classes (Fowler *et al.*, 1987). Class I includes mammalian MTs and polypeptides for other phyla with related primary structure. MTs from human, chicken, trout, crab, and *Neurospora crassa* fall into this class.

MTs from sea urchin, wheat, yeast, and certain prokaryotes display none or only very distant resemblance to the mammalian forms and they are grouped as Class II. Class III MTs contain atypical polypeptides containing γ-glutamylcysteinyl units (Kagi and Schaffer, 1988).

Class I MTs display extensive genetic polymorphism. Mammalian tissues usually contain two major fractions MT-2 and MT-1, differing at neutral pH by a single negative charge. In humans a very complex polymorphism is observed inasmuch as there are ten ISO MT genes which are expressed and some of these are tissue specific.

4.3 Primary Structure

All known class I and II MTs are single-chain proteins. Mammalian forms contain 61 or 62 amino acid residues (Fig. 4-1). Chicken and sea urchin

Figure 4-1. Schematic representation of the amino acid sequence of mammalian MT. [Reprinted with permission from Hunziker, P. E., and Kagi, J. H. R., 1988. Metallothionein: A multigene protein, in *Essential and Toxic Trace Elements in Human Health and Disease* (A. S. Prasad, ed.), Liss, New York, p. 350.]

MTs contain 63 and 64 residues, respectively. In certain fungi and inverte-brates, shorter chains are found. In *N. crassa* MT only 25 residues are present. Class III MTs are often oligomeric structures made up of two or more poly-peptide chains. The most conspicuous feature of all forms is the frequent occurrence of Cys-X-Cys tripeptide sequences, where X is an amino acid residue other than Cys and Cys comprises up to one-third of all residues.

In mammals all 20 Cys residues are invariant and there is also extensive sequence homology with arthropod and certain fungal MTs. Lys and Arg residues are also highly conserved in mammalian MTs which are juxtaposed to Cys and it is believed that they play an auxiliary role in the formation of the metal complexes. Most of the amino acid substitutions are of the conser-vative type and are located in the amino-terminal half of the molecule. No obvious sequence relationships are observed among class II MTs. The class III MTs are homologous, atypical oligo- and polypeptides of the general struc-ture $(\gamma\text{-Glu-Cys})_n X$, where $n = 2\text{--}8$ and X is most often glycine. The individual members of the family of $(\gamma\text{-Glu-Cys})_n \text{Gly}$ are homologues of glutathione and have been designated as phytochelatins (PCn).

4.4 Metal-Binding Sites

The abundance of Cys and their conspicuous arrangement in chelating Cys-Cys, Cys-X-Cys, and Cys-X-Y-Cys, where X and Y are residues other than Cys, predispose MT toward the binding of "soft" metal ions. In mam-malian MTs, these ligands collectively may bind seven bivalent metal ions.

Preparations of pure native mammalian MTs have been shown to be heterogeneous with respect to the proportion of Zn, Cd, Cu, and minor metallic constituents. In human liver MTs, Zn is often the only metal component. It is believed that the metal ion composition is determined by the supply of the metals to the organisms (Kagi and Schaffer, 1988). Zn and Cd are readily released from the protein moiety by acidification.

At neutral pH, the stability constants for zinc, lead and cadmium have been calculated to be $K'_{Zn} = 2 \times 10^{-12} \text{ M}^{-1}$, $K'_{Pb} = 4 \times 10^{-14} \text{ M}^{-1}$, and $K'_{Cd} = 2 \times 10^{-16} \text{ M}^{-1}$. The 10,000-fold higher affinity for Cd(II) over Zn(II) is believed to be the explanation for the tendency of the environmentally much less abundant Cd to be accumulated in MT.

In native mammalian MTs all 20 Cys are deprotonated and participate in metal binding. Spectroscopic studies have shown that complexes are chem-ically and structurally uniform with binding of each bivalent metal to four thiolate ligands arranged in tetrahedrallike symmetry. The specific feature of all classes of MT is the organization of the metal complexes in metal thiolate clusters. This arrangement, which allows sharing of some of the thiolate ligands by adjacent metal ions, reconciles the tetrahedral tetrathiolate coordination with the measured stoichiometry of nearly three thiolate ligands per bivalent metal ion bound. In mammalian MTs the ratio of 20 Cys to 7 metal ions requires that 8 Cys serve as doubly coordinated bridging thiolate ligands and 12 as singly coordinated terminal thiolate ligands.

In mammalian MTs the metal ions are partitioned into two topologically separate metal thiolate clusters (Fig. 4-2). In cluster A, the four metal ions coupled through five bridging thiolate ligands were pictured to form a bicyclo[3.3.1]nonane-like structure made up of two six-membered rings positioned at virtually right angles to each other. It is inferred that in cluster B three metal ions and three bridging thiolate cysteines form a six-membered ring. This partitioning into two completely separate clusters was independently verified by limited enzymatic proteolysis, which resulted in a bisection of the protein into a carboxy- and an amino-terminal portion, designated as α and β domains and containing the Cys ligands of cluster A and cluster B, respectively (Kagi and Schaffer, 1988).

The α domain is located within the carboxy-terminal end of the molecule extending from amino acid 31 to 61 and an amino-terminal β domain extends from residue 1 to 30. Both domains are globular and linked by residues 30 and 31 to give the ellipsoidal shape to the molecule. The β domain contains 9 Cys residues and 3 atoms of Zn or Cd or 6 atoms of Cu. The α domain contains 11 Cys residues and 4 atoms of Zn or Cd or 5 to 6 atoms of Cu. Metal ions are bound exclusively through thiolate-coordination complexes which involve all 20 Cys residues. Native protein contains no disulfide bonds or free SH groups. Both Zn and Cd are bound in 2+ valence state and are tetrahedrally coordinated to 4 Cys thiolate ligands [metal^{2+}(Cys)$_4^-$]$^{2-}$ and are negatively charged, thus imparting overall negative charge to the protein.

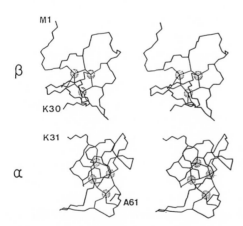

Figure 4-2. Two-dimensional NMR solution structure of β domain (top) and α domain (bottom) of rat liver Cd-MT-2. Stereo view of polypeptide backbone: thick line; Cys side chains: thin lines; metal positions: dotted spheres of radius 0.9 Å. (Reprinted with permission from Kagi, J. H. R., and Schaffer, A., 1988. Biochemistry of metallothionein, *Biochemistry* 27: 8509.)

Copper binds in 1+ valence state. Coordination of Cu binding is probably trigonal with three thiolate ligands involved. Stability constant for Cu is 100-fold greater than that for Cd, which in turn is about 1000-fold greater than that for Zn. Hg and Ag bind with even greater affinity than Cu. Metals are released upon lowering pH. Zn and Cd preferentially fill the α domain first and then the β domain. Cu falls in the reverse order.

4.5 Spatial Structure

Recently, both from 2-D NMR spectroscopic examination in aqueous solution and from x-ray diffraction data obtained on crystals, models for the spatial structure of mammalian MT and the organization of the metal thiolate clusters have been derived (Braun *et al.*, 1986; Kagi and Schaffer, 1988). A stereo view of the polypeptide backbone, Cys side chains, and the metal positions of the two domains of [113]Cd-substituted MT-2 from rat liver as determined by homonuclear and heteronuclear 2-D NMR techniques, is shown in Fig. 4-2. The uniformly sized and nearly spherical α and β domains have a diameter of 15–20 Å and contain in their center the respective metal thiolate clusters as "mineral cores" around each of which the polypeptide chain is wrapped, forming two large helical turns. In the carboxy-terminal α domain the spiral of the peptide fold is left-handed and in the amino-terminal β domain it is right-handed. The domains are connected by a hinge region consisting of the conserved Lys-Lys-Ser segment in the middle of the polypeptide chain. However, owing to the paucity of clearly recognizable interdomain contacts, the mutual orientation of the domains has not been defined as yet. This failure probably reflects a certain flexibility about the hinge region.

The model derived from x-ray crystallography of native rat liver MT-2 resembles the 2-D NMR solution structure in its overall organization including the steric organization of the clusters, the general disposition of the polypeptide chain, and the handedness of the folds (Furey *et al.*, 1986). Nonetheless, there are extensive discrepancies in the detailed spatial arrangements. For instance, there are large differences in the positioning of segments of the chain and consequently only 25% of all Cys agree, with respect to the metal coordination between the two structures. The explanation for this discrepancy is unknown.

The unique character of the polypeptide fold of mammalian MT is its stabilization by a total of 42 intramolecular Cys-metal-Cys cross-links. With 24 of these connections in the α domain and only 18 in the β domain, the conformational stability and the collective affinity for the metal are expected to be lower in the β domain and are more accessible to alkylating agents, and cluster B has a greater tendency to lose metal (Bernhard *et al.*, 1986).

Detection of MTs in tissues is done by isolating the protein via gel filtration chromatography and analyzing the protein for metals by atomic absorption spectrophotometry. MTs characteristically elute at a V_e/V_o ratio of about 1.8, a position where few other metal-binding proteins co-elute. MT isoforms are usually detected by rechromatographing the MT fraction on anion-exchange columns and analyzing the protein for metals. Recovery of MT from ion-exchange columns is much less than 100%. Resolution of isoform is improved with reversed-phase high-performance liquid chromatography. The amount of MT in tissues can be estimated using the above techniques by virtue of known molar ratios of metals to protein provided the complete metal composition is known. Loss of metal from the protein can be prevented by maintaining a pH above 7 and by including a reducing agent such as β-mercaptoethanol in all isolation steps.

MT content of tissues can be determined *in vitro* by saturation of the binding sites of the protein with radioactive Cd (Cousins, 1983). The isotope used is determined by the binding strength of the metals being displaced from this native protein. Quantification is based on known molar ratios of metals to protein and the specific activity of the isotope used to saturate the protein.

The antigenic determinants of vertebrate MT consist of two immunologically dominant regions in the NH_2-terminal domain (residues 1–29). The antigenicity was independent of the degree of folding of the protein favoring a sequential (or continuous) rather than a discontinuous nature of the determinants. The two regions in MT that appear to be important in the interaction of the molecule with the antisera include the NH_2-terminal acetylated methionine and the clusters of lysines in the sequence from residues 20 to 25.

Polyclonal antibodies against several mammalian MTs have been used in radioimmunoassay (RIA) to detect low levels of MT in tissues and plasma (Mehra and Bremner, 1983). An enzyme-linked immunosorbent assay (ELISA) has also been developed (Thomas *et al.*, 1986; Grider *et al.*, 1989). At present, RIA and Cd-saturation methods appear to be preferable. Recent data in rats suggest that assay of plasma MT and zinc may be utilized for assessment of zinc status (Bremner and Morrison, 1988).

4.6 Inducibility of Metallothionein

A remarkable feature of all MTs is that they are inducible. Early observations showed that tissue contents of MT could be increased by the application of Cd salts (Dunn *et al.*, 1987). Table 4-1 lists factors such as metals, hormones, cytotoxic agents, and various pathophysiologic conditions associated with physical or chemical stress, which stimulate biosynthesis of MT in cultured cells or *in vivo*. The induction occurs at the level of transcription

Table 4-1. Factors That Induce Metallothionein
Synthesis in Cultured Cells or *in Vivo*[a]

Metal ions: Cd, Zn, Cu, Hg,	Streptozotocin
Au, Ag, Co, Ni, Bi	2-Propanol
Glucocorticoids	Ethanol
Progesterone	Ethionine
Estrogen	Alkylating agents
Glucagon	Chloroform
Catecholamines	Carbon tetrachloride
Interleukin 1	Starvation
Interferon	Infection
Butyrate	Inflammation
Retinoate	Laparatomy
Phorbol esters	Physical stress
Endotoxin	X-irradiation
Carrageenan	High O_2 tension
Dextran	

[a] Adapted with permission from Kagi, J. H. R., and Schaffer, A., 1988.
Biochemsitry of metallothionein, *Biochemistry* 27:8509.

initiation, leading to a maximum MT concentration within 1–2 days after exposure to an inducer (Kagi and Schaffer, 1988). Actual peak time varies with the type of inducer. In human cells, expression of some of the ISO MT genes appears to be regulated differentially by Cd, Zn, and glucocorticoids and there are indications for tissue-specific expression.

Recent studies have led to the identification of various DNA segments which serve as promoter sites in the 5′ region of various MT genes (Palmiter, 1987). In the mouse MT-1 gene, the functional metal-responsive promoter is composed of a set of four closely related metal-regulatory elements, each made up of eight nucleotides and localized near the TATA box. It is thought to be recognized by one or more MT gene-binding proteins acting as positive transcription factors when activated by the appropriate metal ion (Seguin and Hamer, 1987). Inasmuch as the metal-responsive MT promoter was shown to function in different cell types and independent of species and since it is susceptible to regulation by the metal, the so-called MT-fusion gene technology has found important applications (Palmiter, 1987). In these constructs the promoter is linked to a structural gene of interest, thereby allowing modulation of expression of its gene product in cultured cells by metals. Genes employing the Cu-sensitive promoter of yeast Cu-MT are expected to have applications in the biotechnological and biopharmaceutical syntheses of proteins in yeast (Butt and Ecker, 1987). Fusion genes containing the mouse MT-1 promoter have also been employed in transgenic experiments, allowing regulation of the introduced gene by metal supplementation (Palmiter and Brinster, 1985).

In one study, the promoter or regulatory region of the mouse gene for MT-1 was fused to the structural gene coding for human growth hormone (Palmiter *et al.,* 1983). These fusion genes were introduced into mice by microinjection of fertilized eggs. Twenty-three (70%) of the mice that incorporated the fusion genes showed high concentration of human growth hormone in their serum and grew significantly larger than control mice. Synthesis of human growth hormone was induced further by Cd or Zn, which normally induce MT gene expression.

Heavy metals induce MT synthesis primarily by increasing the rate of MT gene transcription. Zn, Cd, Cu, and Hg increase the transcription rate of the MT-1 gene in mouse kidney and liver several hours before maximal accumulation of MT-1 mRNA and maximal rates of MT biosynthesis occur (Kagi and Schaffer, 1988). In the liver, Cd and Zn were the best inducers, Cu was a good inducer only at high doses, and induction by Hg was weak. Cd, Zn, Cu, Hg, Au, Ni, and Bi have been shown to regulate MT gene transcription. Some of these metals may act indirectly, however, by displacing Zn from MTs and then Zn acts as the direct inducer or via stress associated with administration of pharmacological doses of the metal.

Actinomycin D, cordycepin, and cycloheximide, administered to animals prior to an injection of either Zn or Cd, blocked the induction of synthesis of hepatic MT in rats. When these inhibitors were administered after injection of each metal, however, the induction was not inhibited (Squibb and Cousins, 1974; Squibb *et al.,* 1977). These results have been confirmed in an isolated liver perfusion system. In contrast, Shaikh and Smith (1975a,b, 1977) demonstrated that actinomycin D did not inhibit the biosynthesis of renal MT induced by Cd. These earlier data suggested that Zn and Cd are able to induce the synthesis of MT through changes in the intracellular concentration of MT mRNA.

4.7 Molecular Mechanisms of Metallothionein Induction

The number of structural MT genes in mammal varies from two in the mouse (one of each isoform) to considerably more in primates. The number of metal regulating elements (MRE) for each gene may also vary between species. MREs are upstream from the structural gene and contain closely related base sequences. Having multiple copies of metal regulatory sequences may allow various regulatory factors (possibly specific proteins) generated by different cell types to act either cooperatively or antagonistically with each other in terms of their induction of MT synthesis.

Molecular mechanisms by which metal ions interact with MRE to induce transcription are not well understood. Binding proteins that bind both the

MRE and the inducer metal somehow induce transcription. In bacterial cells a 16-kDa protein has been isolated which binds to the promoter area and confers resistance to Hg (Dunn *et al.*, 1987). This protein acts as a repressor in the absence of Hg and a positive regulator in the presence of Hg. In view of the above observation and the reports that the MT fusion genes are normally inducible in cells that produce little or no thionein, it is unlikely that thionein itself may be a regulatory protein. MT synthesis is also increased by hormones such as glucocorticoids, glucagon, and epinephrine and by other factors such as cAMP, interferon, and interleukin 1. Acute stresses such as food restriction, physical restraint, and tissue injury resulting from inflammatory agents or bacterial toxins, are also known to induce MT synthesis. Glucocorticoids, glucagon, epinephrine, and dibutyryl cAMP (β_2-cAMP) are known to be blocked by actinomycin D, suggesting that these agents act by transcriptional regulation.

The mechanisms by which glucocorticoids stimulate MT gene transcription are unknown. A glucocorticoid receptor complex has been postulated to interact with MT promoter sequences; however, the MT promoter does not appear to be responsive to glucocorticoid hormone in fusion gene experiments (Dunn *et al.*, 1987). Interferon has also been shown to increase the rate of MT gene transcription through an unknown mechanism.

A variety of inflammatory agents and interleukin 1 also induce MT synthesis. It has been postulated that endotoxin and IL-1 induction is mediated or augmented through direct actions of IL-6 on cells. The mode of MT induction by food restriction and physical stress has not been studied.

4.8 Functional Aspects

The functional significance of MTs is not well understood. MT serves as a rather nonspecific metal-buffering ligand to either sequester or dispense metal ions. It may also have specialized function in normal cellular metabolism or development (Fig. 4-3).

MT is responsible for much intracellular sequestration of cadmium. There is also good evidence that MT, owing to its inducibility, provides animals and cells with a mechanism to attenuate at least temporarily the toxicity of cadmium (Dunn *et al.*, 1987).

After a single injection of Cd salts in rats, testicular damage occurs. This effect as well as lethality can be prevented by pretreatment with doses lower than those producing the above effect. In animals protected by pretreatment, a greater portion of testicular Cd as well as of Cd in some other tissues is bound to Cd-binding protein with the same molecular weight as MT. When Cd-containing MT was injected, no testicular damage occurred in mice, but

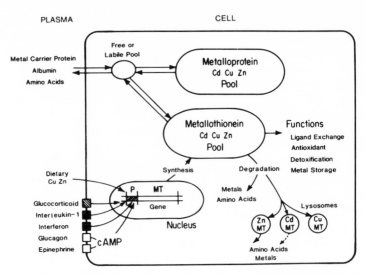

Figure 4-3. Schematic diagram of proposed metabolism and functions of metallothionein. Metals carried in the plasma on proteins or amino acids exchange rapidly with a labile metal pool in the cytoplasm which in turn exchanges metals with MT and other metalloproteins. The size of the MT pool is regulated by MT synthesis and degradation. MT synthesis is regulated by dietary metals, hormones, and other factors which induce transcription of the MT gene via interaction with specific promoters (P) located upstream from the MT gene. cAMP may be the regulatory factor mediating induction by glucagon and epinephrine. Other regulatory factors that interact with the promoter are presumed, but remain unidentified. Degradation of MT may occur in the cytoplasm or within lysosomes. The rate of degradation varies with metal composition (Zn > Cd > Cu). Cu-MTs may accumulate with lysosomes. (Reprinted with permission from Dunn, M. A., Blalock, T. L., and Cousins, R. J., 1987. Minireview, *Metallothionein* 185:107.)

this treatment produced renal damage that was not caused by the same doses of an inorganic Cd salt. Acute renal toxicity of injected Cd-MT is the result of the intracellular release of Cd ions in the lysosomes on degradation of the protein. An alternate hypothesis suggests that Cd-MT is taken up by the renal tubules by pinocytosis and this leads to rupture of the cell membranes and mitochondrial damage.

The activity of certain SH-containing enzymes in the liver was not influenced by Cd in animals exposed repeatedly to this metal, suggesting that MT induced by Cd was protective. On the other hand, the same enzymes were inhibited when Cd was added to liver homogenate *in vitro* (Kagi and Nordberg, 1979).

It is known that acute effects of Cd in testes are preventable by pretreatment with Zn. The dose of Zn must be one or two orders of magnitude higher than that of Cd for its protective effects. It has been suggested that the protection afforded by Zn pretreatment is the result of the induction of MT.

MTs have also been implicated in the sequestration of other metals, such as Hg, Pb, Bi, Ag, Au, and Pt. Such effects have been claimed to be responsible for the development of resistance toward Au- and Pt-containing drugs in cultured cells and for the selective protection of some tissues from such agents in animals following preinduction of MT.

Tumor cell lines with acquired resistance to the antineoplastic agent *cis*-diamminedichloroplatinum(II) overexpressed MT and demonstrated cross-resistance to alkylating agents such as chlorambucil and melphalan. Human carcinoma cells that maintained high levels of MT because of chronic exposure to heavy metals were resistant to *cis*-diamminedichloroplatinum(II), melphalan, and chlorambucil. Thus, overexpression of MT represents one mechanism of resistance to a subset of clinically important anticancer drugs (Kelley *et al.,* 1988). If one could induce MT preferentially in healthy tissues as opposed to malignant cells, the effectiveness of treatment with anticancer agents might be greatly enhanced.

Although Zn, Cu, Hg, Ag, Co, Ni, and Bi induce MT-1 mRNA accumulation in both Cd-resistant and Cd-sensitive cells, the Cd-resistant cells show increased resistance to only a subset of these metals (Zn, Cu, Hg, and Bi). This suggests that not all metals which induce MT mRNA are detoxified by MT and argues against autoregulation of MT genes. MT-1 mRNA is also induced by iodoacetate, suggesting that the regulatory molecule has sensitive sulfhydryl groups.

MT is induced by many forms of chemical and physical stress. In some cases, such as in the exposure to electrophilic agents, i.e., O_2, free radicals, and alkylating agents, the increased supply of MT could provide the "neutralizing" nucleophilic equivalents. It has been suggested that MT behaves as a free radical scavenger and thus protects against UV and x-ray damage.

MTs may have metalloregulatory function in cellular repair processes, growth, and differentiation. This was first suggested by the parallelism of enhanced DNA synthesis with increased Zn-MT formation observed in the liver of rats recovering from partial hepatectomy (Ohtake *et al.,* 1978) and is supported by the programmed regulation of MT mRNA levels and of MT in the course of embryogenesis and in different stages of fetal and perinatal development (Nemer *et al.,* 1984; Andrews *et al.,* 1984). Inasmuch as zinc plays an important role in embryogenesis and participates in a number of DNA and RNA polymerases, and serves as a structural modulator of the zinc finger domains in several DNA-binding proteins, it is possible that Zn-MT plays a part in the storage, transmission, and expression of genetic information. The analogies in the Zn-binding sequence motifs between DNA-binding proteins and MT could imply a similar metal-regulated interaction of the latter with nucleic acids.

It has been proposed that MT could potentially modulate the movement of Cu and Zn within cells either directly through donation of these ions to metal-requiring apoenzymes or metabolic compartments (such as membranes) or indirectly by regulating their free or available concentrations.

The movement of metals through the cellular pool of MT molecules may occur in three ways: (1) release of bound metals concomitantly with the degradation of the protein; (2) exchange of bound metals with free metal ions and metals bound to specific ligands (including MT) without degradation of the protein; and (3) transfer of intact metal–protein complexes out of the cytoplasm into another cellular compartment, such as lysosomes.

Cultured mammalian cells that have lost their ability to make MTs (because of gene hypermethylation) are hypersensitive to cadmium toxicity. Conversely, cell lines with additional MT genes overproduce MTs and become highly resistant to cadmium. Limited evidence suggests that MTs also function to detoxify metals in intact animals.

Not all metals which induce MTs are detoxified by these proteins. Cell lines which overproduce MTs by gene amplification do not show increased resistance to Ag, Au, Co, or Ni, even though these metals induce MT synthesis. It is possible that these metals either do not bind to MTs or produce unstable complexes with the protein that are rapidly degraded.

A function for MT in the regulation of intestinal absorption of copper, zinc, and other metals has been proposed. Richards and Cousins (1975) have obtained experimental evidence suggesting that MT participates in the metabolism of Zn in both intestinal and hepatic cells. In rats pretreated with actinomycin D, MT synthesis was inhibited and there was a concomitant decrease in the hepatic uptake of zinc. In addition, absorption of oral ^{65}Zn was greater in rats pretreated with actinomycin D than in control animals. These data suggested that intestinal MT regulates the efflux of Zn from the intestinal cell to the body while hepatic MT is involved in the uptake of Zn from the blood to the liver.

It has been demonstrated that ^{64}Cu absorption is decreased in rats fed supplementary Cd (Cousins, 1983). The decreased passage of ^{64}Cu was associated with an increased incorporation of ^{64}Cu into a low-molecular-weight Cu-binding protein. These observations suggest that MT functions in the regulation of Cu absorption.

A clear inverse relationship between intestinal MT levels and absorption of copper and zinc has been observed. Zinc exchange rates with MT in intestinal cells is rapid, indicating that this protein could act as a dampening factor in absorption. The lethal-milk (*lm*) mutation in mice has been related to an overproduction of intestinal MT (Dunn *et al.*, 1987). Clinical signs of a systemic zinc deficiency develop and this genotypic expression of *lm* is reversed with zinc supplementation. Similarly, beneficial effects of zinc therapy in Wilson's disease patients which causes negative copper balance have been suggested to result from increased intestinal MT production and a concomitant decrease in copper absorption (Brewer *et al.,* 1983).

Several apoenzymes are known to be activated *in vitro* by Zn- and Cu-MTs (Kagi and Schaffer, 1988). As a homeostatic mediator MT could donate metal ions in the biosynthesis of Zn- and Cu-containing metalloenzymes and metalloproteins. The emergence of Cu_6-MT in *N. crassa* prior to the formation of the Cu-containing enzymes tyrosinase and laccase would be consistent with this role (Huber and Lerch, 1987). Also, *in vitro* experiments have shown that Cu and Zn can be transferred from MT to the apo forms of a number of Cu and Zn proteins, respectively (Beltramini and Lerch, 1982). Conversely, when Cu and Zn accumulate intracellularly, the reactive ions can be sequestered by binding to newly synthesized apo MT. Protection from the effects of excessive ionic Cu by sequestration is thus far the only documented benefit of MT induction in yeast, in *N. crassa,* and in Cu-resistant forms of *Agrostis gigantea* (Thiele *et al.,* 1986).

Activation experiments, however, must be interpreted with caution, inasmuch as an equilibrium could be established between any metal complex and an apoenzyme leading to an activation of the enzyme. This phenomenon has not been demonstrated *in vivo.*

MT gene expression is elevated in several tissues during fetal development (Nemer *et al.,* 1984; Andrews *et al.,* 1984), suggesting a role of MT in controlling metal ion concentrations important to growth and development. This is supported by the finding that growth-arrested human fibroblasts have lower levels of MT mRNA than do actively proliferating cells (Angel *et al.,* 1986). It has been observed, however, that cell lines that have lost their ability to make MTs are fully viable and grow normally, arguing against an essential role for MTs in growth (Compere and Palmiter, 1981; Crawford *et al.,* 1985). Inasmuch as many Zn- and Cu-dependent enzymes are involved in growth, this also argues against any essential role of MT in enzyme activation.

MT has been shown to be an efficient scavenger of free hydroxyl (-OH) ions *in vitro* (Thornalley and Vasak, 1985). Evidence for this centers on the markedly greater reactivity of MT with -OH radicals compared with superoxide radicals. It is believed that following reaction with -OH radicals, metal loss from MT and thiolate oxidation occurs. Regeneration of the metalloprotein could involve reduction with GSH and an appropriate divalent cation. If these reactions occur *in vivo,* scavenging of free radicals released during the acute-phase response could protect tissues from damage. Prevention of lipid peroxidation in hepatocytes in culture by Zn supplementation of culture medium has been correlated with MT induction (Dunn *et al.,* 1987). Free radicals are also responsible for tissue damage resulting from ionizing radiation. It has been observed that cells which overproduce MTs are more resistant to x-ray damage than normal cells (Bakka and Webb, 1981). Both x-ray treatment and ultraviolet light have been shown to induce MTs (Shiraishi *et al.,* 1986). Thus, MT may protect cells by reducing damage from free radicals. Alternatively, MT may serve as a source for zinc for enzymes that repair DNA or other tissue damage.

With respect to human health, it is very well accepted that MTs protect cells from several types of toxic metals. Oral zinc therapy produces a negative copper balance in Wilson's disease (a genetic disorder) patients by inducing MT synthesis and copper retention in intestinal cells. This may be a beneficial treatment for Wilson's disease once copper accumulation has been stabilized by chelation. Also, this treatment may be used as a preventive measure for family members of Wilson's disease patients who may be susceptible to copper toxicity. Pretreatment of animals with zinc to stimulate MT production may afford protection against lethal cadmium toxicity and increased urinary MT observed in factory workers exposed to large amounts of cadmium may be a manifestation of that protective response.

The potential function of MTs as a scavenger of damaging free radicals suggests a role of MTs in defense mechanisms. Involvement of MT in host defense is also suggested by the observation that as a result of infection, zinc from the plasma compartment is transferred to hepatocytes, macrophages, and perhaps the lymphocytes where MT synthesis is induced (Cousins *et al.,* 1986; Patierno *et al.,* 1983; Forre *et al.,* 1982). It has been shown that macrophages are more resistant to endotoxins when MT has been induced (Patierno and Peavy, 1984). Also, natural killer cell activity of lymphocytes is stimulated *in vitro* by treatments that stimulate MT production.

References

Andrews, G. K., Adamson, E. D., and Gedam, L., 1984. The ontogeny of expression of murine metallothionein: Comparison with the alpha-fetoprotein gene, *Dev. Biol.* 103:294.

Angel, P., Poting, A., Mallick, U., Rahmsdor, H. J., Schorpp, M., and Herrlich, P., 1986. Induction of metallothionein and other messenger-RNA species by carcinogens and tumor promoters in primary human skin fibroblasts, *Mol. Cell. Biol.* 6(5):1760.

Bakka, A., and Webb, M., 1981. Metabolism of zinc and copper in neonate: Changes in the concentrations and contents of thionein-bound Zn and Cu with age in the livers of the newborn of various mammalian species, *Biochem. Pharmacol.* 30:721.

Beltramini, M., and Lerch, K., 1982. Copper transfer between Neurospora copper metallothionein and type 3 copper apoprotiens, *FEBS Lett.* 142:219.

Bernhard, W. R., Vasak, M., and Kagi, J. H. R., 1986. Cadmium binding and metal cluster formation in metallothionein: A differential modification study, *Biochemistry* 25:1975.

Braun, W., Wagner, G., Worgotter, E., Vasak, M., Kagi, J. H. R., and Wuthrich, K., 1986. Polypeptide fold in the two metal clusters of metallothionein-2 by nuclear magnetic resonance in solution, *J. Mol. Biol.* 187:125.

Bremner, I., and Morrison, J. N., 1988. Metallothionein as an indicator of zinc status, in *Essential and Toxic Trace Elements in Human Health and Disease* (A. S. Prasad, ed.), Liss, New York, p. 365.

Brewer, G. J., Hill, G. M., Prasad, A. S., Cossack, Z. T., and Rabbani, P., 1983. Oral zinc therapy for Wilson's disease, *Ann. Intern. Med.* 99:314.

Butt, T. R., and Ecker, D. J., 1987. Yeast metallothionein and application in biotechnology, *Microbiol. Rev.* 51:351.

Compere, S. J., and Palmiter, R. D., 1981. DNA methylation controls the inducibility of mouse metallothionein-I gene in lymphoid cells, *Cell* 25:233.

Cousins, R. J., 1983. Metallothionein—Aspects related to copper and zinc metabolism, *J. Inher. Metab. Dis.* 6(Suppl 1):15.

Cousins, R. J., Dunn, M. A., Leinart, A. S., Yedinak, K. C., and DiSilvestro, R. A., 1986. Coordinate regulation of zinc metabolism and metallothionein gene expression in rats, *Am. J. Physiol.* 251:E688.

Crawford, B. D., Enger, M. D., Griffith, B. B., Griffith, J. D., Hanners, J. L., Longmire, J. L., Munk, A. C., Stallings, R. L., Tesmer, J. G., Walkers, R. A., and Hildebrand, C. E., 1985. Coordinate amplification of metallothionein I and II genes in cadmium-resistant Chinese hamster cells: Implications for mechanisms regulating metallothionein gene expression, *Mol. Cell. Biol.* 5:320.

Dunn, M. A., Blalock, T. L., and Cousins, R. J., 1987. Proc. Soc. Exp. Biol. Med. Metallothionein 185:107.

Forre, A. O., Aaseth, J., Dobloug, J., and Ovreb, O., 1982. Induction of a metallothionein-like protein in human lymphocytes, *Scand. J. Immunol.* 15:217.

Fowler, B. A., Hildebrand, C. E., Kojima, Y., and Webb, M., 1987. Nomenclature of metallothionein, *Experientia Suppl.* 52:19.

Furey, W. F., Robbins, A. H., Clancy, L. L., Winge, D. R., Wang, B. C., and Stout, C. D., 1986. Crystal structure of Cd, Zn metallothionein, *Science* 231:704.

Grider, A., Kao, K., Klein, P. A., and Cousins, R. J., 1989. Enzyme-linked immunosorbent assay for human metallothionein: Correlation of induction with infection, *J. Lab. Clin. Med.* 113:221.

Huber, M., and Lerch, K., 1987. The influence of copper on the induction of tyrosinase and laccase in Neurospora crassa, *FEBS Lett.* 219:335.

Hunziker, P. E., and Kagi, J. H. R., 1988. Metallothionein: A multigene protein, in *Essential and Toxic Trace Elements in Human Health and Disease* (A. S. Prasad, ed.), Liss, New York, p. 349.

Kagi, J. H. R., and Nordberg, M., 1979. Metallothionein, *Experientia Suppl.* 34:41.

Kagi, J. H. R., and Schaffer, A., 1988. Biochemistry of metallothionein, *Biochemistry* 27:8509.

Kagi, J. H. R., and Vallee, B. L., 1960. Metallothionein: A cadmium- and zinc-containing protein from equine renal cortex, *J. Biol. Chem.* 235:3460.

Kagi, J. H. R., Himmelhoch, S. R., Whanger, P. D., Bethune, J. L., and Vallee, B. L., 1974. Equine hepatic and renal metallothioneins, *J. Biol. Chem.* 249:3537.

Kelley, S. L., Basu, A., Teicher, B. A., Hacker, M. P., Hamer, D. H., and Lazo, J. S., 1988. Overexpression of metallothionein confers resistance to anticancer drugs, *Science* 241:1813.

Kojima, Y., Berger, C., Vallee, B. L., and Kagi, J. H. R., 1976. Amino-acid sequence of equine renal metallothionein-1B, *Proc. Natl. Acad. Sci. USA* 73:3413.

Margoshes, M., and Vallee, B. L., 1957. A cadmium protein from equine kidney cortex, *J. Am. Chem. Soc.* 79:4813.

Mehra, R. K., and Bremner, I., 1983. Development of a radioimmunoassay for rat liver metallothionein-I and its application to the analysis of rat plasma and kidneys, *Biochem. J.* 213:459.

Nemer, M., Travaglini, E. C., Rondinell, E., and D'Alonzo, J., 1984. Developmental regulation, induction, and embryonic tissue specificity of sea urchin metallothionein gene expression, *Dev. Biol.* 102:471.

Ohtake, H., Haswgawa, K., and Koga, M., 1978. Zinc-binding protein in the livers of neonatal, normal and partially hepatectomized rats, *Biochem. J.* 174:999.

Palmiter, R. D., 1987. Molecular biology of metallothionein gene expression, *Experientia Suppl.* 52:63.

Palmiter, R. D., and Brinster, R. L., 1985. Transgenic mice, *Cell* 41:343.

Palmiter, R. D., Norstedt, G., Gelinas, R. E., Hammer, R. E., and Brinster, R. L., 1983. Metallothionein–human GH fusion genes stimulate growth of mice, *Science* 222:809.

Patierno, S. R., and Peavy, D. L., 1984. Induction of metallothionein in macrophages: A molecular mechanism for protection against LPS-mediated autolysis, *Immunopharmacol. Endotoxicosis* 39.

Patierno, S. R., Costa, M., Lewis, V. M., and Peavy, D. L., 1983. Inhibition of LPS toxicity for macrophages by metallothionein-inducing agents, *J. Immunol.* 130:1924.

Piscator, M., 1964. On cadmium in normal human kidneys together with a report on the isolation of metallothionein from livers of cadmium-exposed rabbits, *Nord. Hyg. Tidskr.* 45:76.

Pulido, P., Kagi, J. H. R., and Vallee, B. L., 1966. Isolation and some properties of human metallothionein, *Biochemistry* 5:1768.

Richards, M. P., and Cousins, R. J., 1975. Mammalian zinc homeostasis: Requirement for RNA and metallothionein synthesis, *Biochem. Biophys. Res. Commun.* 64:1215.

Seguin, C., and Hamer, D. H., 1987. Regulation in vitro of metallothionein gene binding factors, *Science* 235:1383.

Shaikh, Z. A., and Smith, J. C., 1975a. Cadmium induced synthesis of hepatic and renal metal-lothionein, *Fed. Proc.* 34:266.

Shaikh, Z. A., and Smith, J. C., 1975b. Mercury-induced synthesis of renal metallothionein, International Conference on Heavy Metals in the Environment, Abstracts, Toronto, Canada, October, 1975, p. B108.

Shaikh, Z. A., and Smith, J. C., 1977. The mechanisms of hepatic and renal metallothionein biosynthesis in cadmium-exposed rats, *Chem. Biol. Interact.* 19:161.

Shiraishi, N., Yamamoto, H., Takeda, Y., Kondoh, S., Hayashi, H., Hashimoto, K., and Aono, K., 1986. Increased metallothionein content in rat liver and kidney following X-irradiation, *Toxicol. Appl. Pharmacol.* 85:128.

Squibb, K. S., and Cousins, R. J., 1974. Control of cadmium binding protein synthesis in rat liver, *Environ. Physiol. Biochem.* 4:24.

Squibb, K. S., Cousins, R. J., and Feldman, S. L., 1977. Control of zinc-thionein synthesis in rat liver, *Biochem. J.* 164:223.

Thiele, D., Walling, M. J., and Hamer, D., 1986. Mammalian metallothionein is functional in yeast, *Science* 231:854.

Thomas, D. G., Linton, H. J., and Garvey, J. S., 1986. Fluorometric ELISA for the detection and quantification of metallothionein, *J. Immunol. Methods* 89:239.

Thornally, P. J., and Vasak, M., 1985. Possible role for metallothionein in protection against radiation-induced oxidative stress. Kinetics and mechanism of its reaction with superoxide and hydroxyl radicals, *Biochim. Biophys. Acta* 827:36.

Zinc and Hormones   5

Zinc's effects on various hormones will be summarized in this chapter.

5.1 Zinc and Growth

Zinc supplementation studies first performed in Egypt demonstrated reversible growth retardation in zinc-deficient human subjects (Prasad, 1966; Sandstead *et al.*, 1967). Following zinc supplementation, the average increment in height in male subjects between the ages of 14 and 20 (average 17 years) was 5.0 inches per year. Biochemical studies were consistent with the concept that the growth retardation in these subjects was the result of deficiency of zinc.

In a large clinical trial of zinc supplementation at the U.S. Naval Medical Research Unit No. 3, Cairo, Egypt, it was demonstrated that although zinc-supplemented subjects showed a remarkable increase in height, body weight, and gonadal development, they remained anemic. On the other hand, iron-supplemented subjects showed a normal hemoglobin level, but experienced much less effect on growth and gonads than did the zinc-supplemented group.

Later studies by Halsted *et al.* (1972) showed that the development in 19- and 20-year-old subjects receiving a well-balanced diet alone was slow, whereas the effect on height increment and onset of sexual function was strikingly enhanced in those receiving zinc. Additionally, two women described in their report represented the first cases of dwarfism in females caused by zinc deficiency. The 15 men in this study had been rejected by the Iranian Induction Center because of "malnutrition," were all 19 or 20 years of age, and had clinical features similar to those of zinc-deficient dwarfs reported earlier by Prasad *et al.* (1961, 1963a,b). These men were divided into two groups: One of them received a well-balanced diet rich in animal protein plus a placebo capsule for 6 months. A second group received the same diet plus a capsule containing 27 mg of zinc as zinc sulfate for the same period. A third group was given the above diet without additional medication for 6 months, followed by the diet plus zinc for a 6-month period. Zinc supplements resulted in an earlier onset of sexual function, defined by nocturnal emission in males, and menarche in females.

In Turkey, geophagia is common and is a common cause of iron and zinc deficiencies in adolescents. Growth retardation and hypogonadism have been related to zinc deficiency in such cases, and zinc supplementation has resulted in complete correction of these problems (Cavdar *et al.*, 1980). The average increase in height was 4.4 cm after 3 months and 10.4 cm after 6 months of zinc supplementation. The diet of Turkish villagers, consisting mainly of wheat bread that is rich in phytate, which decreases the availability of iron and zinc, may be a major etiologic factor in these deficiencies. Indeed, the deficiency of zinc may be one of the major nutritional problems in Turkey.

Hambidge and Walravens (1976) proposed that marginal zinc deficiency could occur among children in the United States. While surveying zinc concentrations in the hair of normal children in the Denver area, they found that ten subjects between the ages of 4 and 6 years had poor appetite, impaired taste acuity, mild retardation in growth, and decreased concentration of zinc in the hair. These abnormalities were corrected by supplementation with zinc. It was also observed that in the zinc-deficient children, growth rates had typically started to decline in infancy, and weight-for-age percentiles declined before linear growth was affected (Walravens *et al.*, 1989). A recent study found that the zinc-supplemented group of infants who showed nutritional failure to thrive, demonstrated a significant improvement in weight gain relative to placebo-treated controls (Walravens *et al.*, 1989). Thus, it was concluded that a mild zinc deficiency may be one of the etiologic factors in nutritional failure to thrive during infancy in the United States.

Butrimovitz and Purdy (1978) measured plasma zinc concentrations of inner-city children in Baltimore and found them to be lower than the levels of Denver children of middle-income families. This study also demonstrated that the lowest levels of plasma zinc occurred during infancy and puberty, two periods characterized by rapid growth. After curve fitting of the plasma zinc concentration data, these investigators were able to demonstrate an inverse relationship between plasma zinc levels and growth index, an indicator of the relative rate of growth. The authors suggested that depletion of body stores of zinc may occur during periods of rapid growth and that subjects subsisting on a diet low in animal protein would benefit from higher intakes during such periods.

Ghavami-Maibodi *et al.* (1983) studied 13 short children from Nassau County (New York) who had retarded bone age and low hair zinc concentrations, and who were treated with oral zinc supplements for 1 year. The children were divided retrospectively into two groups: those who grew less than 6 cm/year and those who grew more than 6 cm/year while receiving oral zinc supplementation. Both groups of children (ages 9.3 and 10.8 years) had grown poorly during the previous year (3.2 and 2.6 cm, respectively). During the year of zinc supplementation, cumulative growth was 4.1 and 7.5 cm, respectively. The children who grew more than 6 cm/year had significant increases in hair, serum, and urine zinc within 2 months of starting zinc supplementation. The children who grew less than 6 cm/year had insignificant changes in hair, serum, and urine zinc concentrations. In the eight children who grew more than 6 cm/year during 6 months of oral zinc supplementation, serum testosterone increased from 40 to 280 ng/dl ($p < 0.002$), somatomedin C increased from 0.29 to 1.0 μg/ml ($p < 0.001$), and the peak growth hormone response to insulin and to arginine increased from 14 and 19 ng/ml to 25 and 24 ng/ml (insulin peak, $p = 0.0005$; arginine peak, p NS). These findings indicate that a significant number of short children are zinc deficient and that zinc supplementation should be beneficial to them. The interrelationship between growth hormone, somatomedin, and zinc appears to be complex and deserves further investigation.

We and others have reported that patients with adult sickle-cell anemia may suffer from zinc deficiency resulting in growth retardation, hypogonadism in men, hyperammonemia, abnormal dark adaptation, and dysfunction of cell-mediated immunity (Prasad et al., 1975, 1979, 1981; Neill et al., 1979; Prasad, 1981; Prasad and Cossack, 1984; Abbasi et al., 1976; Warth et al., 1981). A limited trial with zinc supplementation found enhancement of secondary sexual characteristics; normalization of the plasma ammonia level; reversal of the dark adaptation abnormality; increased plasma, erythrocyte, and neutrophil zinc levels; and the expected response in the activities of the zinc-dependent enzymes.

The effect of zinc on growth in this disease was corroborated and characterized in a controlled trial of zinc supplementation in a group of 14- to 18-year-old patients with growth retardation (Prasad and Cossack, 1984). Zinc supplementation fully corrected the diminished growth rate in our subjects with sickle-cell anemia.

The administration of bovine growth hormone to nonhypophysectomized, zinc-deficient rats failed to enhance growth, whereas the growth rate in these animals increased after zinc supplementation (Prasad et al., 1969). The growth rate of hypophysectomized rats, however, responded to both hormone and zinc supplementation, regardless of zinc status. It appears that the effects of growth hormone and zinc on the growth of rats are independent and additive. One animal study suggests that somatomedin may be zinc-dependent, but additional experiments are needed to elucidate the interrelationship of somatomedin, growth hormone, and zinc in humans.

A study was conducted on a cohort of 476 women (364 black, 112 white), who attended the Jefferson County Health Department Clinic in Birmingham, Alabama, for their prenatal care, in order to determine the relationship between maternal serum zinc concentration early in pregnancy and birth weight (Neggers *et al.*, 1990). A significant correlation between maternal serum zinc and birth weight was observed in all subjects after various independent determinants of birth weight were controlled for. Pregnant women who had a serum zinc concentration in the lowest quartile had a significantly higher prevalence of low birth weight than did those mothers who had serum zinc concentration in the upper three quartiles during pregnancy. These important findings suggest that maternal serum zinc concentration measured early in pregnancy could be used to identify those women who may be at higher risk of giving birth to a low-birth-weight infant.

A double-blind zinc supplementation trial was conducted recently in low-income, pregnant adolescents thought to be at risk for poor zinc intake (Cherry *et al.*, 1989). Subjects were randomly assigned to receive 30 mg zinc (as gluconate) or placebo. Response to zinc was related to maternal weight. Infants born of normal-weight mothers given zinc had reduced rates of prematurity ($p = 0.05$) and assisted respiration ($p = 0.0006$). Zinc supplementation improved pregnancy outcome in normal-weight women and in underweight multiparas. The nonresponse in underweight primiparas was most likely the result of multiple limiting factors including zinc.

A double-blind, pair-matched study from Canada examined the effects of a zinc supplement of 10 mg for 1 year on linear growth, taste acuity, attention span, biochemical indices, and energy intakes of 60 boys aged 5 to 7 years (Gibson *et al.*, 1989). The inclusion criteria for study were (1) height for age $\leq$ 15th percentile, (2) parent height $>$ 25th percentile, (3) Caucasian, (4) full term with weight at birth appropriate for gestational ages, and (5) apparently healthy with no detectable medical reasons for poor growth.

Boys with initial hair zinc < 1.68 μmol/g had a lower mean ($\pm$ S.D.) weight-for-age Z score (-0.44 ± 0.59 versus -0.08 ± 0.84) and a higher median recognition threshold for salt (15 versus 7.5 mmol/g) than those with hair zinc > 1.68 μmol/g. Only boys with hair zinc < 1.68 μmol/g responded to the zinc supplement with a higher mean change in height-for-age Z score ($p < 0.05$). Taste acuity, energy intakes, and attention span were unaffected. This study documented a syndrome in boys with low weight percentiles, hair zinc levels < 1.68 μmol/g, and impaired taste acuity in southern Ontario, Canada.

Zinc deficiency becomes apparent during nutritional rehabilitation and limits the range of weight gain, according to a study reported from Bangladesh (Simmer *et al.*, 1988). Twenty-five severely malnourished children, who were admitted to the Children's Nutrition unit in Bangladesh, were alternately assigned to two 3.7 mg/day groups. Their mean dietary zinc intake was 3.7 mg/day. One group received a daily zinc supplement of 50 mg for 2 weeks. During the first week, weight gain was similar in both groups, but during the second week, weight gain was 73% higher in the zinc-supplemented group. These results suggest that zinc supplements must be administered to severely malnourished children during nutritional rehabilitation.

Zinc was assayed in plasma and hair in 703 Chinese children aged between 1 and 6 years, and correlated with parameters of growth and development. In the first group of 187 children seen in the Child Health Clinic for routine observation, a positive correlation of hair zinc concentration and height for age was observed and there was an increased prevalence of low hair zinc in children of short stature (Xue-Cun *et al.*, 1985). Another group of 303 children in nurseries and kindergartens in Beijing exhibited a hair zinc level of 92 μg/g, and 34% of these had very low zinc levels (<70 μg/g). In the third group of 213 children who were brought to the outpatient clinic for various reasons such as pica, anorexia, and poor growth, significantly decreased levels of zinc in hair and plasma relative to well-nourished children were noted. These subjects responded to zinc supplementation with improvement in growth and disappearance of pica and anorexia. This study suggests that the diet consumed by the Chinese population may be marginal or inadequate in its content of available zinc.

Zinc deficiency is known to cause growth retardation and delayed sexual development (Prasad *et al.*, 1963a). Zinc supplementation significantly increased growth rates and growth hormone (GH) level of children with both GH and zinc deficiencies (Collip *et al.*, 1982). Some children with constitutional growth delay and delayed adolescence have been observed to demonstrate partial GH deficiency and zinc deficiency (Gourmelen *et al.*, 1979). One boy aged 13 years and 4 months with partial GH deficiency caused by chronic mild zinc deficiency showed a significant increase in growth rate and an increase in GH level following zinc administration (Nishi *et al.*, 1989; Castro-Magana *et al.*, 1981; Ghavami-Maibodi *et al.*, 1983). A state of chronic but mild zinc deficiency in this case was a result of poor dietary zinc intake and possibly renal wasting of zinc. This subject ate only rice and vegetables; animal protein rich in zinc was not included in his diet.

A complex interrelationship exists between zinc, growth hormone, and androgens. Zinc levels of serum and hair in boys beyond Tanner stage 3 of genital development were significantly higher than in stages 1 and 2. A good correlation was also observed between the stage of genital development, serum testosterone level, and zinc levels in serum and hair (Castro-Magana *et al.*, 1981).

The mean concentration of zinc in plasma and urine decreased in patients with GH deficiency after GH injection, whereas after adenoectomy in patients with acromegaly, zinc increased in plasma and decreased in urine (Aihara *et al.*, 1985). These findings may suggest a negative zinc balance and mild zinc deficiency in some patients with GH deficiency on long-term GH replacement or in patients with untreated acromegaly. This is further supported by the observation of Cheruvanky *et al.* (1982) who showed that oral zinc supplements improved the growth rate in patients with GH deficiency on long-term GH therapy. Thus, one must rule out GH deficiency caused by zinc deficiency in patients with growth retardation.

In one study, the zinc-deficient rats showed narrower tibial epiphyseal growth plates than the control rats (Kurtoglu *et al.*, 1987). The number of hypertrophied cells in the epiphyseal plate was decreased in the zinc-deficient rats and GH administration was ineffective in the presence of zinc deficiency.

It is possible that zinc deficiency affects somatogenic binding of GH to liver cells. Another possibility is that receptor unresponsiveness may develop as a result of zinc deficiency. Somatomedin C levels in zinc-deficient rats were reported to be decreased relative to the control group, suggesting that another possible effect of zinc deficiency on growth may be mediated by its effect on somatomedin C.

It is possible that the growth retardation seen in zinc deficiency may be related to either resistance to GH or decreased activity of somatomedin C.

One investigator has shown that plasma zinc varies inversely and urinary zinc varies directly with plasma GH levels (Henkin, 1974b). It has been suggested that treatment of hypopituitary dwarfs with human GH may increase the zinc requirement because of growth. Other reports have not shown any effect of GH therapy on zinc status of subjects. Hypopituitarism resulting in growth retardation is not associated with altered zinc status (Solomons *et al.*, 1976). On the other hand, growth retardation caused by malnutrition is associated with zinc deficiency.

Studies in the 1940s and 1950s demonstrated that crude extracts of pituitary hormones contained zinc and that zinc prolonged the biological activity of these extracts. It appears that as with insulin, the significance of interaction of zinc with steroids and steroid-releasing hormones is partly related to the ability of zinc to prolong the biological activity of these hormones.

In one study $^{35}SO_4$ uptake by glycosaminoglycans (GAG) was significantly lower in zinc-deficient than in either ad libitum-fed control or pair-fed control rat epiphysis (Bolze *et al.*, 1987). Normal growth depends on the regulated influence of both hormonal and nutritional factors. GH is believed to act by increasing circulating levels of somatomedin (Sm), or insulinlike growth factors (IGF) which directly stimulate long bone growth via GAG synthesis (Bolze *et al.*, 1987). In the same study the activity of xylosyltransferase, an enzyme required for GAG synthesis, was observed to be significantly decreased in zinc-deficient rats relative to controls.

Sulfation is the final step in proteoglycan synthesis and the incorporation of $^{35}SO_4$ is a useful marker signaling completion of the GAG chain. Some investigators, using radioimmunoassay technique, have shown that Sm C levels are decreased in zinc-deficient rats relative to controls (pair-fed and ad libitum-fed). Decreased Sm activity has been associated with decreased activity of xylosyltransferase. Thus, decreased sulfation of GAG and decreased xylosyltransferase activity in rat rib epiphyses as a result of zinc deficiency may have been related to Sm levels.

Exposure to bacterial endotoxins is known to stimulate the release of anterior pituitary hormones. Endotoxins are also a potent stimulus for production of interleukin 1 (IL-1) by macrophages and monocytes. Recently, the possibility that IL-1 has a direct effect on the secretion of hormones by rat pituitary cells in a monolayer culture has been investigated by Bernton *et al.* (1987). Recombinant human IL-1β (concentration range 10^{-9} M to 10^{-12} M) stimulated the secretion of adrenocorticotropic hormone, luteinizing hormone (LH), GH, and thyroid-stimulating hormone. Prolactin secretion by the monolayers was inhibited by similar doses of IL-1. The concentrations of IL-1 used in these experiments were within the ranges reported for IL-1 in serum, suggesting that IL-1 generated peripherally by mononuclear cells may affect anterior pituitary cells to modulate hormone secretion *in vivo*. Incubation of IL-1 solutions with antibody to IL-1 neutralized the above effects.

Our recent studies in human volunteers found a decreased production of IL-1 by mononuclear cells *in vitro* when dietary zinc intake was restricted in the volunteers (personal observation). Inasmuch as GH may be decreased and serum prolactin increased in zinc deficiency, one may propose that these hormonal effects are mediated through IL-1.

Recent studies have shown that zinc is required for tight binding of human GH (hGH) to the prolactin receptor, thus demonstrating that a metal ion mediates a direct interaction between a polypeptide hormone and its extracellular receptor (Cunningham *et al.*, 1990). Mutational analysis showed that a cluster of three residues (His-18, His-21, Glu-174) in hGH and His-188 from the human prolactin receptor are most likely zinc ligands. These data provide a possible molecular basis for the association between zinc deficiency and altered GH actions.

5.2 Zinc and Reproductive Functions

Genital and secondary sexual development were retarded in all zinc-deficient male dwarfs from Egypt (Prasad *et al.*, 1961, 1963a,b). Pubic hair was limited to a few fine darkly pigmented hairs at the base of the penis. Facial hair with similar characteristics was present on the upper lip, while axillary hair was absent. Treatment with zinc was followed by surprisingly accelerated sexual maturation. In some individuals, pigmented pubic, extremity, and facial hair appeared within 3 weeks of beginning zinc treatment and was accompanied by maturation changes in the penis and scrotum.

We examined the relationship between zinc deficiency and hypogonadism in 32 adult male patients with sickle-cell anemia (Abbasi *et al.*, 1976). Pubic hair growth was delayed, secondary sex characteristics were abnormal in 29, and eunuchoidal skeletal proportions were present in all except one. Basal serum testosterone, dihydrotestosterone, and androstenedione levels were lower than control values in the 14 patients in whom these assays were performed. Serum LH and follicle-stimulating hormone (FSH) levels before and after stimulation with gonadotropin-releasing (GnRH) hormone were consistent with primary testicular failure and not secondary hypogonadism. Erythrocyte and hair zinc concentrations were significantly decreased, and there was a positive correlation between erythrocyte zinc and serum testosterone ($r = 0.61$, $p < 0.0001$). The androgen deficiency in these patients was corrected with zinc supplements (Prasad *et al.*, 1981). In the first experiment, four subjects received placebo for 12 months, followed by oral zinc supplements (15 mg tid as zinc acetate) for 12 months. In the second experiment, 14 age-matched subjects were divided into two groups, 7 subjects receiving placebo and 7 receiving oral zinc supplements (15 mg of zinc three times as acetate). The first three pairs were treated for 6 months, following which zinc supplementation was discontinued and all of the subjects were observed for an additional 7 months. One pair received the supplementation for 18 months. The periods of supplementation were varied to determine the length of time necessary to obtain maximum effects of serum testosterone levels. The controls for these studies were healthy age-matched subjects.

In the first experiment, plasma zinc, neutrophil zinc, neutrophil alkaline phosphatate activity, and basal serum testosterone levels increased significantly in each subject after zinc but not placebo treatment ($p < 0.01$). Peak serum testosterone and the mean serum testosterone levels following administration of GnRH also increased significantly following supplementation with zinc ($p < 0.05$). In the second experiment, the effect of zinc supplementation appeared to be maximal at the end of 6 months, levels of basal serum testosterone, dihydrotestosterone (DHT), plasma zinc, neutrophil zinc, and neutrophil alkaline phosphatase activity being significantly greater in zinc-treated than in placebo-treated subjects ($p < 0.01$). In the three subjects in whom zinc supplementation was subsequently discontinued for 6 months, there was a significant decline in serum testosterone, neutrophil zinc, and neutrophil alkaline phosphatase activities ($p < 0.01$).

Zinc deficiency may account for the persistence of gonadal dysfunction despite adequate dialysis in uremic men (Mahajan *et al.*, 1982). Twenty patients undergoing hemodialysis three times a week completed a double-blind therapeutic trial using either 50 mg of elemental zinc given orally as zinc acetate (10 patients) or placebo. At the end of the 6-month period, a significant increase in plasma zinc (75 ± 2 μg/dl to 100 ± μg/dl, mean ± S.E., $p < 0.001$), serum testosterone (2.8 ± 0.3 ng/ml, $p < 0.001$), and sperm count (30 ± 3 to 63 ± 5 million per ml, $p < 0.001$) occurred in the zinc-treated group, but not in those receiving placebo. The zinc-treated group also had a significant fall in serum LH (92 ± 10 mIU/ml to 29 ± 26 mIU/ml, $p < 0.05$) and FSH (45 ± 9 mIU/ml to 25 ± 7 mIU/ml, $p < 0.005$), which was not seen in the placebo group. Patients receiving zinc had improved potency, libido, and frequency of intercourse that was not observed in the placebo group. These results suggest that zinc deficiency is a reversible cause of gonadal dysfunction in patients having regular hemodialysis.

A role of zinc in gonadal function is supported by results from animal studies (Lei *et al.*, 1976). Body weight, testicular weight, and testicular zinc content are significantly lower in zinc-deficient rats than in food-restricted (pair-fed) control rats (Tables 5-1 and 5-2). Serum LH and FSH response to intravenous LHRH are higher, and the serum testosterone response is lower in zinc-deficient than in control animals. These findings suggest that zinc has a specific effect on testes and that gonadal function in zinc-deficient rats is affected through some alteration of testicular steroidogenesis. The mechanism by which zinc affects testosterone production is not well understood. It is possible that steroidogenesis may involve one or more zinc-dependent enzymes. Alternatively, testosterone production may decline secondary to the

Table 5-1. Effect of Zinc Deficiency on Body Weight Gain, Wet and Dry Weights and Zinc Content of Testes[a,b]

	Zinc-deficient rats	Control rats	Treatment comparison, p
Body weight gain, g/6 weeks	27.48 ± 3.02 (23)	91.11 ± 2.43 (24)	<0.005
Wet weight of testes, g	1.69 ± 0.12 (23)	3.29 ± 0.07 (24)	<0.005
Dry weight of left testis, g	0.127 ± 0.011 (11)	0.250 ± 0.046 (12)	<0.005
Zinc content of left testis, μg/g dry wt	152.6 ± 6.3 (11)	196.1 ± 2.9 (12)	<0.005

[a] Adapted with permission from Lei, K. Y., Abbasi, A. A., and Prasad, A. S., 1976. Function of pituitary–gonadal axis in zinc-deficient rats, *Am. J. Physiol.* 230:1730.
[b] Values are means ± S.E.; number of rats in parentheses.

Table 5-2. Effect of Zinc Deficiency on Response of Serum LH, FSH, and Testosterone after LHRH Injection in Rat[a,b]

Hormone	Rats	Time after LHRH injection, min					Analysis of variance[c] (treatment comparisons, p)		
		0	5	15	30	60	Zinc deficient vs. control	Time	Interactions
LH, ng/ml	Zinc deficient (n = 11)	31.4 ± 7.7	439.8 ± 42.9	725.0 ± 77.7	763.1 ± 59.7	748.6 ± 79.0	0.005	0.005	NS
	Control (n = 12)	92.3 ± 51.4	270.7 ± 62.0	544.6 ± 103.1	548.0 ± 113.3	518.9 ± 101.8			
FSH, ng/ml	Zinc deficient (n = 6)	766 ± 192	1112 ± 226	1275 ± 285	1575 ± 206	1725 ± 220	0.005	0.005	NS
	Control (n = 6)	260 ± 46	378 ± 50	716 ± 181	673 ± 103	731 ± 101			

Hormone	Rats	Time after LHRH injection, h					Analysis of variance[c] (treatment comparisons, p)		
		0	2	3	4	5	Zinc deficient vs. control	Time	Interactions
Testosterone, ng/ml	Zinc deficient (n = 11)	7.40 ± 2.80	29.05 ± 4.32	23.83 ± 3.87	24.09 ± 4.52	14.37 ± 2.77	0.01	0.005	NS
	Control (n = 12)	6.14 ± 2.04	32.44 ± 2.61	45.79 ± 7.40	35.50 ± 8.69	18.54 ± 2.80			

[a] Reprinted with permission from Lei, K. Y., Abbasi, A. A., and Prasad, A. S., 1976. Function of pituitary–gonadal axis in zinc-deficient rats, Am. J. Physiol. 230:1730.
[b] Values are means ± S.E.
[c] Degree of freedom for zinc effect, 1; degrees of freedom for time effect, 4; degrees of freedom for interactions, 4. NS, not significant.

decrease in testicular size observed in zinc deficiency. Zinc is required for cell division in general, and the testis appears to be very sensitive in this respect. The Sertoli cells, which are target cells for FSH as well as testosterone (both of which are involved in the maintenance and initiation of spermatogenesis), produce an androgen-binding protein (ABP) that plays a role in transporting active androgens from an interstitial compartment through the Sertoli cells in the germinal epithelium. ABP in the testicular fluid also enhances the transport of testosterone from the testis to the epididymis via the efferent testicular fluid. Therefore, ABP provides an androgen concentrating mechanism in the seminiferous tubules as well as in the proximal parts of the epididymis. We examined the possibility that an abnormality in the ABP contributes to the decreased spermatogenesis in zinc deficiency (Meftah *et al.*, 1984). In these experiments, ABP, testosterone, and DHT were measured in the testicular cytosol of zinc-deficient, pair-fed, and ad libitum-fed control rats. Although testosterone, DHT, and DHT-binding protein sites per pair of testes were all reduced in zinc-deficient animals compared with controls ($p < 0.02$), the results were not significantly different when expressed per milligram of protein. Serum testosterone and DHT, testicular weights, and tibial and testicular zinc were also decreased in zinc-deficient rats ($p < 0.02$). These results suggest that the major effect of zinc deficiency is on testicular growth, supporting the postulate that cell division is adversely affected by a lack of zinc.

Although hypogonadism in males resulting from zinc deficiency appears to be more common than menstrual abnormalities in women, zinc does play an important role in pregnancy and lactation. A study on the cytogenetic effects of dietary zinc deficiency on oogenesis and spermatogenesis in mice (Watanabe *et al.*, 1983) demonstrated that the incidence of degenerated oocytes was very high in animals with moderately severe zinc deficiency, and that, even in marginally zinc-deficient mice, hypohaploidy and hyperhaploidy in metaphase II oocytes were significantly increased. In contrast, the incidence of chromosomal aberrations in spermatocytes was unaffected by zinc deficiency. If these results are applicable to humans, they underscore the importance of maintaining adequate zinc status in women before conception.

Another study suggests a very important physiologic role of prostatic zinc as a preserver of an inherent mechanism for head–tail connection of human spermatozoa (Bjorndahl and Kvist, 1982). Finally, the activities of the 3-α and D-3-β steroid dehydrogenases in human prostatic tissue have been found to correlate inversely with zinc concentrations (Leake *et al.*, 1984). This relationship may explain the known increased 5-α-DHT content in prostatic tissues with high zinc concentrations.

Plasma LH is increased but testosterone level is decreased in zinc-deficient adult and immature male rats (Lei *et al.*, 1976; McClain *et al.*, 1984). McClain *et al.* (1984) have related hypogonadism in the rat to Leydig cell failure resulting in decreased testosterone synthesis. According to one study of rats with unilateral maldescended testis, the ectopic testis consistently showed decreased zinc content, while the eutopic testis had a normal zinc level (Chan *et al.,* 1986). Uptake of zinc by the ectopic testis was unaltered. Sephadex gel chromatography showed greatly reduced zinc content of one of the endogenous zinc binding fractions within the cytosol of the ectopic testis in spite of a near-normal protein content. Incorporation of ^{65}Zn into this fraction was also greatly reduced in the ectopic testis. Sodium dodecyl sulfate–polyacrylamide gel electrophoresis demonstrated that a protein of 23 kDa was greatly reduced in quantity. It was suggested that this protein in the ectopic testis may have a role in reduced testicular function (Chan *et al.,* 1986).

The highly S–S cross-linked mammalian sperm nuclear chromatin gives the sperm nucleus its high resistance to different chemical and physical conditions. Evidence has been provided that the nuclear chromatin decondensation (NCD) ability is reversibly inhibited by zinc derived from the prostatic fluid upon ejaculation (Kvist, 1982). It has been observed that the spermatozoa must remain in the female genital tract for several hours to gain its fertilization capacity and that they release zinc during this period. Thus, the role of zinc might be of physiological importance such that it protects sperm in the male by inhibiting NCD ability. NCD ability is reactivated by depletion of zinc from the sperm in the female genital tract. It is generally believed that the ovum provides free-thiol-enzyme systems of high efficiency to cleave the S–S bonds.

Zinc deficiency appears to interfere with secretion of gonadotropins and zinc may be an important component or modulator of hormone receptors. Zinc-deficient pregnant rats have prolonged and delayed parturition (Cunnane, 1981; Cunnane *et al.,* 1983). Delayed progesterone withdrawal is a possible explanation for this phenomenon (Chung *et al.,* 1986). A decreased synthesis of prostaglandin $F_{2\alpha}$ in zinc deficiency may result in delayed activation of the enzyme 20α-hydroxysteroid dehydrogenase which in turn would lead to an elevated serum progesterone level. The number of androgen-binding sites in the cytosol fraction of the zinc-deficient rat prostate was reported to be significantly lower than that of control rats (Chung *et al.,* 1986), although their dissociation constants did not differ. These results suggest that zinc is involved in the androgen-binding process in the target cells.

The androgen receptor (AR) gene has been localized on the human X chromosome between the centromere and q13 (Lubahn *et al.,* 1988). Development of male external genitalia in human embryos and virilization of the pubertal male are dependent on androgen binding to its receptor and subsequent activation of gene expression. Male sex differentiation does not occur if the androgen is missing. In the genetic male that develops female external genitalia, there is androgen insensitivity because the receptor does not function. Thus, androgen, acting through its receptor, functions as a morphogen to direct formation of the male phenotype during early fetal development. At puberty the AR complex functions as a growth and differentiation factor acting in concert with other hormones and growth factors to stimulate reproductive functions that characterize the fully virilized male. Androgen insensitivity is characterized by a lack of target cell response to androgen (testosterone and its 5-α-reduced metabolite, dihydrotestosterone). This disorder is usually associated with abnormal AR androgen-binding activity and absence of nuclear localization of androgen.

The AR belongs to the subfamily of steroid hormone receptors within a larger family of nuclear proteins that most likely evolved from a common ancestral gene. Each contains an amino-terminal region, variable in length, that may have a role in transcriptional activation, a central cysteine-rich DNA-binding domain, and a carboxy-terminal ligand-binding domain (Chang et al., 1988a,b). Highest sequence identity occurs in the DNA-binding domain, including the conserved positioning of cysteines resembling the zinc-binding motif (finger structure) described for *Xenopus laevis* 5 S RNA gene transcription factor IIIA.

Structural analysis of cDNAs for human and rat ARs indicates that the amino-terminal regions of ARs are rich in oligo- and poly-amino acid motifs. A sequence similarity exists among ARs and the receptors for glucocorticoids, progestins, and mineralocorticoids within the steroid-binding domains. The cysteine-rich DNA-binding domains are well conserved. mRNA of AR cDNAs yielded 94- and 76-kDa proteins and smaller forms that bind to DNA and have high affinity toward androgens. The study further revealed that the central prostate and other male accessory organs are rich in AR mRNA and that the production of AR mRNA in the target organs may be autoregulated by androgens.

Zinc is known to potentiate binding of the 4.5 S [^{3}H]dihydrotestosterone–receptor complex to isolated rat prostate Dunning tumor nuclei *in vitro* (Colvard and Wilson, 1984). The receptor–nuclear interaction appears to be selective for zinc; other divalent cations have no effect.

Chang et al. (1988a,b) obtained complementary DNAs (cDNAs) encoding ARs for human testis and rat ventral prostate cDNA libraries. The amino acid sequence deduced from the nucleotide sequences of the cDNAs indicated the presence of a cysteine-rich DNA-binding domain that is highly conserved in all steroid receptors.

The studies of Page et al. (1987) and Page (1988) indicated that zinc-finger Y (ZFY) is testis-determining factor (TDF). Other studies, however, mitigate against this conclusion (Erickson and Verga, 1989).

Most XX males, individuals who are phenotypically completely normal males except for the lack of fertile sperm and who sometimes need testosterone replacements, usually have some Y chromosomal DNA (Stalvey and Erickson, 1987). Some XY females lose a portion of the Y chromosomal sequences (Disteche *et al.,* 1986). Page's group cloned 140-kb DNA, which was present in an XX male and absent in an X,Y:22 translocation female. If indeed this was a simple cytogenetic event, then TDF should lie within this 140-kb interval. The search for a conserved DNA sequence in this region led to the discovery of TDF by Page and his associates.

Isolation and mapping of a mouse cDNA sequence (mouse Y-finger) encoding a multiple, potential zinc-binding, finger protein homologous to the candidate human TDF gene has been reported by Nagamine *et al.* (1989). Four similar sequences were identified in *Hind*III-digested mouse genomic DNA, two (7.2 and 2.0 kb) were mapped in the Y chromosome, and only the 2.0-kb fragment was correlated with testis determination.

5.3 Zinc and Glucose Metabolism

Oral glucose tolerance was assessed in several zinc-deficient Egyptian dwarfs (Sandstead *et al.,* 1967). The majority of them had an apparent delayed absorption of glucose. Disposal of intravenous glucose was normal to slightly accelerated in nearly all patients, implying adequate insulin production. On the other hand, three of four patients showed striking intolerance to intravenous insulin (0.1 unit regular insulin/kg body wt), with severe hypoglycemic reactions (blood glucose < 15 mg/dl), necessitating termination of the test within 20–30 min after insulin injection.

Kinlaw *et al.* (1983) found hyperzincuria (excessive loss of zinc in the urine) in 20 patients with stable type II diabetes mellitus. Urinary zinc loss correlated with the mean serum glucose concentration and was greater in subjects with proteinuria; 25% of these patients had low serum zinc levels, and the oral zinc tolerance test curve was depressed in diabetic subjects. These investigators also demonstrated hyperzincuria following glucose infusion in dogs. They concluded that hyperzincuria, resulting from a nonosmotic glucose-mediated process, in combination with impaired gastrointestinal zinc absorption, produces zinc deficiency in patients with type II diabetes mellitus.

Although hyperzincuria has been reported to be a common finding in most diabetic subjects, plasma zinc concentration has been reported to be normal, increased, or decreased in these patients. We assayed cellular zinc levels of diabetic subjects in order to assess their zinc status (Pai and Prasad, 1988). In our subjects, zinc levels in plasma, lymphocytes, granulocytes, and platelets were decreased significantly relative to controls (Table 5-3). We also

Table 5-3. Zinc in the Plasma, Lymphocytes, Granulocytes, Platelets and Urine and Nucleoside Phosphorylase Activity in the Lymphocytes in Diabetic and Control Subjects[a]

| | Zinc (mean ± S.D.) | | | | Urine | | Nucleoside phosphorylase activity in lymphocytes [ΔOD/h/mg protein (mean ± S.D.)] |
	Plasma (μg/dl)	Lymphocytes (μg/10^10 cells)	Granulocytes (μg/10^10 cells)	Platelets (μg/10^10 cells)	mg/24 h	mg/g creatinine	
Diabetic subjects	108.3 ± 12.9 (16)[b]	44.6 ± 7.42 (16)	35.6 ± 5.86 (16)	1.49 ± 0.29 (16)	0.92 ± 0.61 (14)	0.71 ± 0.64 (14)	3.22 ± 0.93 (11)
Controls	120.2 ± 7.9 (19)	58.0 ± 5.49 (19)	45.0 ± 5.59 (19)	1.93 ± 0.40 (19)	0.26 ± 0.19 (10)	0.17 ± 0.12 (10)	4.1 ± 0.32 (9)
p	<0.01	<0.001	<0.001	<0.001	<0.003	<0.015	<0.01

[a] Adapted with permission from Pai, L. H., and Prasad, A. S., 1988. Cellular zinc in patients with diabetes mellitus, *Nutr. Res.* 8:889.
[b] Number in parentheses indicates number of subjects.

observed hyperzincuria in our subjects. The activity of nucleoside phosphor-ylase, a zinc-dependent enzyme, was decreased in the lymphocytes. Based on these data, we concluded that some diabetic subjects do have zinc deficiency and this must be diagnosed in order to properly manage these cases.

In our recent study in which we investigated the interaction between intestinal absorption of zinc and other solutes using the triple lumen technique, intestinal absorption of zinc was significantly stimulated by the addition of glucose (20 mM) (Lee *et al.*, 1989). Conversely, zinc (0.9 mM) also enhanced the absorption of glucose. The enhanced absorption of zinc or glucose was not accompanied by any increase in absorption of water and sodium. In contrast, increasing the concentration of zinc in the perfusate resulted in de-creased absorption of sodium and water in a dose-related manner. These studies suggest that glucose and zinc interact with a common carrier in the brush border membranes; glucose stimulates the attachment of zinc to the carrier and thereby increases zinc entry into the cell. The same carrier may also enhance intestinal glucose absorption of zinc. Alternatively, glucose may increase the solubility of zinc. The actual mechanism responsible for the in-teraction between zinc and glucose still awaits further study.

It is clear that zinc status and insulin metabolism are interrelated (Arquilla *et al.*, 1978; Engelbart and Kief, 1970; Kirchgessner and Roth, 1980; Levine *et al.*, 1983; Roth and Kirchgessner, 1981; Scott and Fisher, 1938). Not only does zinc promote formation of hexamers from insulin monomers in the beta cell, but it also alters binding of iodoinsulin to the liver and may in fact exert an insulinlike stimulatory effect on lipogenesis. Insulin-resistant but not in-sulinopenic mice have depressed zinc levels in serum and bone (Levine *et al.*, 1983), suggesting either that zinc deficiency plays a role in the pathogenesis of the insulin resistance present in type II maturity-onset diabetes, or that zinc deficiency results from insulin resistance. The finding that genetically obese mice also have decreased zinc levels in plasma, femur, and liver further supports the postulate that "zinc deficiency" is related to tissue sensitivity to insulin.

On the other hand, a study has demonstrated that maternal zinc deficiency is associated with a decreased insulin and glucagon content of the fetal pan-creas, with a decreased number of beta cells and possibly of alpha cells as well (Robinson and Hurley, 1981). The tolerance of zinc-deficient animals to in-traperitoneal injections of glucose appears to be depressed relative to that in pair-fed controls, prompting the suggestion that the rate of insulin secretion in response to glucose stimulus is reduced in zinc deficiency (Arquilla *et al.*, 1978; Kinlaw *et al.*, 1983). Acute stimulation of insulin secretion in rats also reduces the zinc content of the pancreatic beta cells (Engelbart and Kief, 1970). Since zinc participates in the synthesis and storage of insulin in the beta cells, it is possible that the amount of insulin stored during zinc deficiency is decreased. Alternatively, increased degradation of insulin may contribute to the observed decrease in glucose tolerance. It is clear that the interrelationships between glucose, insulin, and zinc metabolism require further investigation.

The crystalline structure of insulin has long been known to contain zinc (Scott, 1934). The conversion of proinsulin to insulin *in vitro* is decreased in zinc-deficient media. Proinsulin binds five times more zinc than insulin, suggesting that the conversion process may be zinc dependent (Grant *et al.*, 1972). Zinc is known to accumulate in the pancreatic islets of Langerhans (Ludvigsen *et al.*, 1979). Zinc increases proinsulin stability and decreases insulin solubility (Gold and Grodsky, 1984).

Circulating insulin has been reported to be either normal or decreased in zinc-deficient rats (Park *et al.*, 1986; Guigliano and Millward, 1987). In one study, impaired glucose tolerance but normal plasma insulin were observed, suggesting peripheral insulin resistance in zinc deficiency (Park *et al.*, 1986). Insulin secretion in response to glucose is decreased in zinc-deficient rats.

Zinc appears to be essential for the release of insulin from the islet of Langerhans. In spite of the close association between zinc and insulin secretion, islet zinc turnover has been reported to be unrelated to insulin release. Insulin is known to prevent increase in zinc excretion caused by somatostatin. It has been observed that an "insulinlike" stimulation of glucose uptake by adipocytes occurs in the presence of 0.1 to 0.3 mM zinc (Coulston and Dandone, 1980). This effect is identical to that demonstrated for prostaglandin (PG) E_1. Although this effect seems to occur in the absence of added insulin, zinc (250 to 1000 μM) does stimulate insulin binding to adipocytes and, at lower concentrations (25 to 50 μM), stimulates insulin binding to liver, lymphocytes, and placenta (Herrington, 1985). In view of the known possible relationship between zinc and PGE, the insulinlike effect of zinc on glucose uptake by adipocytes is most likely mediated by locally synthesized PG.

Zinc and insulin actions are not always complementary, however, and zinc has been reported to inhibit insulin-stimulated uptake of uridine into RNA in mouse mammary gland explants (Rillema, 1979).

Hypozincemia has been noted to occur frequently in non-insulin-dependent diabetics. It appears that insulin resistance may be a problem in zinc deficiency suggesting that insulin receptor function may be zinc dependent, either directly or in association with PGs. It is interesting to note that zinc supplementation helps to reduce clinical obesity.

5.4 Zinc and Adrenals

In zinc-deficient rats, the adrenals are increased in size and the incorporation of cholesterol in the adrenals is also increased. Increased uptake of cholesterol by the adrenals is consistent with the higher circulating cortico-steroid levels observed sometimes in zinc-deficient animals (Quarterman, 1974). Blood levels of adrenal steroids are increased in zinc deficiency. In Cushing's syndrome and following steroid administration, plasma zinc level has been shown to decline (Henkin, 1974a). Conversely, adrenocortical in-sufficiency or following adrenalectomy, plasma zinc concentration is increased (Henkin, 1984). Adrenal steroids induce hyperzincuria and increase zinc up-take by the liver via enhanced metallothionein synthesis (Dunn *et al.,* 1987). Both of these effects may thus may be responsible for decreased level of plasma zinc. In zinc-deficient rats, aldosterone does not increase, even in sodium-depleted rats (Hager, 1986).

In many clinical conditions, such as infections, surgery, and chronic diseases, plasma zinc is decreased. It is also known that in these cases endog-enous steroids may be elevated or else these patients may have received steroid therapy. The clinical significance of hypozincemia in association with increased plasma steroid levels is not well understood at present. In one study, prevention of decrease in plasma zinc level in infected pigs by zinc supplementation was associated with higher mortality than in unsupplemented pigs (Chester, 1978).

IL-1, a monocyte lymphokine secreted after infection, is known to be partly responsible for stimulation of the secretion of glucocorticoids, the ad-renal steroids that mediate important aspects of the response to stress (Bernton *et al.,* 1987). In a recent study in rats, human IL-1 activated the adrenocortical axis by stimulating the release of the controlling hormone corticotropin-releasing factor (CRF) from the hypothalamus (Sapolsky *et al.,* 1987; Ber-kenbush *et al.,* 1987). Infusion of IL-1 induced a significant secretion of CRF into the circulation exiting the hypothalamus, whereas immunoneutralization of CRF blocked the stimulatory effect of IL-1 on glucocorticoid secretion. In this study IL-1 showed no acute direct stimulatory effect on the pituitary or adrenal components of this system. Similar results have been reported by Berkenbush *et al.* (1987). In view of our recent observation that IL-1 production may be decreased in a chronic state of zinc deficiency, one may speculate that some of the effects of zinc deficiency on adrenals may be explained by the above. Reeves *et al.* (1977) have shown that zinc-deficient rats release less corticosterone in response to adrenocorticotropic hormone (ACTH).

5.5 Zinc and Prolactin

Zinc appears to influence the synthesis and secretion of prolactin (Login *et al.,* 1983; Judd *et al.,* 1984; Mahajan *et al.,* 1985). At concentrations between 1 and 10 μM, zinc reduced prolactin secretion and, to a milder extent, synthesis but not release of basal or stimulated GH or LH, *in vitro*. This effect of a physiologic concentration of zinc on prolactin secretion suggests that this trace element may have a role in the regulation of prolactin release *in vivo*. Furthermore, treatment with zinc lowers serum prolactin levels in hemodialyzed uremic patients (Mahajan *et al.*, 1985). Thus, hyperprolactinemia in uremic patients may be related to their zinc-deficient status, and perhaps part of the beneficial effect of zinc on gonadal dysfunction in uremia relates to its ability to lower prolactin.

The uptake of zinc by the dorsolateral prostate is stimulated by prolactin in both animals and human (Gunn *et al.*, 1965; Leake *et al.*, 1984) and the prolactin secretion antagonist, bromocryptine, has been shown to significantly decrease the zinc content of the rat prostate (Harper *et al.*, 1976). The plasma prolactin levels are decreased by zinc supplementation, possibly by its inhibitory effect on prolactin secretion from the pituitary.

In view of the facts that IL-1 production may be decreased because of zinc deficiency (personal observation) and that IL-1 has a direct inhibitory effect on prolactin secretion from the anterior pituitary, it is likely that the observed hyperprolactinemia in zinc deficiency may be mediated by IL-1.

Physiologically relevant concentration of zinc reversibly and selectively inhibits *in vitro* prolactin release. It is unknown whether zinc in systemic or hypothalamic pituitary portal plasma has a role in the control of pituitary prolactin secretion.

5.6 Zinc and Thyroid and Parathyroid Hormones

In normal volunteers, administration of 50 mg elemental zinc given three times a day resulted in an increase in serum zinc and thyroxine levels (Hartome *et al.*, 1979). Serum zinc level below 70 μg/dl was associated with lower triiodothyronine levels (Morley *et al.*, 1981). In zinc deficiency, circulating triiodothyronine and thyroxine have been reported to be decreased (Gordon *et al.*, 1979). Hypothalamic thyroid-releasing hormone is decreased as a result of zinc deficiency (Gordon *et al.*, 1979).

Hypothyroidism has been associated with decreased bidirectional transport of zinc across the gut mucosa, an effect completely reversible by thyroxine administration (McConnell *et al.*, 1977). Thus, low triiodothyronine may be a factor contributing to zinc deficiency in patients with alcoholism and starvation (Morley *et al.*, 1980, 1981).

Hyperzincuria has been observed in patients with thyrotoxicosis, probably as a result of increased protein catabolism (Bremner and Fell, 1977). Erythrocyte zinc level is decreased in thyrotoxicosis and this is correlated with lower plasma albumin and retinol binding protein (Aihara *et al.*, 1984). Untreated patients with hyperparathyroidism also exhibit hyperzincuria and this is accentuated following surgery (Malette and Henkin, 1976).

5.7 Zinc and Brain Hormones

Immunoreactive cholecystokinin is located in areas of the rat hippochcampus and its presence there may be altered by divalent metal chelators (Stengaard-Pederson *et al.,* 1984). Its degradation may require a zinc metalloenzyme.

Zinc is highly localized in the mossy fibers of the rat hippocampus, similar to the localization of enkephalins (Stengaard-Pederson *et al.,* 1981). Zinc is known to block Met-enkephalin binding to opiate receptors in hippocampus synapses (Stengaard-Pederson, 1982). *In vivo,* zinc antagonizes the analgesic effect of morphine (Baraldi *et al.,* 1984).

A deficiency of zinc for 10 days in the rat is known to cause an increase in norepinephrine in the whole brain and in the hypothalamus (Wallwork *et al.,* 1982). In the mouse, pretreatment with zinc increased the survival and liver regeneration and prevented the depletion of brain norepinephrine caused by the toxic effect of the mushroom *Amantia phalloides* (Floersheim *et al.,* 1984).

References

Abbasi, A. A., Prasad, A. S., Ortega, J., Congco, E., and Oberleas, D., 1976. Gonadal function abnormalities in sickle cell anemia: Studies in adult male patients, *Ann. Intern. Med.* 85: 601.

Aihara, K., Nishi, Y., Hatano, S., Kihara, M., Yoshimutsu, K., Takeichi, N., Ito, T., Ezaki, H., and Usui, T., 1984. Zinc, copper, manganese and selenium metabolism in thyroid disease, *Am. J. Clin. Nutr.* 40:26.

Aihara, K., Nishi, Y., Hatano, S., Kihara, M., Ohta, M., Sakoda, K., Uozumi, T., and Usui, T., 1985. Zinc, copper, manganese, and selenium metabolism in patients with human hormone deficiency or acromegaly, *J. Pediatr. Gastroenterol. Nutr.* 4:610.

Arquilla, E., Packer, S., Tarmas, W., and Miyamoto, S., 1978. The effect of zinc on insulin metabolism, *Endocrinology* 103:1440.

Baraldi, M., Caselgrandi, E., and Sauti, M., 1984. Reduction of withdrawal symptoms in morphine-dependent rats by zinc: Behavioral and biochemical studies, *Neurosci. Lett. Suppl.* 18:5371.

Berkenbush, F., Oers, J. V., Rey, A. D., Tilders, F., and Besdedoksy, H., 1987. Corticotropin-releasing factor-producing neurons in the rat activated by interleukin-1, *Science* 238:524.

Bernton, E. S., Beach, J. E., Holaday, J. W., Smallridge, R. C., and Fein, H. G., 1987. Release of multiple hormones by a direct action of interleukin-1 on pituitary cells, *Science* 238:519.

Bjorndahl, L., and Kvist, U., 1982. Importance of zinc for human sperm head–tail connection, *Acta Physiol. Scand.* 116:51.

Bolze, M. S., Reeves, R. D., Lindbeck, F. E., and Elders, M. J., 1987. Influence of zinc on growth, somatomedin, and glycosaminoglycan metabolism in rats, *Am. J. Physiol.* 252:E21.

Bremner, W. F., and Fell, G. S., 1977. Zinc metabolism and thyroid status, *Postgrad. Med. J.* 53:143.

Butrimovitz, G. P., and Purdy, W. C., 1978. Zinc nutrition and growth in a childhood population, *Am. J. Clin. Nutr.* 31:1409.

Castro-Magana, M., Collipp, P. J., Chen, S. Y., Cheruvanky, T., and Maddaiah, V. T., 1981. Zinc nutritional status, androgens, and growth retardation, *Am. J. Clin. Nutr.* 135:322.

Cavdar, A. O., Arcasoy, A., Cin, S., and Gumus, H., 1980. Zinc deficiency in geophagia in Turkish children and response to treatment with zinc sulphate, Haematologica 65:403.

Chan, W., Bates, J. M., Chung, K. W., and Rennert, O. M., 1986. Abnormal zinc metabolism in unilateral maldescended testis of mutant rat strain, *Proc. Soc. Exp. Biol. Med.* 182:549.

Chang, C., Kokontis, J., and Liao, S., 1988a. Molecular cloning of human and rat complementary DNA encoding androgen receptors, *Science* 240:324.

Chang, C., Kokontis, J., and Liao, S., 1988b. Structural analysis of complementary DNA and amino acid sequences of human and rat androgen receptors, *Proc. Natl. Acad. Sci. USA* 85: 7211.

Cherry, F. F., Sandstead, H. H., Rojas, P., Johnson, L. K., Batson, H. K., and Wang, X. B., 1989. Adolescent pregnancy. Associations among body weight, zinc nutriture, and pregnancy outcome, *Am. J. Clin. Nutr.* 50:945.

Cheruvanky, T., Castro-Magana, M., Chen, S. Y., Collipp, P. J., and Ghavami-Maibodi, Z., 1982. Effect of growth hormone on hair, serum, and urine zinc in growth hormone-deficient children, *Am. J. Clin. Nutr.* 35:668.

Chester, J. K., 1978. Biochemical functions of zinc in animals, *World Rev. Nutr. Diet.* 32:135.

Chung, K. W., Kim, S. Y., Chan, W., and Rennert, O. M., 1986. Androgen receptors in ventral prostate glands of zinc deficient rats, *Life Sci.* 38:351.

Collip, P. J., Castro-Magana, M., Petrovic, M., Thomas, J., and Cheruvanky, T., 1982. Zinc deficiency: Improvement in growth and growth hormone levels with oral zinc therapy, *Ann. Nutr. Metab.* 26:287.

Colvard, D. S., and Wilson, E. M., 1984. Zinc potentiation of androgen receptor binding to nuclei in vitro, *Biochemistry* 23:3471.

Coulston, L., and Dandone, P., 1980. Insulin-like effect of zinc on adipocytes, *Diabetes* 29:665.

Cunnane, S. C., 1981, Inhibition of arachidonic acid metabolism in the uteri of zinc deficient parturient rats, *Proc. Nutr. Soc.* 40:78A.

Cunnane, S. C., Majid, E., Senior, J., and Mills, C. F., 1983. Utero-placental dysfunction and prostaglandin metabolism in zinc-deficient pregnant rats, *Life Sci.* 32:2471.

Cunningham, B. C., Bass, S., Fuh, G., and Wells, J. A., 1990. Zinc mediation of the binding of human growth hormone to the human prolactin receptor, *Science* 250:1709.

Disteche, C. M., Casanova, M., Saal, H., Friedman, C., Sybert, V., Graham, J., Thuline, H., Page D. C., Fellovs M., 1986. Small deletions of the short arm of the Y chromosome in 46, XY females, *Proc. Natl. Acad. Sci, USA* 83:7841.

Dunn, M. A., Blalock, T. L., and Cousins, R. J., 1987. Metallothionein, *Proc. Soc. Exp. Biol. Med.* 185:107.

Engelbart, K., and Kief, H., 1970. Uber das funktionelle Verhalten von Zinc und Insulin in den B-Zellen des rathen Pankreas, *Virchows Arch. B* 4:294.

Erickson, R. P., and Verga, V., 1989. Mini review: Is zinc-finger y the sex-determining gene? *Am. J. Hum. Genet.* 45:671.

Floersheim, G. L., Bianchi, L., Probst, A., Chiodetti, N., and Haneggen, C. G., 1984. Influence of zinc-d-penicillamine and oxygen on poisoning with Amantia phalloides: Zinc accelerates liver regeneration and prevents the depletion of brain noradrenaline caused by the mushroom, *Agents Actions* 14:124.

Ghavami-Maibodi, S. Z., Collip, P. J., Castro-Magana, M., Stewart, C., and Chen, S. Y., 1983. Effect of oral zinc supplements on growth, hormonal levels, and zinc in healthy short children, *Ann. Nutr. Metab.* 27:214.

Gibson, R. S., Vanderkooy, P. D. S., MacDonald, A. C., Goldman, A., Ryan, B. A., and Berry, M., 1989. A growth-limiting mild zinc deficiency syndrome in some South Ontario boys with low weight percentiles, *Am. J. Clin. Nutr.* 49:1266.

Gold, G., and Grodsky, G. M., 1984. Kinetic aspects of compartmental storage and secretion of insulin and zinc, *Experientia* 40:1105.

Gordon, J., Morlet, J. E., and Hershman, J. M., 1979. Thyroid function and zinc deficiency, *Clin. Res.* 27:20A.

Gourmelen, M., Pham-Huu-Trung, M. T., and Girard, F., 1979. Transient partial hGH deficiency in prepubertal children with delay of growth, *Pediatr. Res.* 13:221.

Grant, P. T., Coombs, T. L., and Frank, B. H., 1972. Differences in the nature of the interaction of insulin and proinsulin with zinc, *Biochem. J.* 126:433.

Guigliano, R., and Millward, D. J., 1987. The effects of severe zinc deficiency on protein turnover in muscle and thymus, *Br. J. Nutr.* 57:139.

Gunn, S. A., Gould, T. C., and Anderson, W. A. D., 1965. The effect of growth hormone and prolactin preparations on the control by interistitial cell–stimulating hormone of uptake of zinc-65 by the rat dorsolateral prostate, *J. Endocrinol.* 32:205.

Hager, M. H., 1986. Zinc deficiency compromises the adrenal response to sodium deprivation in rats, *Nutr. Rep. Int.* 34:141.

Halsted, J. A., Ronaghy, H. A., Abadi, P., Haghshenass, M., Amirhakemi, G. H., Barakat, R. M., and Reinhold, J. G., 1972. Zinc deficiency in man: The Shiraz experiment, *Am. J. Med.* 53:277.

Hambidge, K. M., and Walravens, P. A., 1976. Zinc deficiency in infants and preadolescent children, in *Trace Elements in Human Health and Disease* (A. S. Prasad, ed.), Academic Press, New York, p. 21.

Harper, M. E., Danutra, V., Chanduer, J. A., and Griffith, K., 1976. The effect of 2-bromo-alpha-ergocryptine (CB 154) administration on the hormone levels, organ weights, prostatic morphology and zinc concentrations in the male rat, *Acta Endocrinol. (Copenhagen)* 83:211.

Hartome, T. R., Sotaniemi, E. A., and Maatanen, J., 1979. Effect of zinc on some biochemical indices of metabolism, *Nutr. Metab.* 23:294.

Henkin, R. I., 1974a. On the role of adrenocorticosteroids in the control of zinc and copper metabolism, in *Trace Element Metabolism in Animals,* Volume 2 (W. G. Hoekstra, J. W., Suttie, H. E. Ganther, and W. Mertz, eds.), University Park Press, Baltimore, p. 647.

Henkin, R. I., 1974b. Growth hormone-dependent changes in zinc and copper metabolism in man, in *Trace Element Metabolism in Animals,* Volume 2 (W. G. Hoekstra, J. W. Suttie, H. E. Ganther, and W. Mertz, eds.), University Park Press, Baltimore, p. 653.

Henkin, R. I., 1984. Zinc in taste function: A critical review, *Biol. Trace Elem. Res.* 6:263.

Herrington, A. C., 1985. Effect of zinc on insulin binding to rat adipocytes and hepatic membranes and to human placental membranes and IM-9 lymphocytes, *Horm. Metab. Res.* 17:328.

Judd, A. M., Macleod, R. M., and Login, I. S., 1984. Zinc may regulate pituitary prolactin secretion, in *The Neurobiology of Zinc, Part A* (C. J. Fredrickson, G. A. Howell, and E. J. Kasarskis, eds.), Liss, New York, p. 91.

Kinlaw, W. B., Levine, A. S., Morley, J. E., Silvis, S. E., and McClain, C. J., 1983. Abnormal zinc metabolism in type II diabetes mellitus, *Am. J. Med.* 75:273.

Kirchgessner, M., and Roth, H. P., 1980. Biochemical changes of hormones and metalloenzymes in zinc deficiency, in *Zinc in the Environment, Part II* (J. O. Nriagu, ed.), Wiley, New York, p. 72.

Kurtoglu, S., Patiroglu, T. E., and Karakas, S. E., 1987. Effect of growth hormone in epiphyseal growth plates in zinc deficiency, *Tokai J. Clin. Med.* 12:325.

Kvist, U., 1982. Spermatozoal thiol–disulphide interaction: A possible event underlying physiological sperm nuclear chromatin decondensation, *Acta Physiol.* 115:503.

Leake, A., Chisholm, G. D., and Habib, F. K., 1984. The effect of zinc on the 5-alpha-reduction of testosterone by the hyperplastic human prostate gland, *J. Steroid Biochem.* 20:651.

Lee, H. H., Prasad, A. S., Brewer, G. J., and Owyang, C., 1989. Zinc absorption in human small intestine, *Am. J. Physiol.* 256:G87.

Lei, K. Y., Abbasi, A. A., and Prasad, A. S., 1976. Function of pituitary–gonadal axis in zinc-deficient rats, *Am. J. Physiol.* 230:1730.

Levine, A. S., McClain, C. J., Handwerger, B. S., Brown, D. M., and Morley, J. E., 1983. Tissue zinc status of genetically diabetic and streptozotocin induced diabetic mice, *Am. J. Clin. Nutr.* 37:382.

Login, I. S., Thorner, M. O., and Macleod, P. M., 1983. Zinc may have a physiological role in regulating pituitary prolactin secretion, *Neuroendocrinology* 37:317.

Lubahn, D. B., Joseph, R. D., Sullivan, P. M., Willard, H. F., French, F. S., and Wilson, E. M., 1988. Cloning of human androgen receptor complementary DNA and localization to the X chromosome, *Science* 240:327.

Ludvigsen, C., McDaniel, M., and Lacy, P. E., 1979. The mechanism of zinc uptake in isolated islets of Langerhans, *Diabetes* 28:570.

McClain, C. J., Gavaler, J. S., and VanThiel, D. H., 1984. Hypogonadism in the zinc-deficient rat: Localization of the functional abnormalities, *J. Lab. Clin. Med.* 104:1007.

McConnell, R. J., Blair-Stanek, C. S., and Rivlin, R. S., 1977. Decreased intestinal transport of zinc in hypothyroidism, *Clin. Res.* 25:658A.

Mahajan, S. K., Abbasi, A. A., Prasad, A. S., Rabbani, P., Briggs, W. A., and McDonald, F. D., 1982. Effect of oral zinc therapy on gonadal function in hemodialysis patients. A double-blind study, *Ann. Intern. Med.* 97:357.

Mahajan, S. K., Hamburger, R. J., Flamenbaum, W., Prasad, A. S., and McDonald, F. D., 1985. Effect of zinc supplementation on hyperprolactinemia in uremic men, *Lancet* 2:750.

Malette, L. E., and Henkin, R. I., 1976. Altered copper and zinc metabolism in primary hyperparathyroidism, *Am. J. Med. Sci.* 272:167.

Meftah, S. P., Prasad, A. S., DuMouchelle, E., Cossack, Z. T., and Rabbani, P., 1984. Testicular androgen binding protein in zinc-deficient rats, *Nutr. Res.* 4:437.

Morley, J. E., Gordon, J., and Hershman, J. M., 1980. Zinc deficiency, chronic starvation and hypothalamic–pituitary–thyroid function, *Am. J. Clin. Nutr.* 33:1767.

Morley, J. E., Russell, R. M., Reed, A., Carney, E. A., and Hershman, J. M., 1981. The interrelationship of thyroid hormones with vitamin A and zinc nutritional status in patients with chronic hepatitis and gastrointestinal disorders, *Am. J. Clin. Nutr.* 34:1489.

Nagamine, C. M., Chan, K., Kozak, C. A., and Lau, Y. F., 1989. Chromosome mapping and expression of a putative testis determining gene in mouse, *Science* 243:80.

Neggers, Y. H., Cutter, R., Acton, R. T., Alvarez, J. O., Bonner, J. L., Goldenberg, R. L., Go, R. C. P., and Roseman, J. M., 1990. A positive association between maternal serum zinc concentration and birth weight, *Am. J. Clin. Nutr.* 51:678.

Neill, H. B., Leach, B. E., and Kraus, A. P., 1979. Zinc metabolism in sickle cell anemia, *J. Am. Med. Assoc.* 242:2686.

Nishi, Y., Hatano, S., Aihara, K., Fujie, A., and Kihara, M., 1989. Transient partial growth hormone deficiency due to zinc deficiency, *J. Am. Coll. Nutr.* 8:93.

Page, D. C., 1988. Is ZFY the sex-determining gene on the human Y chromosome? *Philos. Trans. R. Soc. London* 322:155.

Page, D. C., Mosher, R., Simpson, E. M., Fisher, E. M. C., Mardon, G., Pollack, J., and McGillivray, B., 1987. The sex-determining region of the human Y chromosome encodes a finger protein, *Cell* 51:1091.

Pai, L. H., and Prasad, A. S., 1988. Cellular zinc in patients with diabetes mellitus, *Nutr. Res.* 8: 889.

Park, J. H. Y., Grandjean, C. T., Hart, M. H., Erdman, S. H., Pour, P., and Vanderhof, J. A., 1986. Effect of pure zinc deficiency on glucose tolerance and insulin and glucagon levels, *Am. J. Physiol.* 251:E273.

Prasad, A. S., 1966. Metabolism of zinc and its deficiency in human subjects, in *Zinc Metabolism* (A. S. Prasad, ed.), Thomas, Springfield, Ill., p. 250.

Prasad, A. S., 1981. Zinc deficiency and effects of zinc supplementation on sickle cell anemia subjects, in *The Red Cell: Fifth Ann Arbor Conference* (G. J. Brewer, ed.), Liss, New York, p. 99.

Prasad, A. S., and Cossack, Z. T., 1984. Zinc supplementation and growth in sickle cell disease, *Ann. Intern. Med.* 100:367.

Prasad, A. S., Halsted, J. A., and Nadimi, M., 1961. Syndrome of iron deficiency anemia, hepatosplenomegaly, hypogonadism, dwarfism and geophagia, *Am. J. Med.* 31:532.

Prasad, A. S., Miale, A., Farid, Z., Sandstead, H. H., Schulert, A. R., and Darby, W. J., 1963a. Biochemical studies on dwarfism, hypogonadism and anemia, *Arch. Intern. Med.* 111:407.

Prasad, A. S., Miale, A., Farid, Z., Schulert, A., and Sandstead, H. H., 1963b. Zinc metabolism in patients with the syndrome of iron deficiency anemia, hepatosplenomegaly, dwarfism, and hypogonadism, *J. Lab. Clin. Med.* 61:537.

Prasad, A. S., Oberleas, D., Wolf, P., and Horwitz, J. P., 1969. Effect of growth hormone on nonhypophysectomized rats, *J. Lab. Clin. Med.* 73:486.

Prasad, A. S., Schoomaker, E. B., Ortega, J., Brewer, G. J., Oberleas, D., and Oelschlegel, F. J., 1975. Zinc deficiency in sickle cell disease, *Clin. Chem.* 21: Washington, D.C., 582.

Prasad, A. S., Fernandez-Madrid, F., and Ryan, J. F., 1979. Deoxythymidine kinase activity of human implanted sponge connective tissue in zinc deficiency, *Am. J. Physiol.* 263(3):E272.

Prasad, A. S., Abbasi, A. A., Rabbani, P., and DuMouchelle, E., 1981. Effect of zinc supplementation on serum testosterone level in adult male sickle cell anemia subjects, *Am. J. Hematol.* 19:119.

Quarterman, J., 1974. The effects of zinc deficiency or excess on the adrenals and the thymus of the rat, in *Trace Element Metabolism in Animals,* Volume 2 (W. G. Hoekstra, J. W. Suttie, H. E. Ganther, and W. Mertz, eds.), University Park Press, Baltimore, p. 742.

Reeves, P. G., Frissell, S. G., and O'Dell, B. L., 1977. Response of serum corticosterone to ACTH and stress in the zinc deficient rat, *Proc. Soc. Exp. Biol. Med.* 156:500.

Rillema, J. A., 1979. Effect of zinc ions on the actions of prolactin on RNA and casein synthesis in mouse mammary gland explants, *Proc. Soc. Exp. Biol. Med.* 162:464.

Robinson, L. K., and Hurley, L. S., 1981. Effect of maternal zinc deficiency of food restriction on rat fetal pancreas, *J. Nutr.* 111:869.

Roth, H., and Kirchgessner, M., 1981. Zinc and insulin metabolism, *Biol. Trace Elem. Res.* 3: 13.

Sandstead, H. H., Prasad, A. S., Schulert, A. R., Farid, Z., Miale, A., Bassilly, S., and Darby, W. J., 1967. Human zinc deficiency, endocrine manifestations and response to treatment, *Am. J. Clin. Nutr.* 20:422.

Sapolsky, R., Rivier, C., Yamamoto, G., Plotsky, P., and Vale, W., 1987. Interleukin-1 stimulates the secretion of hypothalamic corticotropin-releasing factor, *Science* 238:522.

Scott, D. A., 1934. Crystalline insulin, *Biochem. J.* 28:1592.

Scott, D. A., and Fisher, A. M., 1938. The insulin and zinc content of normal and diabetic pancreas, *J. Clin. Invest.* 17:725.

Simmer, K., Khanum, S., Carlsson, L., and Thompson, R. P., 1988. Nutritional rehabilitation in Bangladesh: The importance of zinc, *Am. J. Clin. Nutr.* 47:1036.

Solomons, N. W., Rosenfeld, R. L., Jacob, R. A., and Sandstead, H. H., 1976. Growth retardation and zinc nutrition, *Pediatr. Res.* 10:923.

Stalvey, J. R. D., and Erickson, R. P., 1987. Inheritance of the sex-determining factor in the absence of a complete Y chromosome in 46, XX human males, *Ann. N.Y. Acad. Sci.* 513: 505.

Stengaard-Pederson, K., 1982. Inhibition of enkephalin binding to opiate receptors by zinc ions: Possible physiological importance for the brain, *Acta Pharmacol. Toxicol.* 50:213.

Stengaard-Pederson, K., Fredens, K., and Larsson, L. I., 1981. Enkephalin and zinc in the hippocampal mossy fibre system, *Brain Res.* 212:230.

Stengaard-Pederson, K., Larsson, L. I., Fredeno, K., and Rehfeld, J. F., 1984. Modulation of cholecystokinin concentrations in the rat hippocampus by chelation of heavy metals, *Proc. Natl. Acad. Sci. USA* 81:5876.

Wallwork, J. C., Botnen, J. H., and Sandstead, H. H., 1982. Effect of dietary zinc on rat brain catecholamines, *J. Nutr.* 112:514.

Walravens, P. A., Hambidge, K. M., and Koepfer, D. M., 1989. Zinc supplementation in infants with a nutritional pattern of failure to thrive: A double-blind controlled study, *Pediatrics* 83: 532.

Warth, J. A., Prasad, A. S., Zwas, F., and Frank, R. N., 1981. Abnormal dark adaptation in sickle cell anemia, *J. Lab. Clin. Med.* 98:189.

Watanabe, T., Salo, F., and Endo, A., 1983. Cytogenetic effects of zinc deficiency on oogenesis and spermatogenesis in mice, *Yamagata Med. J.* 1:13.

Xue-Cun, C., Tai-An, Y., Jin-Sheng, H., Qiu-Yan, M., Zhi-Min, H., and Li-Xiang, L., 1985. Low levels of zinc in hair and blood pica, anorexia, and poor growth in Chinese preschool children, *Am. J. Clin. Nutr.* 42:694.

Zinc and Lipid Metabolism ⑥

Zinc may have effects on (1) essential fatty acids, (2) cholesterol, (3) phospholipids, (4) triglycerides, (5) lipoproteins, and (6) lipid peroxidation.

6.1 Essential Fatty Acids

Essential fatty acids (EFA) belong to two families, those with the n-6 configuration (first double bond at the sixth carbon from the methyl terminal) and those with the n-3 configuration (first double bond at the third carbon from the methyl terminal) (see Fig. 6-1). The parent fatty acids of these two families (linoleic acid, 18:2n-6; and α-linoleic acid, 18:3n-3) cannot be synthesized *in vivo,* and hence are considered to be essential (Fig. 6-1). Their products are formed by stepwise desaturation and elongation reactions characteristic of the metabolism of long-chain fatty acids.

The clinical features of zinc deficiency in rats are superficially similar to those of EFA deficiency (see Table 6-1). These include growth retardation, skin changes, immunodeficiency, alopecia, male and female infertility, impaired parturition, high perinatal losses, increased capillary permeability, increased epidermal water loss, and exacerbation by calcium and copper.

The effect of zinc deficiency on tissue levels of EFA has been studied (Cunnane, 1988). However, in most studies the total lipid fatty acid composition was analyzed without separating the different classes of lipids. It is preferable to subdivide the lipids into their constituent esters and free fatty acids prior to analysis. Another problem with the available data in the literature is that although the proportional composition and relative concentrations of fatty acids have been reported, the quantitative composition of fatty acids has not been determined.

Abnormalities of both zinc and fatty acid metabolism have been reported to occur in various clinical conditions such as acrodermatitis enteropathica (AE) (Cash and Berger, 1969; White and Montalvo, 1973; Neldner *et al.,*

Figure 6-1. Metabolism of the EFAs. The two families of EFAs are distinguished by the positions of the double bonds in the fatty acid molecules. The linoleic acid family comprises those fatty acids sequentially synthesized from linoleic acid, all of which have the first double bond six carbons from the methyl terminal of the fatty acid. The α-linolenic acid family has fatty acids with the double bonds initially positioned at the third carbon from the methyl terminal. Both families are metabolized alternately by the desaturases and elongases (d-6 D, Δ⁶ desaturase; d-5 D, Δ⁵ desaturase; d-4 D, Δ⁴ desaturase; E, elongase). (Reprinted with permission from Cunnane, S. C., 1988. *Zinc: Clinical and Biochemical Significance,* CRC Press, Boca Raton, Fla., p. 99).

1974; Neldner and Hambidge, 1975; Hambidge *et al.,* 1977), atopic eczema (David *et al.,* 1984), cystic fibrosis (Solomons *et al.,* 1981), Crohn's disease (Cunnane *et al.,* 1986), protein-energy malnutrition (Wolfe *et al.,* 1984), and multiple sclerosis (Dore-Duffy *et al.,* 1983).

An abnormality of EFA metabolism in patients with AE was first reported in 1969 (Cash and Berger, 1969). Many reports have since appeared documenting abnormal levels of EFA in patients with AE. Julius *et al.* (1973) reported that the level of free arachidonic acid was increased in a series of

Table 6-1. Similarities between Deficiencies
of Zinc and Essential Fatty Acids[a]

Growth retardation
Dermal lesions
Immunodeficiency
Alopecia
Male and female infertility
Impaired parturition
High perinatal losses
Increased capillary permeability
Increased epidermal water loss
Exacerbation by calcium and copper

[a] Reprinted with permission from Cunnane, S. C., 1988. *Zinc: Clinical and Biochemical Significance,* CRC Press, Boca Raton, Fla., p. 99.

patients with AE, an observation which may be consistent with the known effects of zinc deficiency on EFA incorporation into lipids. Neldner *et al.* (1974) reported low arachidonic acid levels in AE, suggesting an abnormality of EFA metabolism in this disease.

Previously, diiodoquin was the therapy of choice for the treatment of AE. It has been suggested that diiodoquin's main effect is to enhance absorption of zinc by chelation. Neldner *et al.* (1974), however, suggested that diiodoquin may also enhance absorption of EFA. Some support for this suggestion has been provided by Bettger and O'Dell (1980) who found that administration of diiodoquin to zinc-deficient rats resulted in correction of some of the clinical effects of zinc deficiency but the status of zinc remained unchanged. These observations suggest that in the pathogenesis of skin manifestations of AE, both zinc and EFA may be involved. It is now well documented, however, that zinc supplementation alone to AE patients results in complete reversal of all clinical manifestations and cures this disease.

In general, it appears that AE patients have raised linoleic acid levels, lower levels of metabolites of linoleic acid in plasma phospholipids, and impaired synthesis of arachidonic acid (Cunnane, 1988). In one adolescent subject with AE, low serum zinc and linoleic acid levels were noted, which were reversed by zinc supplementation. In another study of a child manifesting symptoms of both AE and cystic fibrosis, low zinc and EFA levels were found but the serum prostaglandin levels were increased. Zinc supplementation not only corrected zinc and EFA status but also lowered the prostaglandin levels.

EFA abnormalities have also been reported in patients with sickle-cell disease (SCD). Linoleic acid level in erythrocytes was found to be decreased; however, the level of dihomo-γ-linolenic acid (20:3n-6) was increased. In these patients, plasma zinc was decreased but erythrocyte zinc level was normal (Muskiet and Muskiet, 1984). It is not possible to draw any conclusion regarding the zinc status of SCD in this report, inasmuch as detailed studies of zinc metabolism were not carried out, thus making it difficult to relate the changes in EFA to zinc status.

In one study of cystic fibrosis patients, plasma zinc levels correlated positively with the levels of dihomo-γ-linolenic acid and arachidonic acid but did not correlate with linoleic acid (Hamilton *et al.,* 1981). The authors interpreted this to mean that zinc had a role in desaturation of linoleic acid.

Fogerty *et al.* (1985) have also observed a significant positive correlation between zinc and arachidonic acid levels in the liver of infants and young children.

In animal studies, an increase in the ratio of linoleic acid to longer-chain fatty acids (dihomo-γ-linolenic acid and arachidonic acid) was observed as a result of zinc deficiency. In other animal studies, however, arachidonic acid was increased in skin and mammary tissues of zinc-deficient rats. This discrepancy may be explained on the basis that total lipids rather than individual fatty acid profiles were determined and the effect of zinc deficiency may be tissue specific (Cunnane, 1988). It has been demonstrated that skin does not convert linoleic acid to arachidonic acid and it is possible that arachidonic acid in skin was probably transported for physiological healing purposes in zinc deficiency.

Some authors have suggested that the effects of zinc deficiency on EFA metabolism may be secondarily induced by anorexia and reduced food intake which is a consistent finding in zinc-deficient animals. Significant changes in EFA, however, have been observed in zinc-deficient rats compared with their pair-fed controls, thus indicating that zinc may have specific effects on EFA metabolism.

The metabolic steps in the EFA pathway which appear to be most susceptible to the effects of zinc deficiency are the conversion of linoleic acid to longer-chain metabolites, a reaction involving desaturation of linoleic acid (Δ^6) and dihomo-γ-linolenic acid (Δ^5). Studies of the microsomal desaturation of linoleic acid and dihomo-γ-linolenic acid have shown that the effects of zinc deficiency are tissue specific (Cunnane, 1988).

The mechanism by which zinc might affect the desaturation is unknown (Fig. 6-2). Zinc is present in relatively high amounts in liver microsomes, the cellular component also responsible for fatty acid desaturation. Electrons re-

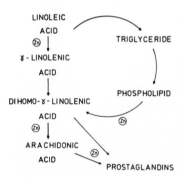

Figure 6-2. Sites at which zinc is known to influence lipid and fatty acid metabolism. Zinc in both humans and animals increases the level of linoleic acid, increases dihomo-γ-linolenic acid, and decreases arachidonic acid, steps involving the Δ^6 and Δ^5 desaturases. Although the desaturases contain iron as the metal cofactor, microsomal zinc levels directly determine electron transfer from nicotinamide adenine dinucleotide hydride (NADH) or nicotinamide adenine dinucleotide phosphate hydride (NADPH) through the cytochromes to the desaturase protein, which is the terminal electron acceptor. The effects of zinc deficiency on this electron transfer pathway are opposite to those of zinc supplementation, suggesting a direct role of zinc in this pathway, rather than an indirect effect of zinc deficiency through, for example, food intake reduction. Zinc also has complex interactions in the synthesis of prostaglandins (PGs), inhibiting some of the enzymes involved, including the phospholipases and thromboxane synthetase. In zinc deficiency, the overwhelming effect is to increase the tissue and serum levels of "2 series" PGs (those derived from arachidonic acid), whereas zinc supplementation decreases PG production. Levels of free (nonesterified) EFA tend to increase in both clinical and experimental zinc deficiency, an effect possibly related to the phospholipases. (Reprinted with permission from Cunnane, S. C., 1988. *Zinc: Clinical and Biochemical Significance,* CRC Press, Boca Raton, Fla., p. 99.)

quired for the desaturation of long-chain fatty acids are provided by NADH and NADPH via cytochromes B-5 and P-450. Cytochrome P-450 activity is significantly decreased in zinc-deficient rats, suggesting a possible mechanism by which zinc deficiency may inhibit the terminal desaturase reaction. Additionally, zinc may be involved in the maintenance of molecular configuration of the enzyme desaturase (Cunnane, 1988).

6.1.1 Lipid Incorporation

The incorporation of EFA into lipid appears to be zinc dependent. EFA are mainly incorporated into triglycerides, phospholipids, and cholesterol esters. Linoleic acid is mainly incorporated in triglycerides, whereas arachidonic acid is preferably incorporated in cholesterol esters and phospholipids. EFA are present as triglycerides in the diet but are absorbed as free fatty acids and later incorporated into appropriate lipids classes.

In zinc-deficient rats, dihomo-γ-linolenic acid accumulates in triglycerides as the free acid. Free arachidonic acid is increased in triglycerides of leukocytes of zinc-deficient pregnant rats. In both Crohn's disease and AE, the free arachidonic acid level is increased in triglycerides. Similar findings have been observed in leukocytes of pregnant women, a condition in which the zinc content of leukocytes may also be decreased. These results suggest that zinc may have a role in the incorporation of EFA into triglycerides (Cunnane, 1988).

Roles for both EFA and prostaglandins (PGs) in zinc absorption have been suggested on the basis of studies in rats but the results are not conclusive. Depending on the type of PG and dose used, zinc absorption may be increased, decreased, or unchanged (Cunnane, 1988). In another study, PGE_1 and zinc levels were correlated in the gut. Whether zinc and PGs bind directly in the gut remains controversial. In another study, the therapeutic use of the PG synthetase inhibitor, indomethacin, given for 3 days did not affect the zinc tolerance test, suggesting that indomethacin had no effect on absorption of zinc during the 4-h test period.

Low concentrations (10 ng to 1 μg/ml) of linoleic acid, γ-linolenic acid, and dihomo-γ-linolenic acid, as well as higher concentrations of arachidonic acid, have been shown to increase zinc absorption (Cunnane, 1988). Zinc deficiency causes increased absorption of zinc and is associated with higher EFA and PG levels in the gut. Human milk is known to have a beneficial effect on zinc absorption in AE. This may be because human milk has a higher EFA content than cow's milk or artificial formulas. Thus, it is reasonable to speculate that EFA and/or PG may have an enhancing effect on zinc absorption, perhaps as a facilitator of mucosal uptake of zinc–ligand complexes.

6.1.2 Prostaglandins

Zinc is known to have physiological and biochemical interactions with PGs (Cunnane, 1988). PG levels (thromboxane B_2 and PGI_2) in the serum of a patient with cystic fibrosis complicated by AE-like dermal lesions were shown to be increased relative to controls and following treatment with zinc PG levels became normal within 10 days.

PGE_1 in gut has been reported to correlate with zinc content (Cunnane, 1988). Increased conversion of labeled arachidonic acid to "2 series" PGs in the placenta of zinc-deficient pregnant rats has been demonstrated, although PG synthesis in the uterus was decreased probably because of excess progesterone in the uterus.

Zinc inhibits the activity of enzymes involved in PG synthesis. Thromboxane synthetase, phospholipase A_2, and phospholipase C are inhibited by zinc (Cunnane, 1988). Zinc also inhibits arachidonic acid release from leukocytes, platelets, and endo/myometrium. Although zinc inhibits aggregation of platelets by 50%, it does not affect thromboxane synthesis. Inhibition of PG synthesis with indomethacin or aspirin has been shown to have no effect on tissue zinc levels.

6.2 Cholesterol

According to Klevay (1975), elevated levels of dietary zinc increased total cholesterol in the serum of rats. Later it was observed that this effect was the result of reduced serum copper induced by high levels of dietary zinc. A significant positive correlation between serum zinc and total cholesterol in animals has been demonstrated by some investigators (Katya-Katya et al., 1984) but not by others (Crouse et al., 1984; Fischer et al., 1980). In these reports, levels of cholesterol in different lipoprotein fractions were not investigated.

Bustamante et al. (1976) observed increased serum zinc/copper ratio and cholesterol levels in patients with atherosclerosis. Okuyama et al. (1982), however, observed that in patients on hemodialysis with atherosclerosis, both serum zinc and high-density-lipoprotein (HDL) cholesterol were low.

In order to evaluate the zinc–cholesterol relationship, one must consider several factors. These are zinc, copper, cholesterol, lipoproteins, sex and species differences. If the dietary intake of zinc was high in relation to copper, an increase in serum total cholesterol was observed in animal experiments (Cunnane, 1988). This was the result of induction of copper deficiency by excess zinc (see Chapter 12). The effect of zinc appeared to be specific for HDL cholesterol, elevation of which is associated with a significantly lower risk of atherosclerosis.

Zinc deficiency is associated with lower adrenal cholesterol in parturient rats but in male rats the cholesterol level in adrenals is increased as a result of zinc deficiency (Cunnane, 1988). In pigs, serum total cholesterol is decreased as a result of zinc deficiency, but in male zinc-deficient rats, serum total cholesterol is increased. Also, the synthesis of cholesterol in the intestinal muscle of zinc-deficient rats has been reported to be increased. It is clear from the above discussion that much more investigation is required in order to understand the precise relationship of zinc and cholesterol in human subjects.

6.3 Phospholipids

According to Petering *et al.* (1977), serum total phospholipids are not affected by increasing dietary zinc from 25 to 40 mg/kg. Sandstead *et al.* (1972), however, observed that the brain of zinc-deficient neonatal rats had increased phosphorus in relation to total lipids as a result of zinc deficiency. Sullivan *et al.* (1980) reported that in zinc-deficient rats, serum zinc was inversely proportional to liver total phospholipid content. This observation is consistent with the observation of Sandstead *et al.* (1972).

The uptake of labeled EFA into phospholipid has been shown to be decreased in zinc-deficient rats. This is most likely related to the inhibitory effect of zinc on phospholipases A_2 and C. Zinc has been shown to form salts with phospholipids in *in vitro* membrane preparations (Cunnane, 1988). Although the physiological importance of this observation remains unclear, it appears that zinc may play a role in maintaining membrane stability. The ability of zinc to increase the resistance of phospholipid monolayers to temperature-induced phase transition suggests that zinc may protect membrane integrity.

6.4 Triglycerides

Carbon tetrachloride causes liver injury which results in lipid accumulation in the liver. Rana and Tayal (1981) have shown that the increase in liver triglycerides caused by carbon tetrachloride can be prevented by zinc administration.

Halevy and Sklan (1986) have isolated a protein from chick liver which stimulates triglyceride lipolysis, has a molecular weight of 6000, but is distinct from metallothionein and contains zinc and copper (4:1).

In zinc-deficient rats, intestinal transport of absorbed triglyceride has been shown to be markedly decreased (Cunnane, 1988). The defect was related to chylomicron formation and may be the result of decreased synthesis or availability of phospholipid or transport proteins.

Free fatty acids in plasma are increased in zinc-deficient rats (Cunnane, 1988). This is most likely caused by increased triglyceride lipolysis rather than by increased phospholipase activity. Zinc at 100 to 500 μM *in vitro* directly stimulates lipogenesis by rat adipocytes, an effect consistent with the known observation of increased lipolysis in zinc deficiency.

6.5 Lipoproteins

According to some investigators, increased dietary zinc is associated with increased cholesterol in plasma HDL. Lefevre *et al.* (1985), however, reported that HDL content of cholesterol was increased in zinc deficiency. Roth and Kirchgessner (1977) found no changes in serum β-lipoproteins as a result of zinc deficiency in rats.

Zinc appears to have a high affinity for serum lipoproteins *in vitro* (Prasad and Oberleas, 1970). Decreased synthesis of the apoprotein component of HDL has been observed in rats fed 100–500 mg zinc/kg per day (Woo *et al.*, 1983).

In one study the effect of zinc status on the distribution of serum cholesterol among the major serum lipoproteins was assessed in adult male rats fed a zinc-deficient diet and in pair-fed and ad libitum-fed zinc-supplemented controls (Koo and Williams, 1981). The zinc-deficient diet produced a significant reduction in total serum cholesterol and this reduction was primarily the result of the selective decrease in HDL cholesterol. Linear regression analysis based on the 36 pairs of serum zinc and HDL cholesterol values showed a significant positive correlation ($r = +0.81$, $p < 0.01$) between these two variables. In this study, serum copper was not affected by zinc deficiency at 4 weeks. These results emphasize a possible adverse effect of inadequate zinc intake on coronary heart disease (CHD), inasmuch as HDL cholesterol is considered to be protective against risk of CHD.

In another experiment, Koo and Ramlet (1983) used two diets, one control containing no cholesterol and the other containing 1% cholesterol which was isocalorically formulated with an equal but adequate level of zinc. Cholesterol feeding produced a significant decrease in the serum level of HDL cholesterol at the 8th week of dietary treatment and a significant increase in very-low-density-lipoprotein cholesterol at the 4th and 8th weeks. Significant decreases in serum zinc levels were observed in cholesterol-fed rats and no changes in the serum levels of other related elements such as copper, calcium, and magnesium were observed as a result of cholesterol feeding. Linear regression analysis of the 44 pairs of serum HDL and zinc values revealed a significant positive correlation ($r = +0.57$, $p < 0.01$) between the two parameters. The mechanism of the effect of cholesterol on serum zinc levels is unknown. Whether the cholesterol effect is the result of altered intestinal absorption of zinc or merely reflects changes in the tissue distribution of the element or alterations in other aspects of zinc metabolism such as turnover and excretion, needs to be studied in the future. It will be of interest to determine what effect dietary cholesterol would have on the serum and body status of zinc and metabolism of HDL, when the diet is marginally (or severely) deficient in zinc, and whether cholesterol feeding would lead to increased requirement of zinc for growing as well as mature animals under these conditions.

Chylomicrons of zinc-deficient rats have been observed to be abnormally large and their levels of apoproteins C and E are reduced (Koo *et al.*, 1986). In one study, the rate of plasma clearance of chylomicron [^{14}C]cholesterol, labeled *in vivo*, was studied in zinc-deficient and control rats (Koo *et al.*, 1986). The ^{14}C clearance curves were nonlinear, consisting of an initial rapid phase followed by a slow phase of clearance. The initial ^{14}C clearance was significantly ($p < 0.05$) delayed regardless of whether the labeled chylomicrons were injected into zinc-deficient or control recipients. The hepatic ^{14}C recovery in extracted lipids was also significantly lower in zinc-deficient rats. These changes were attributed to the molecular alterations of chylomicrons induced by zinc deficiency.

The commonly observed hypocholesterolemic effect of zinc deficiency, however, cannot be attributed to an increase in the rate of turnover or degradation of dietary cholesterol transported by chylomicrons. It appears that the cholesterol-lowering effect of zinc deficiency may be associated with impaired intestinal absorption and/or a decrease in the hepatic synthesis or release of cholesterol. Clearly more studies must be carried out for proper understanding of this phenomenon.

It is obvious from the above discussion that more studies are needed in order to define the relationship between zinc and lipoproteins.

6.6 Lipid Peroxidation

Zinc is associated to a large extent with the lipid component in the membranes. It has been suggested that zinc decreases lipid peroxidation of the membrane and thus behaves as an antioxidant. This topic has been discussed elsewhere (see Chapter 7).

References

Bettger, W. J., and O'Dell, B. L., 1980. Diiodoquin therapy of zinc deficient rats, *Am. J. Clin. Nutr.* 33:2223.

Bustamante, J. B., Mateo, M. C. M., Fernandez, J., DeQuiros, B., and Manchado, O. O., 1976. Zinc, copper, and ceruloplasmin in atherosclerosis, *Biomedicine* 25:244.

Cash, R., and Berger, C. K., 1969. Acrodermatitis enteropathica: Defective metabolism of unsaturated fatty acids, *J. Pediatr.* 74:717.

Crouse, S. F., Hooper, P. L., Atterbom, H. A., and Papenfass, R. L., 1984. Zinc ingestion and lipoprotein values in sedentary and endurance-trained men, *J. Am. Med. Assoc.* 252:785.

Cunnane, S. C., 1988. *Zinc: Clinical and Biochemical Significance,* CRC Press, Boca Raton, Fla., CRC Press, p. 99.

Cunnane, S. C., Ainley, C. C., Keeling, P. W. N., Thompson, R. P. H., and Crawford, M. A., 1986. Metabolism of linoleic and arachidonic acid by peripheral blood leukocytes from patients with Crohn's disease, *J. Am. Coll. Nutr.* 5:451.

David, T. J., Wells, F. E., Sharpe, T. C., and Gibbs, A. C. C., 1984. Low serum zinc in children with atopic eczema, *Br. J. Dermatol.* 111:597.

Dore-Duffy, P., Catalanotto, F., Donaldson, J. O., Ostrom, K. M., and Testa, M. A., 1983. Zinc in multiple sclerosis, *Ann. Neurol.* 14:450.

Fischer, P. W. F., Giroux, A., Belonte, B., and Shah, B. G., 1980. The effect of dietary copper and zinc on cholesterol metabolism, *Am. J. Clin. Nutr.* 33:1019.

Fogerty, A. C., Ford, G. L., Dreosti, I. E., and Tinsley, I. J., 1985. Zinc deficiency and fatty acid composition of tissue lipids, *Nutr. Rep. Int.* 32:1009.

Halevy, O., and Sklan, D., 1986. Effect of copper and zinc depletion on vitamin A and triglyceride metabolism in chick liver, *Nutr. Rep. Int.* 33:723.

Hambidge, K. M., Neldner, K. H., Weston, W. L., Silverman, A., Sabol, J. L., and Brown, R. M., 1977. Zinc and acrodermatitis enteropathica, in *Zinc and Copper in Clinical Medicine* (K. M. Hambidge and B. L. Nichols, eds.), SP Medical and Scientific Books, Jamaica, N.Y., p. 81.

Hamilton, R. M., Gillespie, C. T., and Cook, H. W., 1981. Relationships between levels of essential fatty acids in plasma of cystic fibrosis patients, *Lipids* 16:374.

Julius, R., Schulkind, M., Sprinkle, T., and Rennert, O., 1973. Acrodermatitis enteropathica with immune deficiency, *J. Pediatr.* 83:1007.

Katya-Katya, M., Ensminger, A., Mejean, L., and Debry, G., 1984. The effect of zinc supplementation on plasma cholesterol levels, *Nutr. Res.* 4:633.

Klevay, L. M., 1975. Coronary heart disease: The zinc/copper hypothesis, *Am. J. Clin. Nutr.* 28:764.

Koo, S. J., and Ramlet, J. S., 1983. Dietary cholesterol decreases the serum level of zinc: Further evidence for the positive relationship between serum zinc and high-density lipoproteins, *Am. J. Clin. Nutr.* 37:918.

Koo, S. J., and Williams, D. A., 1981. Relationship between the nutritional status of zinc and cholesterol concentration of serum lipoproteins in adult male rats, *Am. J. Clin. Nutr.* 34:2376.

Koo, S. J., Algilani, K., Norvell, J. E., and Henderson, D. A., 1986. Delayed plasma clearance and hepatic uptake of lymph chylomicron ^{14}C cholesterol in marginally zinc-deficient rats, *Am. J. Clin. Nutr.* 43:429.

Lefevre, M., Keen, C. L., Lonnerdal, B., Hurley, L. S., and Schneeman, B. O., 1985. Different effects of copper and zinc deficiency on composition of plasma high-density lipoproteins in rats, *J. Nutr.* 115:359.

Muskiet, F. D., and Muskiet, F. A. J., 1984. Lipids, fatty acids, and trace elements in plasma and erythrocytes of pediatric patients with homozygous sickle cell disease, *Clin. Chim. Acta* 142:1.

Neldner, K. H., and Hambidge, K. M., 1975. Zinc therapy for acrodermatitis enteropathica, *N. Engl. J. Med.* 292:879.

Neldner, K. H., Hagler, L., Wise, W. R., Stifel, F. B., Lufkin, E. G., and Herman, R. H., 1974. Acrodermatitis enteropathica: A clinical and biochemical survey, *Arch. Dermatol.* 110:711.

Okuyama, S., Mishina, H., Hasegawa, K., Nakeno, N., and Ise, K., 1982. Probable atherogenic role of zinc and copper as studied in chronic hemodialysis, *Tohoku J. Exp. Med.* 138:227.

Petering, H. G., Murthy, L., and O'Flaherty, E., 1977. Influence of dietary copper and zinc on rat lipid metabolism, *J. Agric. Food Chem.* 25:1105.

Prasad, A. S., and Oberleas, D., 1970. Binding of zinc to amino acids and serum proteins in vitro, *J. Lab. Clin. Med.* 76:416.

Rana, S. V., and Tayal, M. K., 1981. Lipotrophic effects of zinc, vitamin B_{12} and glutathione on the fatty liver of the rat. A histochemical study, *Mikroskopie* 38:294.

Roth, H. P., and Kirchgessner, M., 1977. Zum einfloss von zinkmangel auf den Fettstoffwessel, *Int. Z. Vitaminforsch.* 47:277.

Sandstead, H. H., Gillespie, D. D., and Brady, R. N., 1972. Zinc deficiency: Effect on brain of the suckling rat, *Pediatr. Res.* 6:119.

Solomons, N. W., Reiger, C. H. L., Jacob, R. A., Rothberg, R., and Sandstead, H. H., 1981. Zinc nutriture and taste acuity in patients with cystic fibrosis, *Nutr. Res.* 1:13.

Sullivan, J. F., Jetton, M. M., Hahn, H. K. J., and Burch, R. E., 1980. Enhanced lipid peroxidation in liver microsomes of zinc deficient rats, *Am. J. Clin. Nutr.* 33:51.

White, H. B., and Montalvo, J. M., 1973. Serum fatty acids before and after recovery from acrodermatitis enteropathica: Comparison of an infant with her family, *J. Pediatr.* 83:999.

Wolfe, J. A., Margolis, S., Bujdoso-Wolfe, K., Matusick, E., and McLean, W. C., 1984. Plasma and red blood cell fatty acid composition in children with protein-calorie malnutrition, *Pediatr. Res.* 18:162.

Woo, W., Gibbs, D. L., Hooper, P. L., and Garry, P. J., 1983. The effect of dietary zinc on high density lipoprotein synthesis, *Nutr. Rep. Int.* 27:499.

Zinc and Cells 7

7.1 Lymphocytes

Many studies have shown that zinc ions as an integral part of tissues and biologic fluids play an important role in homeostatic mechanisms regulating the reactivity of tissues and cells. In lymphocytes, zinc acts as a nonspecific mitogen (Chvapil, 1976). Within a range of 1.5 to 4.5×10^{-4} M Zn concentration in the culture medium, the blastogenic transformation of lymphocytes as well as their mitosis was significantly increased after 6 days in comparison with the effect of phytohemagglutinin (PHA). Zinc acted as a weak mitogen (Kirchner and Ruhl, 1970; Ruhl *et al.*, 1974). Surprisingly, only zinc and mercury were stimulatory. Calcium and magnesium did not affect DNA synthesis in the culture system. Mn^{2+}, Co^{2+}, Cd^{2+}, Cu^{2+}, and Ni^{2+} at concentrations from 10^{-3} to 10^{-7} M were inhibitory. Inasmuch as DNA-synthesizing enzymes are zinc dependent, one may assume that enhanced mitosis of lymphocytes by zinc may be the result of increased activity of enzymes involved in cell mitosis.

The essential role of Zn^{2+} in DNA replication in PHA-stimulated human lymphocytes was demonstrated by adding a chelator with high affinity for Zn^{2+} to the culture medium, which inhibited the incorporation of thymidine by lymphocytes. Addition of Zn^{2+} to the medium reversed the effect of the chelator (Williams and Loeb, 1973). In another study, lymphocytes isolated from zinc-supplemented animals showed a significantly higher mitotic index when exposed to PHA compared with lymphocytes from animals fed control diets (Chvapil *et al.*, 1974). Zinc transferrin, but not other complexes, showed an increase in DNA synthesis over that seen with PHA alone when it was added to serum-free culture of PHA-stimulated human lymphocytes, suggesting that zinc transferrin complex may have a special role in lymphocyte proliferation (Phillips and Azari, 1974). Although these studies indicate an enzymatic effect of zinc on DNA synthesis, the observation that Zn:8-hydroxyquinoline unsaturated complexes (1:1, 1:2), which do not permeate cell membranes, enhance lymphocyte mitosis as well, indicates the possible membrane-related effect.

7.2 Mast Cells

Although zinc activates lymphocytes *in vitro,* an inhibitory effect of zinc on mast cells, platelets, macrophages, polymorphonuclear leukocytes, and sperm cells has been observed (Chvapil, 1976). Hogberg and Uvnas (1960) first demonstrated the inhibitory effect of Zn^{2+} on the disruption of mast cells of rat mesentery induced by compound 48/80, lecithinase A, or by antigen–antibody reaction. Kazimierczak and Maslinski (1974) administered Zn to rats intraperitoneally for 10 to 20 days (4 mg Zn/kg per day) and found inhibition of the spontaneous release of histamine from lung tissue as well as that induced by a low dose of compound 48/80. The protective effect of Zn on histamine release by mast cells was evident after only a few days of Zn administration to animals. Based on the observation that unsaturated Zn:8-HQ (1:1) complex was an even more potent inhibitor of histamine release from isolated mast cells than Zn alone, it was proposed that Zn acts on the cell membrane (Kazimierczak and Maslinski, 1974). Inasmuch as Zn prevented the histamine release from mast cells only when it was added simultaneously with compound 48/80 to the medium, it appeared that Zn^{2+} competed with compound 48/80 for similar receptor binding sites.

7.3 Platelets

The importance of bivalent cations for the function of platelets is well known. Ca^{2+} is required for contraction and release mechanisms (Harris *et al.,* 1974). Mn^{2+} inhibits *in vitro* platelet aggregation and release mechanisms, possibly by interfering with Ca^{2+} ions. Mg^{2+} is necessary for deaggregation and inhibits aggregation in the presence of Ca^{2+}.

Zn^{2+} was shown to significantly inhibit collagen-induced aggregation of dog platelets and collagen- or epinephrine-induced release of [^{14}C]serotonin from the platelets (Chvapil *et al.,* 1975a). The concentration of Zn required for these functions was 10 to 15 mM. The presence of plasma in the incubating medium was essential for the inhibitory effect of Zn; only fibrinogen, not albumin or globulin, substituted for plasma. The uptake of ^{45}Ca by platelets was also inhibited by Zn^{2+}. Supplementation of zinc *in vivo* in dogs also effectively decreased aggregation of platelets as well as the release of [^{14}C]serotonin from the platelets.

7.4 Macrophages and Polymorphonuclear Granulocytes (PMNs)

Several functions of macrophages or PMNs are inhibited by *in vitro* or *in vivo* Zn supplementation. Macrophages harvested from the peritoneal cavity of animals (rat or guinea pig) fed diets containing a very high level of Zn (2000 ppm) or injected with zinc sulfate remained round for 24 h in a medium with 15% autologous serum, and showed minimal cytoplasmic elongations and extrusions. Their mobility and phagocytosis of *Staphylococcus albus* were inhibited (Karl *et al.,* 1973). The control animals fed 40 ppm Zn showed multiple forms of plasma membrane with elongations, cytoplasmic connections, active migration, and phagocytosis. Macrophages harvested from the peritoneal cavity of zinc-deficient guinea pigs showed enhanced migratory activity. Response to the induction of a sterile inflammatory reaction by, for example, intraperitoneal injection of mineral oil into animals treated with zinc showed less cellular infiltration of the peritoneal cavity by macrophages and PMNs (Chvapil, 1976).

There were several possible effects of Zn at the cell membrane level. Formation of mercaptides with thiol groups of membrane proteins, possible linking to the phosphate moiety of phospholipids, or interaction with carboxyl groups of sialic acid or proteins are some of the sites on the membrane where Zn may interact. This type of Zn reaction may result in change of fluidity of the membrane or may stabilize the membrane as shown for the plasma membrane of fibroblasts and for the lysosomal membrane of hepatocytes (Warren *et al.,* 1966; Chvapil *et al.,* 1972).

There are several enzymes attached to the plasma membrane which control the structure and functions of the membrane. Zn inhibits ATPase and phospholipase A_2 (Chvapil, 1976). Inhibition of either enzyme by Zn may explain immobilization of energy-dependent activity of plasma membrane or increased integrity of the membrane structure.

Histamine and serotonin release agents seem to work through specific receptors at the membrane; thus, masking of such receptor sites by membrane-impermeable Zn:8-HQ complex would explain the inhibition of the release reaction as mentioned above.

7.5 Ecto 5'-Nucleotidase

Ecto 5'-nucleotidase (5'-NT) (EC 3.2.3.5) is an integral plasma membrane enzyme of most mammalian cells and has been used as a membrane marker. The catalytic site of this enzyme faces the outside of the cell (Pilz *et al.,* 1982). Ecto 5'-NT catalyzes the extracellular dephosphorylation of purine and pyrimidine ribonucleotide monophosphate and deoxyribonucleoside 5'-

monophosphate to their corresponding purine and pyrimidine ribonucleo-side and deoxyribonucleoside molecules. Inasmuch as ribonucleoside 5'-monophosphate and deoxyribonucleoside 5'-monophosphate are impermeable to the cell membrane, conversion to purine and pyrimidine ribonucleoside enables them to cross the plasma membrane for utilization in intracellular metabolism and DNA synthesis.

Studies suggest that human lymphoblast plasma membrane 5'-NT can exist as an inactive apoenzyme and zinc plays a unique role in the expression of plasma membrane 5'-NT activity (Pilz et al., 1982). Bacterial 5'-NT is also known to be a zinc metalloenzyme. Thus, it is conceivable that if the activity of ecto 5'-NT is decreased in zinc deficiency, the proliferation of lymphoid cells may be affected adversely.

7.6 Zinc and Free Radicals

Zn inhibits oxidation of NADPH in liver microsomes and in the granular fraction of rabbit lung alveolar macrophages (Chvapil, 1976). NADPH oxidation is crucial to generation of H_2O_2 and thus to bactericidal activity of macrophages and PMNs. Because Zn at low concentrations inhibits NADPH oxidation, this may account for decreased phagocytosis by these cells in the presence of Zn. Zn may also bind to NADPH through phosphate residues, thus impairing the turnover of this nucleotide which is needed for the function of the hexose monophosphate shunt.

Activation of dioxygen to superoxide, H_2O_2, and hydroxyl radical leads to numerous deleterious effects such as lipid peroxidation, and oxidative damage to DNA and chromosomal proteins. In humans, these oxidative in-juries to cells have been implicated in carcinogenesis, rheumatoid arthritis, and aging. A number of factors are known to cause lipid peroxidation. These include high-energy irradiation, ozone, carbon tetrachloride, adriamycin, paraquat, iron, 3-methylindole, and t-butylhydroperoxide. Cellular functions are subsequently compromised as a result of peroxidative damage (Bray and Bettger, 1990).

In zinc-deficient animals, peroxidation of tissue lipids has been noted to occur (Burke and Fenton, 1985; Szeberi et al., 1988; Coppen et al., 1988). The effect of zinc may be related to an inhibition of NADPH-dependent lipid peroxidation. Malondialdehyde formation is increased three- to fourfold as a result of zinc deficiency suggesting that lipid peroxidation is enhanced by zinc deficiency. Zinc may bind to arachidonic acid which may be protective against iron-catalyzed oxidation (Peterson et al., 1981). The level of GSH has been reported to be reduced in plasma of zinc-deficient rats (Fernandez and O'Dell, 1983). Because GSH is an electron donor and free radical scavenger, a reduction of GSH level in zinc deficiency may enhance lipid peroxidation. The beneficial effect of zinc on membrane stabilization may be explained on the basis of inhibition of membrane lipid peroxidation by zinc (Chvapil, 1973, 1976; Chvapil *et al.*, 1972, 1979; Wright *et al.*, 1984). High levels of dietary zinc or the addition of zinc to liver homogenates in rats has been shown to inhibit lipid peroxidation.

Recent data suggest that zinc plays a role in suppression of free radical formation and lipid peroxidation in rat hepatocytes by mechanisms which may involve metallothionein (MT) synthesis, exclusion of extracellular iron, and/or changes in the activities of cytosolic or microsomal enzymes (Coppen *et al.*, 1988). It has been shown that MT may act as a free radical scavenger (Hidalgo *et al.*, 1988; Coppen *et al.*, 1988). Using a Fenton-type reaction (i.e., $Fe^{2+} + H_2O_2$) and the xanthine/xanthine oxidase reaction to generate oxygen radicals, it was demonstrated that MT was effective in quenching hydroxyl radicals compared with superoxide ions. These radicals caused metal loss from the protein and thiolate oxidation. It was suggested that glutathione would regulate reduced SH groups of MT for subsequent rebinding of metals.

Girotti *et al.* (1985) have observed that peroxidation of erythrocyte membranes generated by the xanthine/xanthine oxidase system was inhibited by Zn^{2+} *in vitro*. It was subsequently reported that MT added to this *in vitro* system decreased iron-mediated membrane peroxidation (Coppen *et al.*, 1988). Thus, it appears that the antioxidant activity of MT and reduced uptake of iron into hepatocytes due to Zn, may have contributed to the reduction in free radicals trapped and the reduction of malondialdehyde as reported by Coppen *et al.* (1988).

Zinc inhibits microsomal NADPH oxidation and NADPH-dependent cytochrome c reductase activity (Jeffrey, 1983; Chvapil *et al.*, 1975b,c). One mechanism of inhibition of NADPH-dependent cytochrome c reductase may be that Zn binds to NADPH. By inhibiting microsomal cytochrome reductase, zinc may decrease 3-methylindole radical formation, which subsequently reduces lipid peroxidation (Kubow *et al.*, 1984).

In one study, zinc addition *in vitro* resulted in increased glutathione peroxidase activity, although zinc deficiency does not affect the activity of this enzyme (Coppen *et al.*, 1988). It is clear, therefore, that zinc may play an important role in free radical reactions and preservation of the cell membrane.

Zinc is known to compete with Ca^{2+}. The role of Ca^{2+} in the function of the cell membrane microskeleton has been documented. The contractile elements of this system are responsible for the mobility of microorganelles, transport of granules to the membrane, as well as excitability of the plasma membrane itself. Inhibition of the Ca^{2+} effect by Zn would be another factor involved in the overall effect of Zn on cell membranes.

7.7 Zinc and Erythrocytes

In red blood cells, carbonic anhydrase accounts for a large majority of the erythrocyte zinc and plays an important role in the transport of CO_2 from the tissues to the lungs. In the 1960s, an additional role of zinc was found as a constituent of the enzyme superoxide dismutase (McCord and Fridovich, 1969; Bannister *et al.*, 1971). (This protein was initially named erythrocuprein.) Superoxide dismutase contains two atoms of Zn and two atoms of Cu and catalyzes the reaction:

$$2O_2^- + 2H^+ \rightarrow O_2 + H_2O_2$$

It is thought to play a role in protecting erythrocytes as well as other cells from damage by superoxide ions (McCord *et al.*, 1971).

During the 1970s, interest in other effects of zinc on erythrocytes was generated by the finding that sickle-cell anemia (SCA) frequently correlates with a decrease in the erythrocyte concentration of Zn (Prasad *et al.*, 1977), and the demonstration that zinc affects the oxygenation of hemoglobin (Oelshlegel *et al.*, 1973; Rifkind, 1983). By *in vitro* addition of Zn in excessive amounts, it was shown that zinc binds to the outside surface of the erythrocyte membrane (Passow, 1970). It was also demonstrated that zinc could be transported across the membrane, and inside the cell, zinc bound to the inner leaflet of the membrane (Rifkind, 1983), hemoglobin (Rifkind and Heim, 1977), other proteins, and small molecules (Tsukamoto *et al.*, 1979; Rabenstein and Isab, 1980).

The stabilizing effect of zinc on erythrocyte membrane, as indicated by effects on hemolysis, has been studied extensively. It has been reported that zinc supplementation decreases the osmotic fragility of cells (Chvapil *et al.*, 1974; Settlemire and Matrone, 1967), and that erythrocytes from zinc-deficient animals are susceptible to osmotic stress (Bettger *et al.*, 1978). In one study, addition of zinc *in vitro* to human erythrocytes decreased the osmotic fragility (Kabat *et al.*, 1978). Other studies, however, failed to show a significant decrease in osmotic fragility when Zn was added to erythrocytes *in vitro* (Chvapil *et al.*, 1974; Weismann and Mikkelsen, 1980).

The apparent discrepancy between the results *in vivo*, which indicate that Zn increases osmotic stability, and the bulk of the results *in vitro*, which indicate that Zn has no effect on osmotic stability, may be explained by the known interaction of Zn with Cu *in vivo* (Adams *et al.*, 1979; Bettger *et al.*, 1978; Brewer, 1980a,b; Prasad *et al.*, 1978; Willis, 1965).

Zinc has been shown to protect against hemolysis induced by various bacterial toxins (Avigad and Bernheimer, 1976, 1978; Takeda *et al.*, 1977), complement (Montgomery *et al.*, 1974; Yamamoto and Takahashi, 1975), and other lytic agents (Avigad and Bernheimer, 1976; Takeda *et al.*, 1977). It was subsequently shown that the apparent protection against lysis by vibriolysin, Triton X-100, saponin, and lysolecithin was the result of precipitation by hemoglobin liberated from the cells as well as in the cell (Avigad and Bernheimer, 1976; Takeda *et al.*, 1977). Zinc does, however, directly protect against lysis by other systems. It was shown to prevent the binding of staphylococcal toxin and streptolysin to the membrane (Avigad and Bernheimer, 1978). In the case of *Clostridium perfringens* alpha toxin and perfringolysin as well as complement-induced lysis, steps involving disruption of the membrane subsequent to the binding may be affected by zinc (Boyle *et al.*, 1979).

Zinc produces agglutination of washed human erythrocytes (Passow, 1970). This is in part the result of the binding of Zn to negatively charged groups on the surface of the membrane, such as the carboxyl groups of neuraminic acid and other negatively charged groups on membrane proteins and phospholipids. Studies with liposomes and erythrocyte ghosts have shown that Zn binds to phospholipids in the membrane (McLaughlin *et al.*, 1978). An appreciable amount of Zn binding to the outside surface of the membrane may also involve proteins. Binding of Zn to the outside surface of the membrane causes an increased sodium permeability, probably related to the binding of Zn to external amino groups in the sodium channel (Castranova and Miles, 1977).

7.8 Transport of Zinc across the Erythrocyte Membrane

Zn can penetrate the erythrocyte membrane by a passive process. A relatively small enhancement of Zn influx was observed in the presence of ATP; however, a similar effect was seen in the presence of GTP and CTP, suggesting that Zn may be bound to the nucleotide triphosphate inside the ghosts (Schmetterer, 1978). The transport of Zn across the membrane is dependent on the free Zn concentration in the medium and its uptake is inhibited by complexing agents such as EDTA and histidine (Passow, 1970). Although metal ions such as Co^{2+}, Mn^{2+}, Cd^{2+}, and Cu^{2+} were found to have no effect on Zn^{2+} uptake, Fe^{2+} was inhibitory.

Plasma and serum decrease dramatically the uptake of Zn by erythrocytes as a result of binding of Zn to albumin (Passow, 1970; Karl *et al.*, 1973; Kruckberg *et al.*, 1977; Kruckberg and Brewer, 1978). A 70% increase in Zn uptake was observed when the pH was increased from 7.6 to 8.3 and when the plasma or Krebs phosphate-free bicarbonate buffer (KB) with human serum albumin (HSA) was used. The uptake of Zn was significantly increased at higher temperatures relative to 5°C (see Figs. 7-1 and 7-2). A decrease in

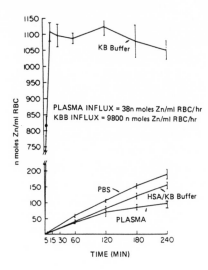

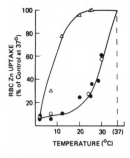

Figure 7-1. Uptake of zinc by erythrocytes in different media. (Reprinted with permission from Kruckberg, W. C., and Brewer, G. J., 1978. The mechanism and control of human erythrocyte zinc uptake, *Med. Biol.* 56:5.)

the uptake of Zn reported for trinitrocresol (TNC) in the absence of plasma or HSA indicates an effect of TNC on membrane transport of Zn (Kruckberg *et al.,* 1977).

Zinc uptake by erythrocytes under certain conditions is very rapid and large quantities of Zn may be taken up by erythrocytes. The efflux of Zn from the erythrocytes is much slower than the influx. The slow efflux and large concentration difference observed between the inside and outside of the erythrocyte indicate that Zn associated with the erythrocyte is bound relatively tightly to the erythrocyte membrane or intracellular molecules.

The mechanism for rapid uptake of zinc by erythrocytes is unknown. Zn uptake is more rapid than the passive transport of calcium (i.e., when the calcium pump is arrested) and furthermore, whereas Zn uptake is inhibited by TNC, the latter accelerates Ca^{2+} uptake, thus indicating different pathways for Ca^{2+} and Zn^{2+} influx.

Figure 7-2. Temperature variation of uptake of zinc by erythrocytes in different media: $\triangle$, KB buffer; O, HSA/KB buffer; ●, plasma. (Reprinted with permission from Kruckberg, W. C., and Brewer, G. J., 1978. The mechanism and control of human erythrocyte zinc uptake, *Med. Biol.* 56:5.)

The rate of transport of Zn is, however, comparable by incorporating into the membrane an artificial carrier such as the ionophore A23187 which is also a calcium ionophore, thus suggesting that the erythrocyte membrane might contain an integral Zn carrier. Zn has been found to be associated with many of the membrane proteins after purification and in particular, a low-molecular-mass membrane protein (<5 kDa) was shown to have a very high affinity for Zn^{2+} (Rifkind, 1983). Further work is needed to determine if there is a carrier protein for Zn^{2+} present in the erythrocyte membrane. Since Zn is known to form relatively stable chloride complexes, transport of the neutral $ZnCl_2$ through the bilayer may be considered as a possible pathway of the transport of Zn^{2+} through the membrane. Torrubia and Garay (1989) have recently provided evidence that zinc may be transported by the anion carrier in the form of the monovalent anion complex $[Zn(HCO_3)_2Cl]^-$.

7.9 Sickle-Cell Membranes and Zn^{2+}

Zinc deficiency in sickle-cell anemia (SCA) patients has been reported. SCA is associated with a modified hemoglobin molecule Hb-SS (β-6 Glu $\rightarrow$ Val), which tends to polymerize when deoxygenated (Bertles et al., 1970; Finch et al., 1973). This polymerization leads to deformation and sickling of the erythrocytes, which lead to occlusion of small vessels, difficulty in oxygen transportation, and pain crisis. It has been shown that the sickling process produces changes in the membrane which can result in cells that retain their abnormal shape even after oxygenation (Dobler and Bertles, 1968). It is believed that these irreversibly sickled cells (ISCs) are responsible for many of the clinical manifestations of SCA.

Administration of zinc in vivo decreases the number of ISCs from 28.0% before treatment to 18.6% during treatment (Brewer et al., 1979b). Since the ISCs have damaged membranes, these results suggest that zinc interacts with the membrane. Zinc at low concentration (0.3 mM) in plasma significantly improved the ability of partially deoxygenated sickle cells at an oxygen pressure of 15 mm Hg to go through the small pores of nucleopore filters (Brewer and Oelshlegel, 1974). This effect of Zn^{2+} was also interpreted as an effect on the membrane.

Arnone and Williams (1977), on the basis of their x-ray evidence for a zinc-binding site on deoxyhemoglobin that links tetramers, have suggested a possible mechanism for zinc at relatively low concentrations to inhibit hemoglobin polymerization (Fig. 7-3). It is possible that zinc's effect on sickle cells relates to both its effect on the membrane and its binding to the hemoglobin molecule.

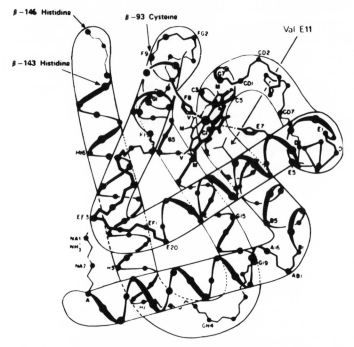

Figure 7-3. Diagrammatic sketch showing the course of the polypeptide chain in the β subunit of hemoglobin. [Reprinted with permission from Rifkind, J. M., 1983. Interaction of zinc with erythrocytes, in *Metal Ions in Biological Systems,* Vol. 15 (H. Sigel, ed.), Dekker, New York, p. 275.]

A large percentage of the control ghosts formed by lysis in 3 mM NaCl and resealed in 10 mM Tris-HCl (pH 7.6) were echinocytes and most of the others were smooth (folded). Lysis in 1.5 mM $ZnSO_4$ very significantly decreased the number of echinocytes, suggesting a direct interaction of zinc with the membranes (Table 1) (Brewer and Oelshlegel, 1974; Dash *et al.,* 1974).

Increase in intracellular calcium produces a number of alterations in the erythrocytes which include decrease in mean cell volume, increase in mean cell hemoglobin concentration, loss of water and potassium, marked changes in morphology, rapid hydrolysis of intracellular ATP, decrease in cell deformability, and stiffening of the erythrocyte membrane (Palek *et al.,* 1971; Eaton *et al.,* 1977; Dunn, 1974; White, 1974, 1976; Weed *et al.,* 1969). Increased calcium concentrations have been found in sickle cells and particularly in ISCs. It has therefore been proposed that the membrane damage in sickling

Table 7-1. Hemoglobin Retention by Ghost Cells[a,b]

Ghosts prepared in the presence of	Hemoglobin in final ghosts (g/100 ml)	Hemoglobin retention (%)
Saline (control)	1.86	7.4
1.5 mM ZnSO$_4$	0.25	1.0
1 mM CaCl$_2$	6.39	24.6
1.5 mM ZnSO$_4$ + 1 mM CaCl$_2$	2.50	10.7

[a] Reprinted with permission from Dash, S., Brewer, G. J., and Oelshlegel, F. J., Jr., 1974. Effect of zinc on haemoglobin binding by red cell membranes, *Nature* 250:251.
[b] Data corrected for change in mean ghost cell volume, average of four experiments.

is the result of increased cellular calcium concentration (Eaton *et al.*, 1977). It has been observed that cell shrinkage, as well as depletion of ATP from Hb-SS cells, is very significantly diminished in the presence of 0.2 mM zinc. On the basis of these results, it has been proposed that the effect of zinc on the number of ISCs in patients with SCA, the filterability at low oxygen pressure of sickle cells, and the decrease in the number of echinocytes found on ghosts prepared from sickle cells can be interpreted in terms of an antagonism between the effect of calcium and zinc, whereby zinc prevents the damage produced by calcium (Dash *et al.*, 1974; Brewer, 1980a).

Calcium has been shown to promote binding of hemoglobin to the inside of the red cell membrane, while zinc inhibits this binding and counteracts the effect of calcium (Dash *et al.*, 1974). Intracellular calcium has been shown to induce limited proteolysis of one of the spectrin bands and at higher concentrations the cross-linking of membrane proteins (Anderson *et al.*, 1977).

Figure 7-4. Zinc inhibition of ATPase activity of erythrocyte ghosts. Assay mixtures contained 3 mM ouabain, 3 mM ATP, 1 mM MgCl$_2$, 50 mM NaCl, 120 mM KCl, and 50 μM CaCl$_2$ and were run at 37°C (pH 7.2). (Reprinted with permission from Brewer, G. J., Aster, J. C., Knutsen, C. A., and Kruckberg, W. C., 1979. Zinc inhibition of calmodulin. A proposed molecular mechanism of zinc action on cellular functions, *Am. J. Hematol.* 7:53.)

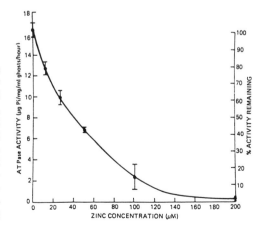

It has been reported that proteolysis is prevented and the activity of transglutaminase, an enzyme involved in cross-linking of membrane protein and activated by calcium, is inhibited by zinc (Anderson *et al.*, 1977). The intracellular level of Ca^{2+} is kept at its usual low level by an ATP-dependent Ca^{2+} pump, which utilizes a Ca^{2+},Mg^{2+}-ATPase activated by intracellular Ca^{2+}. Intracellular Zn^{2+} inhibits calcium-stimulated ATPase activity (Fig. 7-4). On the basis of these results, it has been suggested that the mechanism by which zinc reverses many of calcium's effects may actually be a turning off of ATPase which results in an increase in the intracellular ATP level (Kruckberg *et al.*, 1977).

It has been proposed that in erythrocytes, many of the antagonist effects of zinc on calcium may involve calmodulin, a 17-kDa protein which is found in erythrocytes and many other cells (Brewer, 1980a). Calmodulin is activated by binding calcium, which alters the calmodulin conformation and enables it to stimulate the activity of many enzymes and cellular functions. In the erythrocyte, calmodulin activates the Ca^{2+},Mg^{2+}-Ca–Mg ATPase, induces retention of hemoglobin by erythrocyte ghosts, and induces shrinkage of the erythrocytes (Brewer *et al.*, 1979a). Zinc has been shown to inhibit the cal-

Table 7-2. Test of the Hypothesis That Certain Pharmacological Effects Predict Membrane Expansion and Calmodulin Inhibition

Drug	Known effects	New finding
Zinc	Membrane antisickling, calmodulin inhibition	Expands membranes
Cetiedil	Membrane antisickling, local anesthetic	Inhibits calmodulin, expands membranes
Procaine	Membrane antisickling, local anesthetic, membrane expansion	Inhibits calmodulin
Dibucaine	Local anesthetic, membrane expansion	Inhibits calmodulin
Propranolol	Cardiac antiarrhythmic, membrane expansion	Inhibits calmodulin
Procainamide	Cardiac antiarrhythmic, membrane expansion	Inhibits calmodulin
Quinidine	Cardiac antiarrhythmic	Inhibits calmodulin, expands membranes
U of M experimental drug 272	Cardiac antiarrhythmic	Expands membranes
U of M experimental drug 424	Cardiac antiarrhythmic	Expands membranes
Bretylium	Cardiac antiarrhythmic	Does not expand membranes

[a] Adapted with permission from Brewer, G. J., and Bereza, U. L., 1982. Therapy of sickle cell anemia with membrane expander/calmodulin inhibitor classes of drugs, in *Clinical, Biochemical, and Nutritional Aspects of Trace Elements* (A. E. Prasad, ed.), Liss, New York, p. 211.

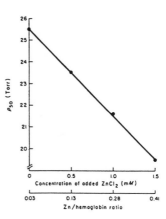

Figure 7-5. Effect of ZnCl₂ on a saline suspension of red blood cells (final hematocrit of 43 vol %). The P_{50} is plotted as a function of the final concentration of added ZnCl₂ on the top line of the horizontal axis; the bottom line gives the molar ratio of Zn to hemoglobin as obtained by assay. (Reprinted with permission from Oelshlegel, F. J., Jr., Brewer, G. J., Knutsen, C., Prasad, A. S., and Schoomaker, E. B., 1974. Studies on the interaction of zinc with human hemoglobin, *Arch. Biochem. Biophys.* 163:742.)

modulin-stimulated hemoglobin retention, as well as several other calmodulin-stimulated enzyme activities in other cells (Brewer, 1980a). On the basis of these results, it has been suggested that the major effect of zinc in erythrocyte membranes involves inhibition of calmodulin's stimulating effect on cellular processes (Table 7-2).

An effect on the oxygenation of hemoglobin by zinc was originally reported by Oelshlegel *et al.* (1973). The investigators found a 9% decrease in P_{50} (i.e., an increase in the oxygen affinity) on incubating normal human erythrocytes as well as erythrocytes of patients with SCA in 1.5×10^{-3} M ZnCl₂ for 2 h (Figs. 7-5 and 7-6). In a subsequent paper (Oelshlegel *et al.*, 1974), they investigated the effect of zinc on partially purified human hemoglobin preparations and found a somewhat larger (24%) decrease in P_{50} from 25.5 to 19.5 mm Hg at a zinc/hemoglobin ratio of 0.41 mol of zinc

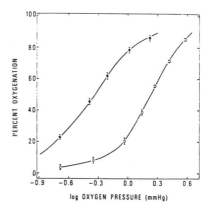

Figure 7-6. Effect of zinc on the oxygenation of stripped human hemoglobin in 0.02 M bistris pH 7.4 at 25°C: O, no zinc added; ●, zinc/heme molar ratio = 0.52. (Reprinted with permission from Rifkind, J. M., and Heim, J. M., 1977. Interaction of zinc and hemoglobin: Binding of zinc and the oxygen affinity, *Biochemistry* 16:4438.)

bound per mol of hemoglobin tetramer. They also demonstrated that zinc interacted directly with the hemoglobin.

The high affinity of hemoglobin for zinc indicates that a large majority of zinc binds to hemoglobin in the concentration range that produces the increase in oxygen affinity. Gilman and Brewer (1978) and Rifkind and Heim (1977) have utilized their data on the zinc-binding site to propose how zinc stabilizes the liganded conformation and thereby increases the oxygen affinity. Gilman and Brewer (1978) suggest that zinc destabilizes the deoxy conformation by binding to histidine β-146, which in deoxyhemoglobin is involved in an intrachain salt bridge with aspartic acid β-94. Furthermore, the decrease in the effect of zinc on the oxygen affinity of hemoglobin with the β-93 SH blocked is explained by the disruption of the salt bridge involving β-146 even in deoxyhemoglobin when the β-93 SH is blocked.

Rifkind and Heim (1977) have proposed that a hydrogen bond between the glutamine at β-143 and asparagine β-139 of the other beta chain would stabilize the liganded conformation and that Zn could coordinate between the asparagine and histidine. Such an interaction would decrease the dissociation constant of tetramers to dimers, which seems to increase in the presence of Zn.

It has been shown that under certain conditions Zn causes dissociation of hemoglobin to dimers and perhaps even to monomers (Rifkind, 1983). Since dimers have a higher oxygen affinity than tetramers, it is possible that the effect on Zn of the oxygenation could at least partly be caused by dissociation of hemoglobin. Obviously more studies are needed in this area in order to fully understand the mechanism of zinc's effects on hemoglobin.

References

Adams, K. F., Johnson, G., Jr., Hornowski, K., and Lineberger, T., 1979. The effect of copper on erythrocyte deformability: A possible mechanism of hemolysis in acute copper intoxication, *Biochim. Biophys. Acta* 550:279.

Anderson, D. R., Davis, J. L., and Carraway, K. L., 1977. Calcium-promoted changes of the human erythrocyte membrane. Involvement of spectrin, transglutaminase and a membrane-bound protease, *J. Biol. Chem.* 252:6617.

Arnone, A., and Williams, D., 1977. The binding of zinc to human deoxyhemoglobin and its possible relevance to the anti-sickling effect of zinc, in *Zinc Metabolism: Current Aspects in Health and Disease* (G. J. Brewer and A. S. Prasad, eds.), Liss, New York, p. 217.

Avigad, L. S., and Bernheimer, A. W., 1976. Inhibition by zinc of hemolysis induced by bacterial and other cytolytic agents, *Infect. Immun.* 13:1378.

Avigad, L. S., and Bernheimer, A. W., 1978. Inhibition of hemolysis by zinc and its reversal by L-histidine, *Infect. Immun.* 19:1101.

Bannister, J., Bannister, W., and Wood, E., 1971. Bovine erythrocyte cupro-zinc protein. 1. Isolation and several characterizations, *Eur. J. Biochem.* 18:178.

Bertles, J. F., Rabinowitz, R., and Dobler, J., 1970. Hemoglobin interaction modification of solid phase composition in the sickling phenomenon, *Science* 169:375.

Bettger, W. J., Fish, T. J., and O'Dell, B. L., 1978. Effects of copper and zinc status of rats on erythrocyte stability and superoxide dismutase activity, *Proc. Soc. Exp. Biol. Med.* 158:279.

Boyle, M. D. P., Langone, J. J., and Borsos, T., 1979. Studies on the terminal stages of immune hemolysis, *J. Immunol.* 122:1209.

Bray, T., and Bettger, W. J., 1990. The physiological role of zinc as an antioxidant, *Free Radical Biol. Med.* 8:281.

Brewer, G. J., 1980a. Calmodulin, zinc, and calcium in cellular and membrane regulation, *Am. J. Hematol.* 8:231.

Brewer, G. J., 1980b. Zinc and copper in hematology, in *Zinc and Copper in Medicine* (Z. A. Karcioglu and R. M. Sarper, eds.), Thomas, Springfield, Ill., p. 347.

Brewer, G. J., and Bereza, U. L., 1982. Therapy of sickle cell anemia with membrane expander/calmodulin inhibitor classes of drugs, in *Clinical, Biochemical, and Nutritional Aspects of Trace Elements* (A. S. Prasad, ed.), Liss, New York, p. 211.

Brewer, G. J., and Oelshlegel, F. J., Jr., 1974. Anti-sickling effect of zinc, *Biochem. Biophys. Res. Commun.* 58:854.

Brewer, G. J., Aster, J. C., Knutsen, C. A., and Kruckberg, W. C., 1979a. Zinc inhibition of calmodulin: A proposed molecular mechanism of zinc action on cellular functions, *Am. J. Hematol.* 7:53.

Brewer, G. J., Brewer, L. F., and Prasad, A. S., 1979b. Suppression of irreversibly sickled erythrocytes by zinc therapy in sickle cell anemia, *J. Lab. Clin. Med.* 90:549.

Burke, J. P., and Fenton, M. R., 1985. Effect of zinc deficient diet on lipid peroxidation in liver and tumor subcellular membranes, *Proc. Soc. Exp. Biol. Med.* 179:187.

Castranova, Y., and Miles, P. R., 1977. The effect of zinc and other metals on the stability of lysosomes, *J. Membr. Biol.* 33:263.

Chvapil, M., 1973. New aspects in the biological role of zinc: A stabilizer of macromolecules and biological membranes, *Life Sci.* 13:1041.

Chvapil, M., 1976. Effect of zinc on cells and biomembranes, *Med. Clin. North Am.* 60(4):799.

Chvapil, M., Ryan, J. N., and Zukoski, C. F., 1972. The effect of zinc and other metals on the stability of lysosomes, *Proc. Soc. Exp. Biol. Med.* 140:642.

Chvapil, M., Zukoski, C. F., Hattler, B. G., Stankova, L., Montgomery, D., Carlson, E. C., and Ludwig, J. C., 1974. Zinc and cells, in *Trace Elements in Human Health and Disease* (A. S. Prasad, ed.), Academic Press, New York, p. 269.

Chvapil, M., Weldy, P. L., Stankova, L., Clark, D. S., and Zukoski, C. F., 1975a. Inhibitory effect of zinc ions on platelet aggregation and serotonin release reaction, *Life Sci.* 16:561.

Chvapil, M., Ludwig, J. C., Sipes, G., and Halladay, S. C., 1975b. Inhibition of NADPH oxidation and oxidative metabolism of drugs in liver microsomes by zinc, *Biochem. Pharmacol.* 24:1.

Chvapil, M., Ludwig, J. C., Sipes, G., and Misiorowski, R., 1975c. Inhibition of NADPH oxidation and related drug oxidation in liver microsomes by zinc, *Biochem. Pharmacol.* 25:1787.

Chvapil, M., Montgomery, D., Ludwig, J. C., and Zukoski, C., 1979. Zinc in erythrocyte ghosts, *Proc. Soc. Exp. Biol. Med.* 162:480.

Coppen, D. E., Richardson, D. E., and Cousins, R. J., 1988. Zinc suppression of free radicals induced in cultures of rat hepatocytes by iron, t-butyl hydroperoxide, and 3 methylindole, *Proc. Soc. Exp. Biol. Med.* 189:100.

Dash, S., Brewer, G. J., and Oelshlegel, F. J., Jr., 1974. Effect of zinc on haemoglobin binding by red cell membranes, *Nature* 250:251.

Dobler, J., and Bertles, J. F., 1968. The physical state of hemoglobin in sickle cell anemia erythrocytes in vivo, *J. Exp. Med.* 127:711.

Dunn, M. F., 1974. Red blood cell calcium and magnesium: Effects upon sodium and potassium transport and cellular morphology, *Biochim. Biophys. Acta* 352:97.

Eaton, J. W., Berger, E., White, J. G., and Jacob, H. S., 1977. Metabolic and morphologic effect of intraerythrocyte calcium: Implications for the pathogenesis of sickle cell disease, in *Metabolism: Current Aspects in Health and Disease* (G. J. Brewer and A. S. Prasad, eds.), Liss, New York, p. 275.

Fernandez, M. A., and O'Dell, B. L., 1983. Effect of zinc deficiency on plasma glutathione in the rat, *Proc. Soc. Exp. Biol. Med.* 173:564.

Finch, J. T., Perutz, M. J., Bertles, J. F., and Dobler, J., 1973. Structure of sickle erythrocyte and of sickle-cell hemoglobin fibers, *Proc. Natl. Acad. Sci. USA* 70:718.

Gilman, J. G., and Brewer, G. J., 1978. The oxygen-linked binding site of human haemoglobin, *Biochem. J.* 169:625.

Girotti, A. W., Thomas, J. P., and Jordon, J. E., 1985. Inhibitory effect of zinc on free radical lipid peroxidation in erythrocyte membranes, *J. Free Radicals Biol. Med.* 1:395.

Harris, G. L. A., Cove, D. H., and Crawford, N., 1974. Effect of divalent cations and chelating agents on the ATPase activity of platelet contractile protein, "thrombosthenin," *Biochem. Med.* 11:10.

Hidalgo, J., Campmany, L., Borras, M., Garvey, J. S., and Armario, A., 1988. Metallothionein response to stress in rats, *Am. J. Physiol.* 255:E518.

Hogberg, B., and Uvnas, B., 1960. Further observations on the disruption of rat mesentery mast cells caused by compound 48/80, antigen–antibody reaction, lethciinase A and decylamine, *Acta Physiol. Scand.* 48:133.

Jeffrey, E. H., 1983. The effect of zinc on NADPH oxidation and monooxygenase activity in rat hepatic microsome, *Cell Pharmacol.* 23:467.

Kabat, I. A., Niewworok, J., and Blaszcyk, J., 1978. Der einfluss von zinkionen auf ausgewahlte osmo tische parameter zentralbl bakteriol, *Zentralbl. Bakteriol. Parasitenkd. Infektionskr. Hyg. Abt. I Orig. B* 166:375.

Karl, L., Chvapil, M., and Zukoski, C. F., 1973. Effect of zinc on the viability and phagocytic capacity of peritoneal macrophages, *Proc. Soc. Exp. Biol. Med.* 142:1123.

Kazimierczak, W., and Maslinski, C., 1974. The mechanism of the inhibitory action of zinc on histamine release from mast cells by compound 48/80, *Agents Actions* 4:203.

Kirchner, H., and Ruhl, H., 1970. Stimulation of human peripheral lymphocytes by Zn^{2+} in vitro, *Exp. Cell Res.* 61:229.

Kruckberg, W. C., and Brewer, G. J., 1978. The mechanism and control of human erythrocyte zinc uptake, *Med. Biol.* 56:5.

Kruckberg, W., Knutsen, C. A., and Brewer, G. J., 1977. Mechanisms of red cell zinc uptake with a note on zinc and red cell metabolism, in *Zinc Metabolism: Current Aspects in Health and Disease* (C. J. Brewer and A. S. Prasad, eds.), Liss, New York, p. 259.

Kubow, S., Janzen, E. G., and Bray, T. M., 1984. Spin-trapping of free radicals formed during in vitro and in vivo metabolism of 3-methylindole, *J. Biol. Chem.* 259:4447.

McCord, J. M., and Fridovich, I., 1969. Superoxide dismutase. An enzyme function for erythrocuprein (hemocuprein), *J. Biol. Chem.* 244:6049.

McCord, J. M., Keele, B. B., Jr., and Fridovich, I., 1971. An enzyme-based theory of obligate anaerobiosis: The physiological function of superoxide dismutase, *Proc. Natl. Acad. Sci. USA* 68:1024.

McLaughlin, A., Greatwohl, C., and McLaughlin, S., 1978. The absorption of divalent cations to phosphatidylcholine bilayer membranes, *Biochim. Biophys. Acta* 513:2778.

Montgomery, D. W., Don, L. K., Zukoski, C. F., and Chvapil, M., 1974. The effect of zinc and other metals on complement hemolysis of sheep red blood cell in vitro, *Proc. Soc. Exp. Biol. Med.* 145:263.

Oelshlegel, F. J., Jr., Brewer, G. J., Prasad, A. S., Knutsen, C., and Schoomaker, E. B., 1973. Effect of zinc on increasing oxygen affinity of sickle and normal red blood cells, *Biochem. Biophys. Res. Commun.* 53:560.

Oelshlegel, F. J., Jr., Brewer, G. J., Knutsen, C., Prasad, A. S., and Schoomaker, E. B., 1974. Studies on the interaction of zinc with human hemoglobin, *Arch. Biochem. Biophys.* 163: 742.

Palek, J., Curby, W. A., and Lionetti, F. J., 1971. Effect of calcium and adenosine triphosphate on volume of human red cell ghosts, *Am. J. Physiol.* 220:19.

Passow, H., 1970. The use of pharmacological doses of zinc in the treatment of sickle cell anemia, in *Effect of Metals on Cells, Subcellular Elements and Macromolecules* (G. J. Brewer and A. S. Prasad, eds.), Thomas, Springfield, Ill., p. 291.

Peterson, D. A., Gerrand, G. M., Peller, J., Rao, G. H. R., and White, J. C., 1981. Interactions of zinc and arachidonic acid, *Prostaglandins Med.* 6:91.

Phillips, J. L., and Azari, P., 1974. Enhancement of nucleic acid synthesis in phytohemagglutinin-stimulated human lymphocytes, *Cell. Immunol.* 10:31.

Pilz, R. B., Willis, R. C., and Seegmiller, J. E., 1982. Regulation of human lymphoblast plasma membrane 5'-nucleotidase by zinc, *J. Biol. Chem.* 257:13544.

Prasad, A. S., Abbasi, A., and Ortega, J., 1977. Zinc deficiency in man: Studies in sickle cell disease, in *Current Aspects in Health and Disease* (G. J. Brewer and A. S. Prasad, eds.), Liss, New York, p. 211.

Prasad, A. S., Brewer, G. J., Schoomaker, E. B., and Rabbani, P., 1978. Hypocupremia induced by zinc therapy in adults, *J. Am. Med. Assoc.* 240:2166.

Rabenstein, D. L., and Isab, A. A., 1980. The complexation of zinc in intact human erythrocytes studied by Ih spin-echo NMR, *FEBS Lett.* 221:6.

Rifkind, J. M., 1983. Interaction of zinc with erythrocytes, in *Metal Ions in Biological Systems,* Vol. 15 (H. Sigel, ed.), Dekker, New York, p. 275.

Rifkind, J. R., and Heim, J. M., 1977. Interaction of zinc and hemoglobin: Binding of zinc and the oxygen affinity, *Biochemistry* 16:4438.

Ruhl, H., Kirchner, H., and Bochert, G., 1974. Kinetics of the Zn^{2+} stimulation of human peripheral lymphocytes in vitro, *Proc. Soc. Exp. Biol. Med.* 137:1089.

Schmetterer, G., 1978. ATP dependent uptake of zinc by human erythrocyte ghosts, *Z. Naturforsch.* 33:210.

Settlemire, C. T., and Matrone, G., 1967. In vivo interference of zinc with ferritin iron in the rat, *J. Nutr.* 92:1959.

Szeberi, S., Eskelson, C. D., and Chvapil, M., 1988. The effect of zinc on iron-induced lipid peroxidation in different lipid systems including liposome and micelles, *Physiol. Chem. Phys. Med. NMR* 20:205.

Takeda, Y., Ogiso, Y., and Miwatani, T., 1977. Effect of zinc ion on the hemolytic activity of thermostable direct hemolysin from Vibrio parahaemolyticus streptolysin O, and Triton X-100, *Infect. Immun.* 17:239.

Torrubia, J. O. A., and Garay, R., 1989. Evidence for a major route for zinc uptake in human red blood cells: $(Zn(HCO_3)_2Cl)^-$ influx through the (Cl^-/HCO_3) anion exchanger, *J. Cell. Physiol.* 138:316.

Tsukamoto, T., Yoshinaga, T., and Sano, S., 1979. The role of zinc with special reference to the essential thiol groups in 8-aminolevulinic acid dehydratase of bovine liver, *Biochim. Biophys. Acta* 570:167.

Warren, L., Glick, M. C., and Nass, M. K., 1966. Membranes of animal cells I. Method of isolation of the surface membrane, *J. Cell. Physiol.* 68:269.

Weed, R. I., LaCelle, P. L., and Merrill, E. W., 1969. Metabolic dependence of red cell deformability, *J. Clin. Invest.* 48:795.

Weismann, K., and Mikkelsen, H. I., 1980. Osmotic lysis of erythrocytes in relation to the zinc concentration of the medium, *Arch. Dermatol. Res.* 269:105.

White, J. G., 1974. Effects of ionophore, A23187, on the surface morphology of normal erythrocytes, *Am. J. Pathol.* 77:507.

White, J. G., 1976. Scanning electron microscopy of erythrocyte deformation: The influence of a calcium ionophore, A23187, *Semin. Hematol.* 13:121.

Williams, R. O., and Loeb, L. A., 1973. Zinc requirement for DNA replication in stimulated human lymphocytes, *J. Cell Biol.* 58:594.

Wills, E. D., 1965. Mechanism of lipid peroxide formation in tissues. Role of metals and haematin proteins in the catalysis of the oxidation of unsaturated fatty acids, *Biochim. Biophys. Acta* 98:238.

Wright, C. E., Gaull, G. E., and Pasentes-Morales, H., 1984. Protective effects of taurine, zinc and vitamin E on human cell membranes: Possible relevance to retina, *J. Am. Coll. Nutr.* 3:248.

Yamamoto, K., and Takahashi, M., 1975. Inhibition of the terminal stage of complement-mediated lysis (reactive lysis) by zinc and copper ions, *Int. Arch. Allergy Appl. Immunol.* 48:653.

Zinc and Neurobiology

8.1 Zinc and Cell Division

Zinc deficiency is known to affect the development of many derivatives of the primitive neural tube. Defects such as agenesis and dysmorphogenesis of the brain, spinal cord, eyes, and olfactory tract have been reported in the offsprings of zinc-deficient female rats (Hurley and Shrader, 1972). Hydrocephalus caused by closure of the aqueducts of Sylvius has also been noted as a result of zinc deficiency (Hurley, 1974). In general, the pattern of early brain malformations appears to be consistent with impaired mitosis during embryonic development and the involvement of zinc in DNA synthesis and cell division offers a plausible explanation of these observations (see Figs. 8-1, 8-2, and 8-3). It has also been suggested that the developing brain is more sensitive to zinc deficiency with respect to cell division than other organs (Eckert and Hurley, 1977).

Biochemically, the teratogenicity of zinc is widely ascribed to impaired nucleic acid synthesis during embryonic development and to the resulting asynchrony in histogenesis and organogenesis, which would distort the differential rate of growth necessary for normal morphogenesis (Swenerton *et al.,* 1969; Dreosti *et al.,* 1972). Zinc is required for a number of enzymes involved in the process of transcription and translation, but the primary role of zinc in cell division remains to be established. Considerable evidence suggests that diminished activity of the zinc-dependent enzyme thymidine kinase may be an important factor responsible for reduced mitotic activity, inasmuch as this enzyme is widely recognized to represent a rate-limiting step in DNA biosynthesis (Fujioka and Lieberman, 1964; Prasad and Oberleas, 1974; Duncan and Dreosti, 1975).

Studies by Record and Dreosti (1979) showed that the activity of thymidine kinase fell more in the brain (53%) than in the liver (34%) of zinc-deficient 20-day-old fetuses compared with restricted-fed controls. Thymidine kinase pathway for DNA synthesis represents a mechanism for salvaging ex-

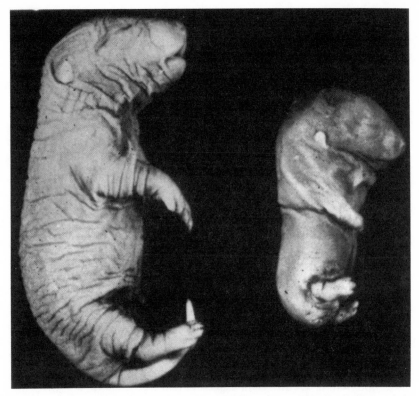

Figure 8-1. Abnormal fetus following *in utero* zinc deficiency in rats. [Reprinted with permission from Dreosti, I. E., 1984. Zinc in the central nervous system. The emerging interactions, in *The Neurobiology of Zinc* (C. J. Frederickson, G. A. Howell, and E. J. Kasarskis, eds.), Part A, Liss, New York, p. 2.].

isting thymidine, while the second pathway utilizes the enzyme thymidylate synthetase and permits synthesis of metabolite *de novo.*

In one study (Dreosti, 1984), under normal conditions the salvage pathway contributed more to the supply of thymidine phosphate (TMP) in the fetal brain than it did in the fetal liver (43 versus 18%). Although both pathways were reduced by zinc deficiency, as was total DNA synthesis, it appears that the brain has a much greater reliance on the supply of preformed nucleotides for DNA synthesis than does the liver (Dreosti, 1984).

Sever and Emanuel (1973) proposed that widespread zinc deficiency in the Middle East may account for the greater incidence of teratology in the region. This hypothesis was based on the observations that deficiency of zinc was prevalent in the Middle East (Prasad *et al.,* 1963) and that zinc deficiency

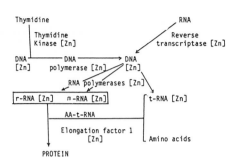

Figure 8-2. Zinc in nucleic acid and protein metabolism. [Reprinted with permission from Dreosti, I. E., 1984. Zinc in the central nervous system. The emerging interactions, in *The Neurobiology of Zinc* (C. J. Frederickson, G. A. Howell, and E. J. Kasarskis, eds.), Part A, Liss, New York, p. 4.]

was teratogenic in chicks and rats (Blamberg *et al.*, 1960; Hurley and Swenerton, 1966). A report from Turkey (Cavdar *et al.*, 1980) of an association between poor maternal zinc nutriture and fetal anencephaly supported this hypothesis (see Tables 8-1 and 8-2).

8.2 Zinc and Tubulin

Zinc plays a role in the polymerization of tubulin into microtubules. Microtubule reassembly is reduced in brain extracts from zinc-deficient animals because of impaired ability of tubulin to repolymerize (Torre *et al.*, 1981). Thus, in zinc deficiency, formation of the neural fold may be defective, inasmuch as polymerization of tubulin is essential for neural fold formation.

8.3 Zinc and Brain Function

Behavioral changes in zinc-deficient animals include impaired long-term memory, some reduction in short-term memory and learning ability, coupled with diminished activity, less emotional control, and higher susceptibility to stress (Caldwell *et al.*, 1970; Sandstead *et al.*, 1975). The neurological lesions

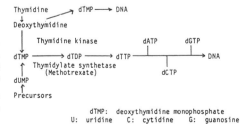

Figure 8-3. Pathways of DNA synthesis. [Reprinted with permission from Dreosti, I. E., 1984. Zinc in the central nervous system. The emerging interactions, in *The Neurobiology of Zinc*, (C. J. Frederickson, G. A. Howell, and E. J. Kasarskis, eds.), Part A, Liss, New York, p. 5.]

Table 8-1. Brain Malformations and Zinc Deficiency in Humans[a]

Terata	Country/condition	Reference
Anencephalus	Middle East	Damyanov and Dutz (1971), Sever and Emanuel (1973)
	Turkey	Cavdar et al. (1980)
	USA	Stewart et al. (1982)
	Britain	Soltan and Jenkins (1982)
Myelomeningocoele	Sweden	Jameson (1976)
Spina bifida	West Germany	Bergmann et al. (1981)
Anencephalus	Acrodermatitis enteropathica	Hambidge et al. (1975)
Exencephalus microcephalus	Valproate usage	Hurd et al. (1983)

[a] Adapted with permission from Dreosti, I. E., 1984. Zinc in the central nervous system: The emerging interactions, in *The Neurobiology of Zinc* (C. J. Frederickson, G. A. Howell, and E. J. Kasarskis, eds.), Part A, Liss, New York, p. 1

responsible for the behavioral changes are not well defined. Reduced cell division, impaired synaptogenesis and myelination have been suggested as possible explanations but future studies are needed in order to understand this more completely (Sandstead *et al.*, 1975; Duerre *et al.*, 1977; Record and Dreosti, 1979).

Another study revealed neuroanatomical and behavioral deficits in the adult offsprings of rats fed a mildly zinc-deficient diet during pregnancy and lactation (Dreosti *et al.*, 1981; Dreosti, 1984). These deficits included impaired learning and/or working memory, and significant trends toward reduced areas of neuropil and neuronal layers with the hippocampus and dentate gyrus, as well as increased neuronal density in the hippocampal CA3 stratum pyramidale and the dentate stratum granulosum (Dreosti, 1984). Growth retar-

Table 8-2. Zinc Status Associated with Neurological Disorders in Humans[a]

Condition	Zinc status	Reference
Dementia	Diminished	Burnett (1981)
Depression	Diminished	McLardy (1975), Moynahan (1976)
Epilepsy	Increased	Barbeau and Donaldson (1974)
Fifth-day fits	Diminished	Goldberg and Sheehy (1982)
Mental retardation	Diminished	Pihl and Parkes (1977), Krischer (1978)
Pick's disease	Increased	Constantinidis and Tissot (1981)
Schizophrenia	Diminished	Pfeiffer and Iliev (1972)

[a] Adapted with permission from Dreosti, I. E., 1984. Zinc in the central nervous system: The emerging interactions, in *The Neurobiology of Zinc* (C. J. Frederickson, G. A. Howell, and E. J. Kasarskis, eds.), Part A, Liss, New York, p. 1.

dation, mental lethargy, and decreased sexual activity were mentioned as part of the overall clinical description of zinc deficiency. Results of the first systematic study of the relationship of zinc to behavior were reported by Caldwell *et al.* (1970). They reported behavioral changes in zinc-deficient rats. Learning and memory deficits and increased aggression resulting from zinc deficiency have now been demonstrated in rats, mice, and monkeys (Caldwell *et al.,* 1973; Sandstead *et al.,* 1977). Behavioral changes have been observed both as a result of zinc deficiency occurring during prenatal or early postnatal development.

Mild zinc deficiency impaired working memory and caused a small but persistent injury across all eight regions of the hippocampus according to one study (Dreosti, 1984).

8.4 Zinc Homeostasis

Regulation of zinc homeostasis in the brain has been studied using ^{65}Zn (Kasarskis, 1984). Based on these studies, it appears that zinc enters the brain via transport sites which are as yet anatomically undefined. The net transport is increased in response to lowered circulating levels of plasma zinc. Most likely, zinc is taken up by neurons (and glia as well) and a proportion of zinc is transported distally by slow axonal transport (Knull and Wells, 1975). Inasmuch as zinc is involved in the assembly of microtubules, it has been speculated that zinc may also be transported in association with tubulin. Zinc may be incorporated into the metalloenzymes or other ligands either in the neuronal soma, the axon, or the synapse.

Excess zinc appears to be eliminated via cerebrospinal fluid pathways and probably reenters plasma by transport throughout the choroid plexus. The possibility that zinc egress may occur through other barrier tissues such as cerebral capillaries cannot be ruled out.

Zinc deficiency does not alter the zinc content of most regions of rat brain. Other elements, however, such as copper, manganese, sodium, and potassium, were altered in various regions of rat brain. These changes in brain elemental composition may be responsible for increased norepinephrine levels observed in zinc-deficient rat brains, particularly in hippocampus and the brain remnants (Dreosti and Record, 1984).

8.5 Zinc-Related Enzymes in the Brain

Table 8-3 shows the principal enzymes that have been studied in the brain in zinc deficiency. Observations of the decreased activities of myelin

Table 8-3. Effect of Zinc Status on the Activity of Several Enzymes
in the Rat Brain[a]

Enzyme	Zinc status[b]	Enzyme activity[b]	Brain region[c]	Reference
2′,3′-cyclic nucleotide	−	−	Ce	Prohaska et al. (1974)
3′-Phosphohydrolase	−	−	Ce, Hi	Dreosti et al. (1981)
Alkaline phosphatase	−	−	Fetal brain	Dreosti et al. (1980)
Glutamate	−	−	Ce, Hi	Dreosti et al. (1981)
dehydrogenase				
In vitro	−	−	Hi, Nc	Wolf and Schmidt (1982)
In vitro	−	−	Hi, Nc	Wolf and Schmidt (1982)
Dopamine-β-	−	−	Co, Ce, Hi	Wenk and Stemmer (1982)
hydroxylase				
Phenylethanolamine-N-	−	−	Co, Ce, Hi	Wenk and Stemmer (1982)
methyltransferase				
Glutamate	+	−	Hi	Itoh and Ebadi (1981)
decarboxylase	+	0	A, CN, Ht	
(intercerebral) Zn				
Thymidine kinase	−	−	Fetal brain	Record and Dreosti (1979)

[a] Adapted with permission from Dreosti, I. E., 1984. Zinc in the central nervous system: The emerging interactions, in *The Neurobiology of Zinc* (C. J. Frederickson, G. A. Howell, and E. J. Kasarskis, eds.), Part A, Liss, New York, p. 1.
[b] −, diminished; +, increased; 0, unaffected.
[c] A, amygdala; Ce, cerebellum; CN, caudate nucleus; Co, cortex; Hi, hippocampus; Ht, hypothalamus; Nc, neocortex.

marker enzyme 2′,3′-cyclic nucleotide phosphohydrolase and brain alkaline phosphatase as a result of zinc deficiency are important inasmuch as these enzymes play a role in the process of myelination and brain maturation (Dreosti, 1984). In a 6-month-old human infant with acrodermatitis enteropathica, improved myelination was observed following zinc therapy. Changes in the activity of glutamate dehydrogenase as a result of zinc deficiency may have an important effect on the metabolism of the neurotransmitter glutamic acid and on glutamergic neural systems (Dreosti, 1984).

Another enzyme, cuprozinc superoxide dismutase, which protects against damage by superoxide radicals, may be of some importance, as the activity of this enzyme and that of the mangano form of superoxide dismutase increase dramatically in the first 2 months postnatally in rat brain. In aging rat brains, the activity of superoxide dismutase declines and it has been observed that this may be accompanied by membrane injury. Superoxide dismutase may protect catecholamines from being oxidized by superoxide and indeed during development its activity is much higher in catecholamine-rich areas of the brain. The activity of superoxide dismutase in brain, however, is known to decrease as a result of copper deficiency and zinc deficiency does not affect its activity in either rat liver or rat erythrocytes.

Wallwork and Sandstead (1981) observed that the levels of norepinephrine and dopamine were higher in the brains of zinc-deficient weanling rats and they suggested that behavioral changes in zinc-deficient animals may be partly related to catecholamine alterations. Wenk and Stemmer (1982), however, reported that the activity of the enzymes dopamine-β-hydroxylase and phenylethanolamine-N-methyltransferase, which respectively convert dopamine to norepinephrine and norepinephrine to epinephrine, were decreased in certain regions of the zinc-deficient rat brain. These results contradict those of Wallwork and Sandstead (1981) and more studies are needed in order to reconcile the results.

Of all brain regions, the hippocampus accumulates the most zinc and hippocampal mossy fibers are among the most densely stained using heavy metal histochemical methods and dynorphin antisera (McGinty et al., 1984). The mere colocalization of zinc and opioid peptides in mossy fibers has suggested that there may be a functional interaction between them. Zinc ions block [^{3}H]-D-Ala^2-Met^5-enkephalinamide binding to rat brain membranes, notably in the hippocampus, suggesting that zinc ions may be important regulators of opioid peptide action in brain regions where they are colocalized. Thiol reducing agents can restore the binding capacities of ZnCl$_2$-treated membranes, suggesting that oxidation of opioid receptor SH groups by zinc ions may be the mechanism through which zinc blocks opioid binding.

The following evidence suggest interactions between zinc and opioid peptides in the hippocampus: (1) zinc and enkephalin-containing opioid peptides are selectively localized in hippocampal mossy fibers, (2) zinc blocks opioid binding, (3) zinc complexes with enkephalins *in vitro,* and (4) zinc and opioid peptides administered intraventricularly induce limbic seizures.

These findings suggest that opioid peptides and zinc in the mossy fibers may interact in the physiological regulation of hippocampal excitability. These endogenous metabolites may act synergistically with an excitatory amino acid during hippocampal synaptic transmission. Together they could effectively regulate the normal excitability and seizure threshold in limbic circuits, possibly by altering the degree of GABAergic inhibition in the hippocampus.

Anorexia is a major symptom of zinc deficiency. Recent studies have suggested an important role for endogenous opiate peptides in appetite regulation. Dynorphin, a leucine-enkephalin-containing opiate peptide, is a potent inducer of spontaneous feeding. Zinc-deficient animals were found to be resistant to dynorphin-induced feeding (Essatara et al., 1984a,b). Zinc-deficient animals had lower levels of dynorphin in the hypothalamus and [^{3}H]naloxone binding was significantly increased in the isolated brain membranes from zinc-deficient animals. The effects of zinc deficiency on endogenous opiate action appear to include alterations in receptor affinity, a postreceptor defect, and alterations in the synthesis and/or release of dynorphin.

According to Stengaard-Pedersen *et al.* (1981), both zinc and enkephalin are present in the hippocampal mossy fiber pathway. Since it is known that opioid peptides exert an effect on hippocampal electrophysiology and that metal ions inhibit binding to opiate-receptor sites, it has been suggested that zinc may act neurochemically by reducing enkephalin binding through a modulation of opiate receptor binding. Alternatively, cleavage of an enkephalin peptide may occur following activation by zinc of an aminopeptidase or by enkephalinase, a zinc metalloenzyme.

Several experiments have reported altered metabolism of the adrenocorticoid hormones, adrenal hypertrophy, hypersensitivity to adrenocorticotropic hormones, and alterations in the serum levels of circulating corticosteroids as a result of zinc deficiency in the rats (Quarterman, 1972, 1974; Reeves *et al.,* 1977). Interestingly, Hesse *et al.* (1979) have reported that behavioral characteristics in chronically zinc-deficient rats were consistent with excess glucocorticoids.

8.6 Zinc and the Hippocampus

Trace element analysis and histochemistry of the brain have shown consistently that zinc is especially concentrated in hippocampus and that it accumulates in the intrahippocampal mossy fiber pathway during the first 2–4 weeks postnatally, also the period when most development of the region is known to occur (Crawford and Connor, 1972). In chronic zinc deficiency, hippocampal electrophysiology in the rat has been shown to be disturbed, suggesting that the presence of zinc in the mossy fiber boutons may be important for synaptic transmission (Dreosti, 1984).

The role of zinc in the mossy fibers may be associated with the metabolism of the neurotransmitter glutamic acid, either through the zinc metalloenzyme L-glutamic acid dehydrogenase, or as a stable zinc glutamate storage complex in the giant boutons (Crawford and Connor, 1975). The levels of glutamic acid and the activity of the enzyme glutamic acid dehydrogenase are both higher in the region of the horn of Ammon containing the giant boutons, than in other hippocampal areas. Also, glutamic acid appears to be released following electrical stimulation of the mossy fibers, and binding sites with a high affinity for glutamic acid have been found to occur specifically on the membranes of the giant mossy fiber boutons. Thus, the findings in relation to the activity of glutamic acid dehydrogenase are especially pertinent as they may provide a neurochemical basis for the disturbed hippocampal function in zinc-deficient rats.

γ-Aminobutyric acid (GABA) exerts its inhibitory action on synaptic function by stimulating membrane receptors coupled to chloride channels (Baraldi *et al.*, 1984). The hyperpolarizing response to GABA is related to the inflow of chloride ions into the neuron, while the depolarizing response seems to be related to the reverse process.

In the brain, there are two different types of GABA receptors, named $GABA_A$ and $GABA_B$, the former coupled with chloride channels and the latter linked with divalent cation channels. Zinc is known to regulate the activity of glutamic acid decarboxylase (GAD), the target enzyme of GABA synthesis in nerve terminals. At low levels, zinc stimulates the activity of GAD, while at higher concentrations it inhibits the activity of this enzyme. Zinc, by increasing or decreasing GAD activity, may modulate the functional activity of the GABA receptor complex and in turn derange the system from physiological to pathological state under some circumstances. It has been observed that intracerebroventricular administration of zinc may cause convulsions in rats, which has been attributed to an inhibition of GAD activity in hippocampus and to an inhibition of Na^+,K^+-ATPase in hippocampus and hypothalamus by other investigators. On the other hand, Baraldi *et al.* (1984) have reported a decrease in the amount of zinc in several brain areas associated with changes in [^{3}H]GABA binding characteristics in cases of experimentally induced hepatic coma. These findings indicate that both an increase and a decrease in zinc content of the brain may be associated with pathological states.

Recent reports indicate that zinc plays a physiological role for normal binding of GABA to its receptors in the brain. At low concentrations, zinc seems to be essential, and at higher concentrations it progressively inhibits Na^+-independent [^{3}H]GABA binding. The biphasic activity of zinc on Na^+-independent GABA binding correlates very well with its known activity on GAD. It has also been suggested that zinc, by combining with the phosphate moiety, binds to intrinsic components of the membrane, forming strong complexes. This type of reaction may result in changes of membrane fluidity capable in turn of affecting the binding capacity of GABA receptors.

Some of the memory and learning defects associated with zinc deficiency resemble partly the behavioral syndrome resulting from ablation of the hippocampus. More work is required in order to elucidate the form in which zinc is stored in the mossy fibers and the respective functions of the metal complexes present in the mossy fibers.

In zinc-deficient weanling rat brains, norepinephrine was consistently increased and in one experiment brain serotonin was increased while 5-hydroxyindoleacetic acid was decreased (Essatara *et al.*, 1984a,b). These alterations were reversed on zinc supplementation. Interestingly, the zinc level in the brain was not affected by zinc deficiency but the copper level was increased.

Prenatal zinc deprivation from the 14th to the 20th day of gestation resulted in depressed fetal weight, brain weight, and total brain DNA (McKenzie *et al.,* 1975). In postnatally zinc-deprived pups, the levels of brain DNA, RNA, and protein were lower at all ages compared with control pups (Fosmire *et al.,* 1975). Forebrain polysomal RNA was decreased after only 5 days of Zn deprivation compared with pups nursed by pair-fed dams. Thymidine incorporation into DNA, sulfur into the trichloroacetic acid-precipitable fraction of brain, and lysine into histone proteins were depressed by zinc deficiency from birth to postnatal day 12 (Sandstead *et al.,* 1972; Duerre *et al.,* 1977). Total lipid in cerebellum was also decreased. Effects of zinc deficiency on morphology of the brain were characteristics of immaturity in general (Sandstead, 1984).

8.7 Zinc and Human Brain

In humans, unlike in animals, a direct causal relationship has not yet been established between zinc deficiency and brain dysmorphogenesis. None-theless, current evidence suggests that the developing human fetus is no less vulnerable to zinc depletion than are the offspring of other species.

Behavioral consequences of zinc imbalance in humans are less well documented than in animals, but alterations in zinc status have been reported in several disorders of the central nervous system (Table 8-2). Mental lethargy has been reported to be a feature of the zinc deficiency syndrome observed in the Middle East (Prasad *et al.,* 1961), and psychological disturbances occur regularly as a zinc-responsive symptom associated with acrodermatitis enteropathica. In cases of iatrogenically induced severe zinc deficiency in infants and adults, psychological disturbances include jitteriness, impaired concentration, depression, and mood lability. Patients with cirrhosis of the liver and chronic renal disease often have low plasma zinc levels and accompanying taste disorders, anorexia, and mental disturbances (Lindeman *et al.,* 1978).

Chronic alcoholics are frequently noted to be chronically zinc deficient. Zinc levels are much reduced in the hippocampus from alcoholic patients and the granular cell layer of the dentate gyrus is abnormally thin, suggesting that zinc deficiency may be involved pathogenically in alcohol-related mental deterioration (McLardy, 1975). Walker *et al.* (1980) reported that in adult rats, chronic ethanol consumption led to a loss of hippocampal pyramidal cells and granule cells in the dentate gyrus. The organization of mossy fibers in the hippocampus was significantly altered in rats exposed to alcohol prenatally (West *et al.,* 1981).

One of the manifestations of the alcohol abstinence syndrome in chronic alcoholics is grand mal convulsive seizures which sometimes precede the tremulousness. Chronic alcoholics are frequently noted to be zinc deficient (Prasad, 1979). Whether or not zinc deficiency plays a biochemical role in alcoholic seizures remains to be determined.

8.8 Zinc and Epilepsy

Hair analysis for trace elements in epileptic subjects has revealed high magnesium and low zinc content (Shrestha and Oswaldo, 1987). The significance of this observation is not understood.

High level of dietary zinc protects against the toxic effect of lead. This may relate to an inhibitory effect on absorption of lead. Lead intoxication in humans and other species causes seizures. In addition, the exposure to sub-convulsive doses of lead enhances the convulsant activity of picrotoxin, iso-niazid, mercaptopropionic acid, and strychnine. Chronic lead exposure produces cholinergic hypoactivity and catecholaminergic hyperactivity (Ebadi and Pfeiffer, 1984). In addition, lead-induced seizures may be produced by the inhibition of GABAergic transmission.

Acute zinc deficiency has been implicated in a syndrome of unexplained neonatal convulsions (Pryor *et al.*, 1981; Goldberg and Sheehy, 1982) which occur 4 to 6 days after birth. These seizures are self-limited. The concentration of zinc in cerebrospinal fluid is considerably reduced in such cases.

Low serum zinc concentrations have been demonstrated in untreated male epileptics (Palm and Hallmans, 1982). Phenytoin, phenobarbital, or carbamazepine, individually or in combination, have been shown to increase the plasma zinc concentration (Pippenger *et al.*, 1980).

In rats, the intraventricular administration of zinc or copper produced epileptic seizures and it was proposed that the divalent ions inhibited Na^+,K^+-ATPase and hence had ouabainlike actions (Tokuoka, 1967; Donaldson *et al.*, 1971). The intracerebroventricular administration of GABA (0.4 μmol in 10 μl) prevented the zinc-induced epileptic seizures, suggesting that zinc may interfere with GABAergic transmission.

GABA is synthesized by α-decarboxylation of L-glutamate under the catalytic activity of GAD, which has an absolute requirement for pyridoxal phosphate (PLP). A positive correlation has been observed between the activity of GAD and the concentration of GABA to the extent that assay of GAD activity rather than GABA concentration is a more accurate measurement of GABAergic transmission (Ebadi and Pfeiffer, 1984).

When GAD activity was assayed in the presence of 0.2 mM PLP, no differences were noted in tissues obtained from saline-treated or zinc sulfate-treated animals. On the other hand, in the absence of PLP, GAD activity was significantly reduced in the zinc-treated sample only in the hippocampus area (Itoh and Ebadi, 1982). Interestingly, the hippocampus had the highest concentration of zinc and the lowest level of PLP. In another experiment, zinc did not reduce the concentration of PLP. It was therefore suggested that zinc may have inhibited the binding of PLP to GAD apoenzyme in the hippocampus (Miller *et al.,* 1978).

Intravenously or intraperitoneally administered zinc, even in large doses (up to 100 mg/kg), did not cause convulsive seizures, in contrast to intracerebroventricular administration of zinc, which induced epileptic seizures in rats (Itoh and Ebadi, 1982). It was assumed that the peripherally administered Zn became bound to various proteins in the plasma and was unavailable to the central nervous system. Although under normal conditions the level of zinc in the hippocampus is high, it is possible that the free zinc concentration is low and therefore GAD activity is not affected adversely. Recently, the existence of three zinc-binding proteins with apparent estimated molecular weights of 15,000, 25,000, and 210,000 in the brain has been documented (Itoh and Ebadi, 1982; Ebadi *et al.,* 1982; Itoh *et al.,* 1983). Also, it has been shown that the intracerebroventricular but not intraperitoneal administration of zinc stimulates the synthesis of zinc-binding protein in the brain. It has been proposed that GABA is synthesized by α-decarboxylation of L-glutamate under the catalytic activity of GAD, which has an absolute requirement for pyridoxal phosphate PLP. Conversion of pyridoxal to pyridoxal phosphate requires pyridoxal kinase, which is a zinc-dependent enzyme.

Growing guinea pigs and chicks seem to be most prone to stiffness, abnormal gait, and hypersensitivity to touch as a result of zinc deficiency (O'Dell *et al.,* 1989). Because guinea pigs vocalize readily, it is easier to assess discomfort caused by neuromuscular disorder resulting from zinc deficiency. O'Dell *et al.* (1989) used vocalization and posture as indices of the neuromuscular pathology that develops in zinc-deficient guinea pigs and determined the rate of depletion before and of repletion after intraperitoneal zinc therapy. The first signs of vocalization relating to handling, abnormal posture, and skin lesions developed after approximately 4 weeks on the zinc-deficient diet and severe signs were observed after 5 to 6 weeks. A single intraperitoneal dose of $ZnSO_4$ (50 μmol/kg) caused remission of signs within 4–5 days following which all signs regressed within 7 days. Zinc level decreased only in the plasma and bone in severely deficient guinea pigs. On the other hand, major soft tissues, such as muscle, brain, liver, and skin, showed no changes. These findings suggest that major soft tissues do not serve as mobilizable stores of zinc for other critical metabolic functions. Although the bone zinc slowly mobilized during zinc deficiency, the rate of mobilization is insufficient to maintain health or even life of the animals.

The neuromuscular signs relating to zinc deficiency may arise from a defect in muscle, nerve, the neuromuscular junction, or a combination of these sites. The low pain threshold most likely resides in the abnormal nerve metabolism, the biochemical nature of which is unknown.

References

Baraldi, M., Caselgrandi, E., and Santi, M., 1984. Effect of zinc on specific binding of GABA to rat brain membranes, in *Neurobiology of Zinc, Part A* (C. J. Fredrickson, G. A. Howell, and E. J. Kasarskis, eds.), Liss, New York, p. 73.

Barbeau, A., and Donaldson, J., 1974. Zinc, taurine and epilepsy, *Arch. Neurol.* 30:52.

Bergmann, K. E., Makosch, E., and Tows, K. H., 1980. Abnormalities of hair zinc concentration in mothers of newborn infants with spina bifida, *Am. J. Clin. Nutr.* 33:2145.

Blamberg, D. L., Blackwood, U. B., Supplee, W. C., and Combs, G. F., 1960. Effect of zinc deficiency in hens on hatchability and embryonic development, *Proc. Soc. Exp. Biol. Med.* 104:217.

Burnett, F. M., 1981. A possible role of zinc in the pathology of dementia, *Lancet* 1:186.

Caldwell, D. F., Oberleas, D., Clancy, J. J., and Prasad, A. S., 1970. Behavioral impairment in adult rats following acute zinc deficiency, *Proc. Soc. Exp. Biol. Med.* 133:1417.

Caldwell, D. F., Oberleas, D., and Prasad, A. S., 1973. Reproductive performance of chronic mildly zinc deficient rats and the effects on behavior of their offspring, *Nutr. Rep. Int.* 7:309.

Cavdar, A. O., Arcasoy, A., Baycu, T., and Himmetoglu, O., 1980. Zinc deficiency and anencephaly in Turkey, *Teratology* 23:141.

Constantinidis, J., and Tissot, R., 1981. Role of glutamate and zinc in hippocampal lesions of Pick's disease, in *Glutamate as a Neurotransmitter* (G. Dichiaa and G. L. Gessa, eds.), Raven Press, New York, p. 413.

Crawford, I. L., and Connor, J. D., 1972. Zinc in maturing rat brain: Hippocampal concentration and localization, *J. Neurochem.* 19:1451.

Crawford, I. L., and Connor, J. D., 1975. Zinc and hippocampal function, *Orthomol. Psychiatry* 4:34.

Damyanov, I., and Dutz, W., 1971. Anencephaly in Shiraz, Iran, *Lancet* 1:82.

Donaldson, J., St. Pierre, T., Minnich, J. L., and Barbeau, A., 1971. Seizures in rats associated with divalent cation inhibition of Na^+K^+ ATPase, *Can. J. Biochem.* 49:1217.

Dreosti, I. E., 1984. Zinc in the central nervous system: The emerging interactions, in *The Neurobiology of Zinc, Part A* (C. J. Fredrickson, G. A. Howell, and E. J. Kasarskis, eds.), Liss, New York, p. 1.

Dreosti, I. E., and Record, I. R., 1984. Accumulation of zinc in the hippocampus of neonatal rats, in *The Neurobiology of Zinc, Part A* (C. J. Fredrickson, G. A. Howell, and E. J. Kasarskis, eds.), Liss, New York, p. 119.

Dreosti, I. E., Grey, P. C., and Wilkins, P. J., 1972. Deoxyribonucleic acid synthesis, protein synthesis and teratogenesis in zinc deficient rats, *S. Afr. Med. J.* 46:1585.

Dreosti, I. E., Record, I. R., and Manuel, S. J., 1980. Incorporation of ^{3}H-thymidine into DNA and the activity of alkaline phosphatase in zinc deficient fetal rat brains, *Biol. Trace Element Res.* 2:21.

Dreosti, I. E., Manuel, S. J., Buckley, R. A., Fraser, F. J., and Record, I. R., 1981. The effect of late prenatal and/or early postnatal zinc deficiency on the development and some biochemical aspects of the cerebullum and hippocampus in rats, *Life Sci.* 28:2133.

Duerre, J. A., Ford, K. M., and Sandstead, H. H., 1977. Effect of zinc deficiency on protein synthesis in brain and liver of suckling rats, *J. Nutr.* 107:1082.

Duncan, J. R., and Dreosti, I. E., 1975. A proposed site of action for zinc in DNA synthesis, *J. Comp. Pathol.* 86:81.

Ebadi, M., and Pfeiffer, R. F., 1984. Zinc in neurological disorders and in experimentally induced epileptiform seizures, in *The Neurobiology of Zinc, Part B* (C. J. Fredrickson, G. A. Howell, and E. J. Kasarskis, eds.), Liss, New York, p. 307.

Ebadi, M., Itoh, M., and Swanson, S., 1982. The nature and mechanism of zinc-induced epileptic seizures, *Trans. Soc. Neurosci.* 12:139.

Eckert, C. D., and Hurley, L. S., 1977. Reduced DNA synthesis in zinc deficiency: Regional differences in embryonic rats, *J. Nutr.* 107:855.

Essatara, M. B., McClain, C. J., Levine, A. S., and Morley, J. E., 1984a. Zinc deficiency and anorexia in rats: The effect of central administration of norepinephrine, muscimol and brom-ergocryptine, *Physiol. Behav.* 32:479.

Essatara, M. B., Morley, J. E., Levine, A. S., Elson, M. K., Shafer, R. B., and McClain, C. J., 1984b. The role of endogenous opiates in zinc anorexia, *Physiol. Behav.* 32:475.

Fosmire, G. J., Al-Ubaidi, Y. Y., Halas, E., and Sandstead, H. H., 1975. Some effects of postnatal zinc deficiency on developing rat brain, *Pediatr. Res.* 9:89.

Fujioka, M., and Lieberman, I., 1964. A zinc requirement for the synthesis of DNA by rat liver, *J. Biol. Chem.* 239:1164.

Goldberg, H. J., and Sheehy, E. M., 1982. Fifth day fits: An acute zinc deficiency syndrome? *Arch. Dis. Child.* 57:632.

Hambidge, K. M., Neldner, K. H., and Walravens, P. A., 1975. Zinc, acrodermatitis enteropathica and congenital malformations, *Lancet* 1:577.

Hesse, G. W., Frank-Hess, K. A., and Catalanotto, F. A., 1979. Behavioral characteristics in rats experiencing chronic zinc deficiency, *Physiol. Behav.* 22:211.

Hurd, R. W., Wilder, B. J., and Van Rinsvelt, H. A., 1983. Valproate, birth defects and zinc, *Lancet* 1:181.

Hurley, L. S., 1974. Zinc and its influence on development in the rat, in *Clinical Applications of Zinc Metabolism* (W. J. Pories, W. H. Strain, J. M. Hsu, and R. L. Woosley, eds.), Thomas, Springfield, Ill., p. 57.

Hurley, L. S., and Shrader, R. E., 1972. Congenital malformation of the nervous system in zinc deficient rats, in *Neurobiology of the Trace Elements* (C. C. Pfeiffer, ed.), Academic Press, New York, p. 7.

Hurley, L. S., and Swenerton, H., 1966. Congenital malformations resulting from zinc deficiency in rats, *Proc. Soc. Exp. Biol. Med.* 123:692.

Itoh, M., and Ebadi, M., 1981. The selective inhibition of glutamic acid decarboxylase (GAD) in hippocampus by Zn^{++}, *Pharmacologist* 23:243.

Itoh, M., and Ebadi, M., 1982. The selective inhibition of hippocampal glutamic acid decarboxylase in zinc-induced epileptic seizures, *Neurochem. Res.* 7:1287.

Itoh, M., Ebadi, M., and Swanson, S., 1983. The presence of zinc binding proteins in brain, *J. Neurochem.* 41:823.

Jameson, S., 1976. Variations in maternal serum zinc during pregnancy and correlation to congenital malformation, dysmaturity and abnormal parturition, *Acta Med. Scand. Suppl.* 593: 21.

Kasarskis, E. J., 1984. Regulation of zinc homeostasis in rat brain, in *Neurobiology of Zinc, Part A* (C. J. Fredrickson, G. A. Howell, and E. J. Kasarskis, eds.), Liss, New York, p. 27.

Knull, H. R., and Wells, W. W., 1975. Axonal transport of cations in the chick optic system, *Brain Res.* 100:121.

Krischer, K. N., 1978. Copper and zinc in childhood behavior, *Psychopharmacol. Bull.* 14:58.

Lindeman, R. D., Baxten, D. J., Yunice, A. A., and Kraikitpanitch, S., 1978. Serum concentration and urinary excretions of zinc in cirrhosis, nephrotic syndrome and renal insufficiency, *Am. J. Med. Sci.* 275:17.

McGinty, J. F., Henriksen, S. J., and Chavkin, C., 1984. Is there an interaction between zinc and opioid peptides in hippocampal neurons? in *Neurobiology of Zinc, Part A* (C. J. Fredrickson, G. A. Howell, and E. J. Kasarskis, eds.), Liss, New York, p. 73.

McKenzie, J. M., Fosmire, G. J., and Sandstead, H. H., 1975. Zinc deficiency during the latter third of pregnancy: Effects on fetal rat brain, liver and placenta, *J. Nutr.* 105:1466.

McLardy, T., 1975. Hippocampal zinc and structural deficit in brains from chronic alcoholics and some schizophrenics, *Orthomol. Psychiatry* 4:32.

Miller, L. P., Martin, D. L., Mazumdar, A., and Waiters, J. P., 1978. Studies on the regulation of GABA synthesis: Substrate-promoted dissociation of pyridoxal-5'-phosphate from GAD, *J. Neurochem.* 30:361.

Moynahan, E. J., 1976. Zinc deficiency and disturbances of mood and visual behaviors, Lancet 1:91.

O'Dell, B. L., Becker, J. K., Emery, M. P., and Browning, J. D., 1989. Production and reversal of the neuromuscular pathology and related signs of zinc deficiency in guinea pigs, *J. Nutr.* 119:196.

Palm, R., and Hallmans, G., 1982. Zinc and copper metabolism in phenytoin therapy, *Epilepsia* 23:453.

Pfeiffer, C. C., and Iliev, V., 1972. A study of zinc deficiency and copper excess in the schizophrenia, in *Neurobiology of the Trace Metals Zinc and Copper* (C. C. Pfeiffer, ed.), Academic Press, New York, p. 141.

Pihl, R. O., and Parkes, M., 1977. Hair element content in learning disabled children, *Science* 198:204.

Pippenger, C. E., Garlock, C., Fernandez, F., Slavin, W., and Iannarone, J., 1980. Effect of antiepileptic drugs on manganese, zinc and copper concentrations in whole blood, RBC, and plasma of epileptics, in *Advances in Epileptology: XIth Epilepsy International Symposium* (R. Canger, ed.), Raven Press, New York, p. 435.

Prasad, A. S., 1979. *Zinc in Human Nutrition,* CRC Press, Boca Raton, Fla. p. 17.

Prasad, A. S., and Oberleas, D., 1974. Thymidine kinase activity and incorporation of thymidine into DNA in zinc deficient tissue, *J. Lab. Clin. Med.* 83:634.

Prasad, A. S., Halsted, J. A., and Nadimi, M., 1961. Syndrome of iron deficiency anemia, hepatosplenomegaly, hypogonadism, dwarfism and geophagia, *Am. J. Med.* 31:532.

Prasad, A. S., Miale, A., Farid, Z., Schulert, A., and Sandstead, H. H., 1963. Zinc metabolism in patients with the syndrome of iron deficiency anemia, hypogonadism, and dwarfism, *J. Lab. Clin. Med.* 61:537.

Prohaska, J. R., Luecke, R. W., and Jasinski, R., 1974. Effect of zinc deficiency from day 18 of gestation and/or during lactation on the development of some rat brain enzymes, *J. Nutr.* 104:1525.

Pryor, D. S., Don, N., and Macourt, D. C., 1981. Fifth day fits: A syndrome of neonatal convulsions, *Arch. Dis. Child.* 56:753.

Quarterman, J., 1972. The effect of zinc deficiency on the activity of the adrenal glands, *Proc. Nutr. Soc.* 31:74A.

Quarterman, J., 1974. The effect of zinc deficiency or excess on the adrenals and the thymus in the rat, in *Trace Elements in Animal Metabolism,* Volume 2 (W. G. Hoekstra, J. W. Suttie, A. E. Ganther, and W. Mettz, eds.), University Park Press, Baltimore, p. 742.

Record, I. R., and Dreosti, I. E., 1979. Effects of zinc deficiency on the liver and brain thymidine kinase activity in the fetal rat, *Nutr. Rep. Int.* 20:749.

Reeves, P. G., Frissell, S. G., and O'Dell, B. L., 1977. Response of serum corticosterone to ACTH and stress in the zinc deficient rat, *Proc. Soc. Exp. Biol. Med.* 156:500.

Sandstead, H. H., 1984. Neurobiology of zinc, in *Neurobiology of Zinc, Part B* (C. J. Fredrickson, G. A. Howell, and E. J. Kasarskis, eds.), Liss, New York, p. 73.

Sandstead, H. H., Gillespie, D. D., and Brady, R. N., 1972. Zinc deficiency: Effect on brain of the suckling rat, *Pediatr. Res.* 6:119.

Sandstead, H. H., Fosmire, G. J., McKenzie, J. M., and Halas, E. S., 1975. Zinc deficiency and brain development in the rat, *Fed. Proc.* 34:86.

Sandstead, H. H., Fosmire, G. J., Halas, E. S., Strobel, D., and Duerre, J., 1977. Zinc: Brain and behavioral development, in *Trace Element Metabolism in Man and Animals,* 3rd ed. (M. Kirchgessner, ed.), Freising University, Munich, p. 203.

Sever, L. E., and Emanuel, I., 1973. Is there a connection between maternal zinc deficiency and congenital malformations of the central nervous system? *Teratology* 7:117.

Shrestha, K. P., and Oswaldo, A., 1987. Trace elements in hair of epileptic and normal subjects, *Sci. Total Environ.* 67:215.

Soltan, M. H., and Jenkins, D. M., 1982. Maternal and fetal plasma zinc concentration and fetal abnormality, *Br. J. Obstet. Gynaecol.* 89:56.

Stengaard-Pederson, K., Fredens, K., and Larson, L. I., 1981. Enkephalin and zinc in the mossy fiber system, *Brain Res.* 212:230.

Stewart, C., Katchan, B., Collip, P. J., Clegan, S., Pudalov, S., and Chen, S. Y., 1981. Zinc and birth defects, *Pediatr. Res.* 15:515.

Swenerton, H., Shrader, R. E., and Hurley, R. L., 1969. Zinc deficient embryos: Reduced thymidine incorporation, *Science* 166:1014.

Tokuoka, S., 1967. Neurochemical considerations on the alleviating effect of caudal resection of the pancreas on epileptic seizures: Relationship of zinc metabolism to brain excitability, *Bull. Yamaguchi Med. Sch.* 14:1.

Torre, J., Villasante, A., Corral, J., and Avila, J., 1981. Factors implicated in determining the structure of zinc tubulin-sheets: Lateral tubulin–tubulin interaction is promoted by the presence of zinc, *J. Supramol. Struct. Cell Biochem.* 17:183.

Walker, E. W., Barnes, D. E., Zornatzer, S. F., Hunter, B. E., and Kubanic, P., 1980. Neuronal loss in hippocampus induced by prolonged ethanol consumption in rats, *Science* 209:711.

Wallwork, J., and Sandstead, H. H., 1981. Effect of zinc deficiency on brain catecholamine concentrations in the rat, *Fed. Proc.* 40:939.

Wenk, G. L., and Stemmer, K. L., 1982. Activity of the enzymes dopamine-beta-hydroxylase and phenylethanolamine-N-methyltransferase in discrete brain regions of the copper-zinc deficient rat following aluminum ingestion, *Neuro Tox* 3:93.

West, J. R., Hodges, C. A., and Black, A. C., 1981. Prenatal exposure to ethanol alters the organization of the hippocampal mossy fibres, *Science* 211:957.

Wolf, G., and Schmidt, W., 1982. Zinc as a putative regulatory factor of glutamate dehydrogenase activity in glutamergic systems, in *Neuronal Plasticity and Memory Formation* (C. Ajmone Marson and H. Matthies, eds.), Raven Press, New York, p. 437.

Zinc and Immunity

9.1 Studies in Animal Models

During the past decade considerable knowledge has accumulated concerning the role of zinc in cellular immunity. It has been known for many years that zinc deficiency in experimental animals results in atrophy of thymic and lymphoid tissue (Prasad and Oberleas, 1971). These changes are associated with a variety of functional abnormalities.

Young adult zinc-deficient mice showed thymic atrophy, reductions in the absolute number of splenocytes, and greatly depressed responses to both T-cell-dependent (TD) and T-cell-independent (TI) antigens (Fraker *et al.,* 1986; Fernandes *et al.,* 1979; Fraker, 1983; Zwickl and Fraker, 1980). In response to sheep red blood cells (SRBC), a TD antigen, the zinc-deficient mice produced 40% as many IgM and IgG plaque-forming cells (PFC) per spleen as did the zinc-adequate mice. Fluorescent cell cytometry revealed that although the ratio of T cells to B cells was unaltered, the deficient mice had nearly double the proportion of B cells bearing high amounts of surface IgM as did the control mice. Further studies suggested that a greater proportion of immature B cells were accumulating in the spleens of zinc-deficient mice (Fraker, 1983). Other investigators observed that greater numbers of immature T cells were present in zinc-deficient mice relative to controls (Nash *et al.,* 1979).

Both primary and secondary antibody responses have been reported to be depressed in zinc-deficient mice (Fraker *et al.,* 1977). The influence of suboptimal zinc nutriture on the memory cells has been investigated in mice (Fraker *et al.,* 1986). Normal adult mice were primed with SRBC. Two weeks later, groups of mice were placed on diets containing various amounts of zinc. After a 4-week period, the mice were challenged with a second injection of SRBC. The mice fed suboptimal zinc produced only 43% as many PFC per spleen as did mice fed adequate zinc. Further, cells of the zinc-deficient mice responded less well to antigenic stimulation when adoptively transferred to normal, irradiated syngeneic hosts than did cells of mice fed adequate zinc. An extensive 4-week period of nutritional repletion restored some but not all of the immunologic memory response, which suggested that some memory cells had been destroyed by zinc deficiency. This finding indicates that after a period of malnutrition it may be necessary to revaccinate the patients to restore immunity that existed before the nutritional deficiency.

A decline in *in vivo*-generated cytotoxic T-killer activity to allogeneic tumor cells in zinc-deficient mice has been observed (Fernandes *et al.*, 1979; Good and Fernandes, 1979). Natural killer (NK) cell activity was reduced and grossly deficient responses to cutaneous sensitization have been reported in zinc-deficient mice (Fraker *et al.*, 1985).

Dietary-induced zinc deficiency resulted in an impaired cell-mediated immune response to non-H_2 allogeneic tumor cells in mice (Frost *et al.*, 1981). Animals maintained on a zinc-deficient diet for as little as 2 weeks developed a severe impairment in their ability to generate a cytotoxic response in the face of the tumor challenge. This impairment was totally reversible by returning zinc-deficient mice to normal dietary zinc intake. Such animals demonstrated a normal cytotoxic response to tumor challenge. If mice were treated with toxic doses of dietary zinc, a similar impairment of the cell-mediated cytotoxic response occurred. This suggests that either a deficiency or toxic level of zinc impairs the immune response to all allogeneic tumor cells.

The *in vitro* production of antibody to TD antigens was depressed as a function of age in mice (Fraker *et al.*, 1985). The addition of supplemental zinc to cultures of cells from young adult, middle-aged, and immunodepressed aged mice enhanced antibody production. The responses of young mice were enhanced 30 to 40% but the responses in cultures for aged mice were increased 5- to 12-fold and the maximum response achieved was often equivalent to that of young, fully competent mice. Zinc affected the early events in the activation of antibody-forming cells suggesting that zinc may play a role in the interactions between the antigen-presenting cell and T cells or between T cells and the precursors of antibody-forming cells.

9.1.1 Effect of Zinc Deficiency on Development of Immune Response

Zinc deficiency is harmful to the development of the immune response if it occurs during the critical periods of ontogeny. The effects of zinc deprivation on the immunologic development of outbred N:NIH(S) mice were investigated (Beach *et al.*, 1979, 1980a). At 4 weeks of age, direct splenic PFC responses to SRBC were dramatically diminished in mice that were moderately or severely deprived of zinc during the postnatal period. The zinc-deficient mice also exhibited a highly disordered serum immunological profile, with absence of detectable IgM, IgG2a, and IgA, along with greatly elevated serum levels of IgG1. These findings suggest that zinc deficiency during the period of rapid growth and development may predispose higher animals to a severe acquired immune deficiency and set the stage for opportunistic infection (Beach *et al.*, 1980b).

In another experiment, progeny of zinc-deprived dams showed significant growth retardation, and preferentially decreased growth of spleen and thymus relative to controls (Beach *et al.*, 1982a,b). Cross-fostering of zinc-deprived pups on control dams improved the growth of organs but failed to improve the growth of spleen and thymus. These findings were more striking when intergestational effects were studied. Pregnant Swiss–Webster mice fed a diet moderately deficient in zinc from day 7 of gestation until parturition had offspring with depressed immune function through 6 months of age. In addition, the second and third filial generation, each fed a normal control diet, continued to manifest reduced immunocompetence, although not to the same degree as the first generation.

Recent studies indicate that B-cell development *in vitro* is significantly altered by a marginal deficiency of zinc and supplemental zinc restores antibody formation in cultures of aged spleen cells (Golub *et al.*, 1984; Haynes *et al.*, 1985).

9.1.2 Zinc and Autoimmunity

The effect of zinc nutriture on the progression of autoimmune disease in New Zealand Black (NZB), NZB/W, and MRL/I mice was investigated (Beach *et al.*, 1981, 1982c,d). Beginning at either 6 weeks or 6 months of age, NZB mice consumed diets containing 199 ppm zinc (control), or 9, 5, or 2.5 ppm zinc. In addition, a group of mice were pair-fed the control diet in amounts equal to the intake of the mice fed 5 ppm zinc. NZB mice fed 9 ppm zinc from the age of 6 weeks showed decreased autoimmune hemolytic anemia and lower antierythrocyte autoantibody titers, and they lived longer. Mice fed 5 ppm zinc also had higher packed cell volumes and lower antierythrocyte titers than their pair-fed controls. These findings indicate that deprivation of zinc may account for a significant influence on the development of autoimmunity. NZB mice started on the zinc-restricted diets at 6 months of age also showed a reduction in autoimmune hemolytic anemia. Similar results were obtained in NZB/W and MRL/I mice.

9.1.3 Effects of Zinc on Monocytes and Macrophages

Monocytes and macrophages from zinc-deficient mice exhibit comparable or enhanced Fc receptor (FcR) and complement receptor (CR) expression and phagocytic activity relative to cells from zinc-adequate mice. Recent data, however, show that macrophages from zinc-deficient mice were unable to kill the parasite *Trypanosoma cruzi* after they were engulfed (Fraker *et al.,* 1985).

The increased activities and/or rate of monocyte differential observed in these experiments may lead to increased antigen catabolism and decreased immune responsiveness as seen in the Biozzi low-responder mice.

Animals on low-protein and low-zinc diets showed significantly impaired delayed hypersensitivity to dinitrochlorobenzene and tuberculin compared with animals consuming the control and commercial chow regimen. The ability of peritoneal exudate cells to produce macrophage migration inhibition factor (MIF) when stimulated with PPD *in vitro* was diminished in animals maintained on the low-zinc diet for more than 5 weeks, but was unaffected by protein deficiency. Reduced capacity to produce functional lymphokines may contribute to skin test anergy in chronic zinc deficiency.

9.2 Zinc Deficiency and Immune Functions in Humans

Abnormalities of cellular immunity have been observed in zinc-deficient humans. An extreme example of the effects of zinc deficiency on the human immune system is acrodermatitis enteropathica, a genetic disorder of zinc malabsorption (Oleske *et al.,* 1979). This condition is characterized by mucocutaneous lesions, diarrhea, failure to thrive, and frequent severe infections with fungi, viruses, and bacteria. Affected subjects have thymic atrophy, anergy, reduced lymphocyte proliferation response to mitogens, a selective decrease in T4$^+$ helper cells, and deficient thymic hormone activity (Chandra and Dayton, 1982). All of these changes are corrected by zinc supplementation (Chandra and Dayton, 1982; Good *et al.,* 1982). Less severe cellular immune defects have been observed in patients who become zinc deficient while receiving total parenteral nutrition. These abnormalities, which include lymphopenia, decreased ratios of T-helper and T-suppressor cells, decreased NK activity, and increased monocyte cytotoxicity, are readily corrected by zinc supplementation (Allen *et al.,* 1983).

Further evidence of a relationship between zinc deficiency and immune dysfunction in humans comes from studies of two groups known to have a high incidence of mild to moderate zinc deficiency: uremic patients on hemodialysis and the elderly. In one study, delayed hypersensitivity skin tests to mumps antigen were carried out in 25 apparently well-nourished men who gave a prior history of mumps infection and were receiving regular hemodialysis because of end-stage renal disease (Antoniou *et al.,* 1981). Nine patients were receiving zinc in their dialysis bath for the treatment of hypogonadism. Only one patient in the zinc-treated group was anergic to mumps. In contrast, anergy to mumps and other antigens was observed in 11 of 16 untreated patients. Skin sensitivity test was restored to normal in three of four anergic patients treated with zinc.

In another study, 15 institutionalized, apparently healthy persons aged 81 ± 5 years were given 100 mg zinc daily for 1 month, while 15 control subjects aged 79.6 ± 4.2 years were given placebo (Duchateau et al., 1981). The group receiving zinc displayed an increased percentage of circulating T lymphocytes, an increased frequency and magnitude of delayed hypersensitivity skin reactions to purified proteins, and a greater IgG antibody response to tetanus toxoid. Although the results clearly suggest that some defects in cellular immunity in these two groups were related to zinc deficiency, the findings were not definitive since zinc status was not evaluated prior to the treatment.

Results of our studies on human zinc deficiency are summarized in Tables 9-1 to 9-4 and Figs. 9-1 to 9-10.

9.2.1 Sickle-Cell Anemia

Infection is the most common cause of death in children with SCA, according to some investigators. In one study it was noted that the risk of

Table 9-1. Cellular Zinc in Zinc-Deficient SCA and Non-SCA Subjects before and after Zinc Supplementation

	Zinc		
	Lymphocytes	Granulocytes	Platelets
	$\mu g/10^{10}$ cells		
SCA ($n = 6$)			
Before	51.8 ± 4.6	35.7 ± 4.2	1.35 ± 0.26
After Zn supplementation[b]	61.4 ± 5.5	42.8 ± 3.37	2.24 ± 0.71
p	<0.01	<0.01	<0.03
Non-SCA ($n = 6$)			
Before	47.7 ± 8.4	37.8 ± 1.95	1.6 ± 0.17
After Zn supplementation[c]	62.6 ± 5.7	44.0 ± 2.9	1.9 ± 0.15
p	<0.005	<0.002	<0.01
Controls ($n = 28$)	54.4 ± 6.6	49.8 ± 7.5	2.2 ± 0.5

[a] Adapted with permission from Prasad, A. S., Meftah, S., Abdallah, J., Kaplan, J., Brewer, G. J., Bach, J. F., and Dardenne, M., 1988. Serum thymulin in human zinc deficiency, J. Clin. Invest. 82:1202.
[b] SCA subjects were supplemented with 50 mg of zinc (as acetate) orally in two divided doses daily for 3–6 months.
[c] Non-SCA subjects were supplemented with 50 mg of zinc (as acetate) orally in two divided doses daily for 3 months.

bacterial infection reduced markedly after age 3 and proven pneumococcal infection was not seen unless they had experienced previous pneumococcal infection (Barrett-Conner, 1971). Survival past the age of 3 without prior recognized bacterial infection implied a definite reduced risk of subsequent bacterial infections except for infections by coliform bacteria. Although the study reported by Barrett-Conner (1971) did not find an increased incidence of salmonella infection or increased mortality and morbidity resulting from tuberculosis in patients with sickle-cell disease, other reports have observed such associations.

The risk of salmonella osteomyelitis is several hundred times greater in patients with SCA than in the normal population (Golding et al., 1959). An increased incidence of urinary tract infection, with clinical pyelonephritis in up to 25% of adults with SCA caused by E. coli, has been observed (Barrett-Conner, 1971). These patients also appear to be susceptible to Enterobacter–Klebsiella infection. Our own clinical impression is that a large number (three out of four subjects) of our adult SCA patients who are admitted to the hospital give suggestive evidence for upper respiratory tract infection, probably viral in etiology, and/or urinary tract infection prior to the onset of pain crisis.

A recent review suggests that an overwhelming infection caused by encapsulated bacteria—Salmonella spp. and Plasmodium falciparum (in malarious areas)—is an important cause of morbidity and death in patients with SCA (Onwubalili, 1983). Although contributing factors to increased susceptibility to infections in patients with SCA may include a state of functional asplenia, an opsonophagocytic defect caused by an abnormality of the alternative complement pathway, and a deficiency of specific circulating antibodies, the role of lymphocytes in cell-mediated immunity in this disease remains to be established.

Some investigators have observed that patients with SCA are unusually susceptible to mycoplasma pneumonia infection, often manifesting unusual clinical features (Shulman et al., 1972; Mann et al., 1975). Although proven examples of viral hepatitis in SCA patients do not suggest increased incidence of this infection, 20–40% incidence of macronodular cirrhosis among adult subjects has been reported. It has been shown that primary infection with a novel parvovirus-like agent (PVLA) is the major cause of aplastic crises in children with SCA (Sergeant et al., 1981). Cryptococcal pneumonia and pneumonia caused by cytomegalovirus in patients with SCA have also been reported (Hardy et al., 1986; Haddad et al., 1984).

Cellular immune responses of which delayed hypersensitivity reactions are a prototype, are of paramount importance in the defense against a number of obligate or facultative intracellular parasites such as viruses, rickettsiae, mycobacteria, Listeria monocytogenes, Brucella abortus, Salmonella typhosa, and certain protozoa. In patients with acquired immunodeficiency syndrome

(AIDS)—an extreme example of impaired cellular immune function affecting T cells—*Pneumocystis carinii* pneumonia, candidiasis, cytomegalovirus (CMV) infection, perirectal herpes simplex virus (HSV) infection, *Mycobacterium avium* intracellular complex infection, cryptococcosis, cryptosporidiosis, toxoplasmosis, salmonella bacterium, and aspergillosis were observed (Douglas *et al.*, 1984). Multiple infections were common and overall mortality was 97.3% in 12 months. Clearly, we are not dealing with a severe T-cell disorder in SCA, nonetheless it appears that these patients do have problems of cell-mediated immunity.

9.2.2 Anergy (Decreased Delayed Response to Common Skin Antigens)

In order to examine the relationship between zinc status and cell-mediated immunity in SCA, we measured *in vivo* delayed skin test responsiveness in 26 adult SCA patients (Ballester and Prasad, 1983)—16 men and 10 women, with a mean age of 25.8 years. All patients were evaluated when they were without pain. No patient was admitted to the study within 3 weeks of a crisis or infection, or within 3 months of a blood transfusion. Six patients who had proven zinc deficiency had been receiving an oral zinc supplement (zinc acetate, 45 mg/day) for a minimum of 1 year. Eleven healthy, age- and sex-matched persons—seven blacks and four whites—were recruited as controls. The zinc levels and results of skin tests with four antigens (PPD, SKSD, candida, and mumps) were recorded. An area of induration of 5 mm or more was recorded as a positive reaction.

Six patients were anergic, with negative responses to all four antigens. Four patients had only one positive test result. The remaining 16 patients had two or more positive reactions. Ten of the eleven controls also had two or more positive reactions. For the purpose of comparison, patients were grouped according to the delayed hypersensitivity reaction responses. Group A included patients who were anergic or had only one positive test, whereas all patients in group B had two or more positive tests. Patients in group C, those receiving oral zinc supplements, were analyzed separately. When compared with groups B and C and the controls, patients in group A were found to have significantly lower neutrophil zinc concentrations and purine nucleoside phosphorylase (PNP) activity ($p < 0.001$) in erythrocytes, and decreased levels of plasma and erythrocyte zinc ($p < 0.01$). No significant differences in zinc levels or PNP activity were found in group A between anergic patients and patients with one positive test. There were no significant differences in zinc levels or PNP activity in groups B and C and the controls, except for higher plasma zinc levels in patients receiving an oral zinc supplement (group C). There were no significant differences in the activity of adenosine deaminase in the various groups.

The correlation between neutrophil zinc (one of the best indicators of tissue zinc status) and the activity of erythrocyte PNP in 21 patients, with both measurements done simultaneously, was highly significant ($r = 0.716$, $p < 0.0001$). All patients in group A (with one exception) had lower levels of neutrophil zinc and PNP activity than the controls.

Three anergic patients from group A entered a trial of oral zinc supplementation (zinc acetate, 45 mg/day) and were reevaluated 6 months later. All three patients showed a correction of neutrophil zinc levels and normal or significantly improved PNP activity. On repeat skin tests, one patient had a positive response to two tests, and the remaining two patients had positive responses to one test each.

Taken together, these findings indicate that in patients with SCA, anergy is associated with zinc deficiency. The improvement in skin test responses of some but not all zinc-deficient patients following zinc treatment suggests that in these subjects zinc deficiency may be directly responsible for their anergy. Longer periods of supplementation or higher doses of zinc may be necessary to restore normal tissue zinc stores in others.

The correlation of decreased zinc levels with decreased activity of PNP is of interest because the congenital deficiency of this enzyme in humans induces severe defects in cell-mediated immunity (Giblett *et al.,* 1975). Our patients with SCA, with zinc deficiency and impaired delayed hypersensitivity reactions had a mean PNP activity 59% that of normal controls. Heterozygotes for congenital PNP deficiency with comparable decreases in PNP activity have been reported to be clinically unaffected, suggesting that the occurrence of decreased PNP activity may not be sufficient to explain our findings. The role played by decreased nucleoside phosphorylase activity, however, needs further evaluation inasmuch as a chronic accumulation of toxic nucleotides as a result of PNP deficiency in lymphocytes may ultimately be harmful.

9.2.3 Decreased NK Cell Activity in SCA Patients with Zinc Deficiency

Because NK cells are known to be important in host defense against viruses, we compared the NK activity of 8 zinc-deficient SCA patients (group 1) with that of 8 SCA patients with normal zinc status (group 2), 5 SCA patients who were receiving zinc therapy because of previously diagnosed zinc deficiency (group 3), and 12 healthy age- and sex-matched adults (9 men and 3 women, 7 blacks and 5 whites) who served as controls (Tapazoglou *et al.,* 1985). None of the SCA patients participated within 3 weeks of a crisis or infection, or within 3 months of a blood transfusion. The three groups of

SCA patients had no significant differences in hemoglobin levels, hematocrit values, mean corpuscular volume, and erythrocyte, reticulocyte, and leukocyte counts. Hemoglobinopathy was ruled out in black controls by hemoglobin electrophoresis. We also studied two male volunteers, aged 48 (patient 1) and 33 (patient 2) years, who were inpatients on a metabolic ward where their dietary zinc intake was restricted to 3.0 mg/day for 20 weeks. These volunteers were screened thoroughly before inclusion. They were free of any disease, and their zinc status was normal as judged by zinc assay in plasma, erythrocytes, hair, neutrophils, and urine, and by metabolic zinc balance data. These subjects were tested for NK activity and plasma zinc concentration at baseline, during dietary zinc restriction, and after zinc repletion with a dietary zinc intake of 30 mg/day for 14 weeks. NK activity decreased in these volunteers as a result of dietary zinc restriction and was corrected after repletion with zinc.

Zinc-deficient SCA patients (group 1) had significantly lower NK activity than the patients in groups 2 and 3 and the normal controls. The three groups of patients with SCA had no significant differences in hemoglobin levels, hematocrit values, mean corpuscular volume, and erythrocyte, reticulocyte, and leukocyte counts.

These findings suggest that zinc deficiency in SCA patients is associated with diminished NK activity. This defect in NK activity is likely related to zinc deficiency *per se* rather than to some associated condition, since a decline in NK activity was observed in zinc-restricted volunteers.

9.2.4 Interleukin-2 (IL-2) Production

In addition to its critical role in T-cell proliferation, IL-2 is a major *in vivo* activator of NK activity. Could the impairments of both T-cell proliferation and NK activity associated with zinc deficiency result from an underlying defect in IL-2 production? The results of our preliminary study are consistent with this possibility. Supernatants of peripheral blood lymphocytes stimulated for 48 h with phytohemagglutinin (PHA) were assayed for IL-2 activity in a murine thymocyte proliferation assay.

IL-2 production by zinc-deficient SCA patients (plasma zinc < 100 μg/dl) was significantly lower ($p < 0.05$) than that of SCA patients with normal zinc levels and control subjects. Our results suggest that diminished IL-2 production may be responsible for impaired T-cell proliferation and NK activity in zinc-deficient subjects (Prasad *et al.,* 1986).

Our data in the experimental human model showed that IL-2 activity declined as a result of zinc restriction (Prasad *et al.,* 1988). IL-2 activity was promptly corrected during the zinc repletion period (see Fig. 9-6).

9.2.5 Lymphocyte Subpopulation Abnormalities in Zinc Deficiency

In this study, we evaluated lymphocyte subpopulations in 23 adults with SCA (Ballester *et al.*, 1986). When compared with controls, SCA patients had higher lymphocyte counts with normal numbers of T101$^+$ cells (T lymphocytes) and T4$^+$ cells. T8$^+$ cells were significantly increased in SCA patients relative to controls (1684 ± 243 versus 980 ± 367, $p < 0.001$). This increment was largely dependent on a T101$^-$, T8$^+$ cell population. The SCA patients as a group had a significantly decreased T4/T8 ratio ($p < 0.0001$). The SCA patients with a history of blood transfusions had higher T4$^+$ cell numbers and a higher T4/T8 ratio, but no other significant differences from nontransfused patients were noted. Thus, in our study, a distinct pattern of abnormalities was seen in lymphocyte subpopulations of adult SCA patients.

The significant changes in lymphocyte subpopulations in the experimental human model as a result of zinc depletion and zinc repletion are shown in Fig. 9-5. The T4/T8 ratio declined on the zinc-restricted diet, and this was corrected by zinc repletion. Significant increases in T101$^+$, sIg$^+$, T4$^+$, and T8$^+$ cells were noted. The changes in the T4/T8 ratio during zinc depletion were related to a slight decrease in T4$^+$ cells and a similar slight increase in T8$^+$ cells that were reversed by zinc supplementation (Table 9-2). Similar changes in lymphocyte subpopulations in adult SCA subjects before and after zinc supplementation are shown in Table 9-3.

Table 9-2. Lymphocyte Subpopulations in Experimental Human Model[a,b]

Subjects	Total lymphocytes	T cells (T101$^+$)	B cells (sIg$^+$)	T101$^-$ (sIg$^-$)	T helper (T4$^+$)	T suppressor (T8$^+$)	T4/T8 ratio
			No. cells/mm^3 blood				
A							
Baseline	2956	2263	455	238	1138	1116	1.02
Zn depletion	3313	2460	425	428	926	1279	0.72
Zn repletion	3079	2238	527	314	1465	1246	1.18
B							
Baseline	2063	1609	263	191	827	604	1.37
Zn depletion	2089	1517	230	334	798	752	1.06
Zn repletion	1831	1406	240	203	851	616	1.39

[a] Adapted with permission from Prasad, A. S., Meftah, S., Abdallah, J., Kaplan, J., Brewer, G. J., Bach, J. F., and Dardenne, M., 1988. Serum thymulin in human zinc deficiency, *J. Clin. Invest.* 82:1202.
[b] The results are averages of two determinations at baseline, two determinations during the fifth and sixth months of the Zn depletion phase, and two determinations during Zn repletion in the eighth and ninth months of the study.

Table 9-3. Lymphocyte: Subpopulations in Adult SCA Patients[a,b]

	Total lymphocytes	T cells (T101+)	B cells (sIg+)	T101- (sIg-)	T helper (T4+)	T suppressor (T8+)	T4/T8 ratio
			No. cells/mm³ blood				
SCA							
1. Before	2754	1052	772	759	674	919	0.73
After	3023	1254	1149	604	1254	665	1.88
2. Before	4207	2545	883	778	1703	1346	1.26
After	5779	3900	1271	577	3409	1618	2.10
3. Before	4083	2653	816	612	1449	1163	1.20
After	4772	2958	1264	524	2529	1097	2.30
4. Before	4230	2538	782	909	1226	1099	1.10
After	6818	5011	988	818	3818	1804	2.04
Mean ± S.D.							
Before	3819 ± 713	2197 ± 756	813 ± 50	765 ± 122	1263 ± 438	1132 ± 176	1.07 ± 0.24
After	5098 ± 1616	3281 ± 1591	1168 ± 132	631 ± 129	2753 ± 1135	1296 ± 516	2.08 ± 0.17
Paired t test							
p	0.088	0.13	0.006	0.01	0.04	0.49	0.0008
Controls ($n = 21$)	2779 ± 819	2122 ± 684	332 ± 160	372 ± 149	1428 ± 472	980 ± 367	1.54 ± 0.40

[a] Adapted with permission from Prasad, A. S., Meftah, S., Abdallah, J., Kaplan, J., Brewer, G. J., Bach, J. F., and Dardenne, M., 1988. Serum thymulin in human zinc deficiency, J. Clin. Invest. 82:1202.
[b] The results are averages of two determinations before and two determinations after zinc supplementation.

9.2.6 Serum Thymulin and Zinc Deficiency

Thymulin (formerly called serum thymic factor) is a well-defined zinc-dependent thymic hormone with the following amino acid sequence: <Glu-Ala-Lys-Ser-Gln-Gly-Gly-Ser-Asn-OH (Bach *et al.,* 1977). Since its initial isolation from pig serum, a number of actions of thymulin on the immune system have been recognized. Dardenne *et al.* (1982) showed that zinc was required for thymulin to express its biological activity. Recent studies indicate the existence of two forms of thymulin, the first one deprived of zinc and biologically inactive, the second one containing zinc and biologically active.

Antithymulin monoclonal antibodies have been produced against either the synthetic (ASI) or the natural intraepithelial (AEI) molecule. By using biological and immunofluorescence assays, the two antibodies were shown to recognize exclusively the zinc-coupled thymulin molecule. Based on these studies, it was suggested that a zinc-specific conformation of the thymulin molecule existed.

The zinc–thymulin relationship was previously investigated using two models of *in vivo* zinc deficiency (Dardenne *et al.,* 1984; Bensman *et al.,* 1985). In the first model, active thymulin levels in sera from mice subjected to a long-term marginally zinc-deficient diet were studied. In spite of the absence of thymic atrophy, the serum levels of thymulin decreased as early as 2 months after beginning the diet. However, these levels could be consistently restored after *in vitro* addition of zinc.

Similar observations were made with sera from children suffering from nephrotic syndrome with zinc deficiency (the second model), a disease in which a low level of thymulin activity was observed that could be restored to normal after *in vitro* serum chelation and incubation with zinc. These results confirm the presence of the inactive hormone in the serum of zinc-deficient individuals and its potential activation following zinc addition. The specificity of these results was confirmed by the lack of activation in experiments performed with sera from thymectomized mice or patients with DiGeorge's syndrome in whom the hormone is nonexistent.

Six zinc-deficient homozygous male SCA subjects were included in our recent study (Prasad *et al.,* 1988). Their ages ranged from 17 to 33 years. They were ambulatory, had no neurological or psychiatric deficits, were not dependent on drugs, and no other hemoglobinopathies such as alpha-thalassemia, beta-thalassemia, or Hb C disease were present in these cases. These subjects have been followed regularly in the adult sickle cell clinic of the University Health Center of Wayne State University, Detroit Medical Center. They had not received blood transfusion during the 6 months prior to this study. Prior to selection of these subjects for our study, zinc was assayed in lymphocytes, granulocytes, and platelets by flameless atomic absorption spectrophotometry. In this study, care was taken to remove platelets from granulocyte and lymphocyte fractions prior to assay. Plasma zinc levels ranged from 98 to 118 μg/dl and were considered to be within the normal range. The serum level of biologically active thymulin was evaluated by a rosette assay described in detail elsewhere and shown by us and several other investigators to be strictly thymus specific. The assay analyzes the conversion of relatively azathioprine (Az)-resistant spleen of adult thymectomized mice to the positive rosette-forming cells that are more sensitive to Az. In the presence of thymulin-containing sera, rosette formation was inhibited by Az. The results were expressed as the log 2 of reciprocal highest serum dilution conferring sensitivity to Az inhibition upon spleen cells from adult thymectomized mice. To confirm the specificity of the biological activity measured, all of the determinations were repeated after preincubation of the sera under study with an antithymulin monoclonal antibody or a specific antithymulin immunoabsorbent.

Sera (100 μl) from patients and control subjects were incubated for 30 min at room temperature with an equal volume of Chelex 100 at 50 mg/ml in distilled water. At the end of the incubation, the mixture was centrifuged at 12,000g for 2 min to eliminate the chelating resin and the biological activity in the supernatant of Chelex 100-treated serum was measured. Ten micrograms of $ZnCl_2$ was then added to 100 μl of Chelex-treated serum. This mixture was incubated for 15 min at room temperature and its biological activity measured by the rosette assay.

Abnormally low levels of active thymulin in the sera were found in zinc-deficient SCA patients compared with age-matched healthy individuals (Figs. 9-1 and 9-2). Following supplementation with 50 mg of zinc (as acetate) orally in two divided doses daily for 3 to 8 months, the serum thymulin activity (mean ± S.D.) became normal (before, 2.49 ± 0.81 versus after, 4.44 ± 0.65 log 2, $p < 0.0001$). Zinc levels (mean ± S.D.) in granulocytes, lymphocytes, and platelets ($\mu g/10^{10}$ cells) before and after supplementation were: granulocytes, 35.7 ± 4.2 versus 42.8 ± 3.37, $p < 0.01$; lymphocytes, 51.8 ± 4.6

Figure 9-1. Effect of zinc therapy on the levels of thymulin activity in (A) SCA subjects and (B) non-SCA deficient patients. Results are expressed as log-2 reciprocal titers. Each point represents the mean of three determinations. *p < 0.001; **p < 0.001. (Reprinted with permission from Prasad, A. S., Meftah, S., Abdallah, J., Kaplan, J., Brewer, G. J., Bach, J. F., and Dardenne, M., 1988. Serum thymulin in human zinc deficiency, *J. Clin. Invest.* 82:1202.)

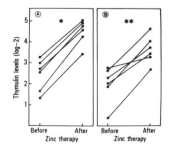

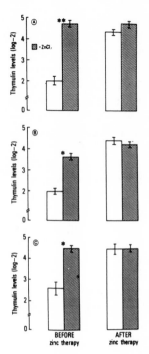

Figure 9-2. Restoration of normal thymulin activity in sera of zinc-deficient patients after *in vitro* addition of $ZnCl_2$. After chelation, 200 μl of serum was incubated for 1 h at 37°C with 10 ng of $ZnCl_2$. Thymulin activity determination was performed before and after *in vitro* addition on individual samples. *p < 0.01; **p < 0.001. (A) Healthy volunteers submitted to zinc restriction. (B) Non-SCA zinc-deficient subjects. (C) SCA zinc-deficient subjects. (Reprinted with permission from Prasad, A. S., Meftah, S., Abdallah, J., Kaplan, J., Brewer, G. J., Bach, J. F., and Dardenne, M., 1988. Serum thymulin in human zinc deficiency, *J. Clin. Invest.* 82:1202.)

versus 61.4 ± 5.5, p < 0.01; platelets, 1.35 ± 0.26 versus 2.24 ± 0.71, p < 0.03 (Table 9-1).

Recently we have established an experimental human model for the study of a mild state of zinc deficiency. Two adult male volunteers, ages 23 and 25, were hospitalized at the Clinical Research Center of the University of Michigan Medical School Hospital. The protocol for all human studies was reviewed and approved by the Committee on Research in Human Subjects at Wayne State University School of Medicine, Detroit, and the University of Michigan Medical School, Ann Arbor. A semipurified diet that supplied 3.0 mg of zinc on a daily basis was used to produce zinc deficiency. The details of the methodology have been published elsewhere (Rabbani *et al.*, 1987).

They were given a hospital diet containing animal protein daily for 4 weeks. This diet averaged 12 mg zinc/day, consistent with the recommended dietary allowance of the National Research Council, National Academy of Sciences. After this, they received 3.0 mg of zinc daily while consuming a soy protein-based experimental diet. This regime was continued for 28 weeks, at the end of which two cookies containing 27 mg of zinc supplement were added to the experimental diet. The supplementation was continued for 12 weeks.

Activity of circulating thymulin observed in the two healthy young volunteers before the onset of dietary zinc restriction was normal compared with a group of age-matched healthy controls. Thymulin activity began to decrease 3 months after initiation of zinc deprivation, and was undetectable after 6 months (Figs. 9-3 and 9-4). *In vivo* zinc supplementation induced a rapid return to a normal level of thymulin activity within 1 month. The final level of activity was even higher than those observed before zinc restriction (Fig. 9-3). Recent data suggested that the natural thymulin molecule could exist in two forms, one biologically active, containing zinc, and the other inactive and lacking this metal ion (Gasintinel *et al.,* 1984; Laussac *et al.,* 1985). To confirm the possibility that the peptide could be present in the serum from zinc-deprived individuals in an inactive form as previously shown in zinc-deficient mice and patients with nephrotic syndrome (Dardenne *et al.,* 1984; Bensman *et al.,* 1985), we performed supplementary experiments, adding $ZnCl_2$ to the serum under study. This addition was done after serum chelation to eliminate other metals, which could compete with zinc for the binding site in the peptide, present in the serum (Gasintinel *et al.,* 1984).

As observed in Fig. 9-2, after chelation and incubation with $ZnCl_2$, the activity of thymulin in the serum of zinc-deficient individuals reached normal levels similar to those observed after *in vivo* zinc supplementation. In contrast, no change in the activity of the hormone was observed after incubation of serum with $ZnCl_2$ when these experiments were done with the serum obtained from patients after zinc supplementation, i.e., at a time when thymulin activity had already reached normal levels, or in healthy volunteers before the beginning of zinc restriction at a time when thymulin was fully active.

We also identified six human volunteers who showed a mild state of zinc deficiency. Low levels of active thymulin were found in these subjects which was corrected by zinc supplementation both *in vivo* and *in vitro* (Prasad *et al.,* 1988).

Although several polypeptides have been extracted from the thymus, and a number of these have been characterized chemically, from the physiological point of view, only two of these are true thymic hormones, actually produced

Figure 9-3. Levels of thymulin activity in sequential study of young human volunteers submitted to a zinc-restricted diet for 6 months followed by zinc supplementation. Results are expressed as log-2 reciprocal titers (mean ± S.E.). Each determination was performed in triplicate. (Reprinted with permission from Prasad, A. S., Meftah, S., Abdallah, J., Kaplan, J., Brewer, G. J., Bach, J. F., and Dardenne, M., 1988. Serum thymulin in human zinc deficiency, *J. Clin. Invest.* 82:1202.)

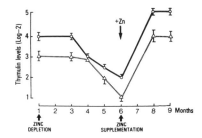

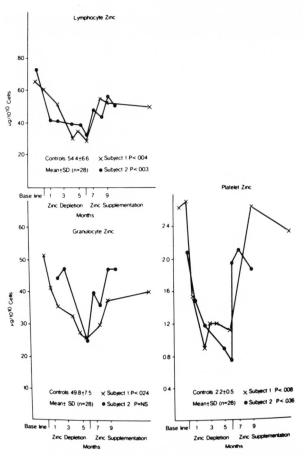

Figure 9-4. Cellular zinc concentrations in a sequential study of young human volunteers submitted to a zinc-restricted diet for 6 months followed by zinc supplementation. (Reprinted with permission from Prasad, A. S., Meftah, S., Abdallah, J., Kaplan, J., Brewer, G. J., Bach, J. F., and Dardenne, M., 1988. Serum thymulin in human zinc deficiency, *J. Clin. Invest.* 82:1202.)

by the thymus and biologically active on T cells (Bach *et al.,* 1977; Bach, 1983; Savino *et al.,* 1982). These are thymopoietin and thymulin, both of which are produced by the thymic epithelium (Bach *et al.,* 1977; Bach, 1983). Recent data demonstrate that thymulin binds to high-affinity receptors, induces several T-cell markers, and promotes T-cell function including allogeneic cytotoxicity, suppressor function, and IL-2 production. For the first time, our study has provided evidence that even a mild deficiency of zinc may affect serum thymulin level.

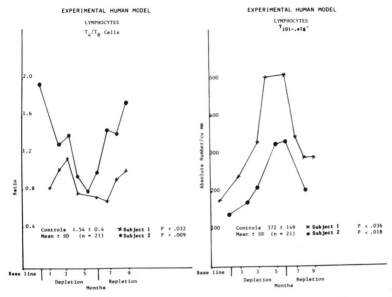

Figure 9-5. Changes in lymphocyte subpopulations (T4/T8 and T101−, sIg− cells) as a result of zinc restriction and zinc repletion in the experimental human model subjects. Each data point represents the average of two separate determinations. (Reprinted with permission from Prasad, A. S., Meftah, S., Abdallah, J., Kaplan, J., Brewer, G. J., Bach, J. F., and Dardenne, M., 1988. Serum thymulin in human zinc deficiency, *J. Clin. Invest.* 82:1202.)

Interestingly, not only did oral zinc supplementation *in vivo* correct serum thymulin activity in all zinc-deficient subjects in this study, but thymulin activity was also corrected by the addition of zinc *in vitro*. It is likely that some of the T-cell dysfunctions observed as a result of zinc deficiency in humans may indeed be caused by decreased thymulin activity. In the experimental human model, we observed changes in lymphocyte subpopulations and decreased IL-2 activity during the zinc-restricted period that were corrected by zinc supplementation. Our studies thus provide suggestive but not conclusive evidence that the changes in lymphocyte subpopulations and IL-2 activity may be related to thymulin activity.

The relationship between cellular immunity, thymic hormones, and serum zinc levels in 19 patients with common varied immunodeficiency was investigated by Cunningham-Rundles *et al.* (1981). Five (26%) had serum zinc levels that were 2 S.D. below normal and 11 (58%) had abnormally low lymphocyte proliferation to phytohemagglutinin and concanavalin A. Forty-two percent had abnormally low levels of serum thymulin and 74% had low levels of thymopoietin. Three patients with the most profound zinc deficiency

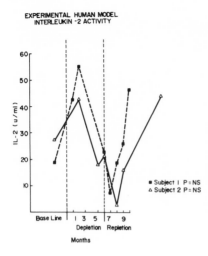

EXPERIMENTAL HUMAN MODEL
INTERLEUKIN -2 ACTIVITY

Figure 9-6. Changes in IL-2 activity as a result of zinc restriction and zinc repletion in the experimental human model subjects. Each data point represents the average of two separate determinations. (Reprinted with permission from Prasad, A. S., Meftah, S., Abdallah, J., Kaplan, J., Brewer, G. J., Bach, J. F., and Dardenne, M., 1988. Serum thymulin in human zinc deficiency, *J. Clin. Invest.* 82:1202.)

had substantial increases in thymic hormones after zinc repletion, and two had complete resolution of intractable diarrhea. A therapeutic potential of zinc for certain patients with hypogammaglobulinemia should be considered.

9.2.7 Zinc and Chemotaxis

Impaired leukocyte chemotaxis was observed as a result of zinc-restricted diet in normal human volunteers (Baer *et al.*, 1985). Abnormal chemotaxis correctable with zinc supplementation has also been observed in patients with acrodermatitis enteropathica who became severely zinc deficient.

9.2.8 Probable Mechanisms of Zinc Action on Cell-Mediated Immunity

Although the mechanism underlying zinc deficiency-related abnormalities of cellular immunity remains to be determined, some are almost certainly related to the deleterious effects of zinc deficiency on lymphocyte proliferative responses. Zinc probably operates at several different levels to influence lymphocyte proliferation. Zinc deficiency can impair DNA synthesis. Deoxythymidine kinase (TK) is an example of a zinc-dependent enzyme which is extremely sensitive such that even a mild state of zinc deficiency may significantly affect its activity adversely. Our recent studies (unpublished) indicate that the level of mRNA of TK is decreased when Jurkat cells (malignant T lymphoblastoid cells) are grown in zinc-deficient medium. Whether or not zinc is involved in TK gene expression needs to be established and further studies are required to show that this mechanism of zinc action is also applicable to human T cells.

A congenital deficiency of the purine enzyme PNP is associated with a severe T-cell immune deficiency and a congenital deficiency of adenosine deaminase results in impaired development of T and B cells (Giblett *et al.,* 1975). Our recent studies in experimental animals and humans indicate that PNP may be zinc dependent (Ballester and Prasad, 1983; Prasad and Rabbani, 1981; Meftah and Prasad, 1989). Thus, a decreased activity of PNP may additionally account for T-cell dysfunction in zinc deficiency. The regulation of expression of three purine-metabolizing enzymes—adenosine deaminase (ADA), PNP, and ecto 5'-nucleotidase (5'-NT)—and TdT (terminal deoxyribonucleotidyl transferase) are considered to be important for homeostatic T-cell development (Blazsek and Mathe, 1984). Ma *et al.* (1983) have suggested that the high ADA versus low PNP and 5'-NT and a sudden fall of TdT activities in cortical thymocytes is a lethal combination. The preferential substrate of TdT is dGTP and dATP, so TdT may build up "non-sense" DNA polymers which in turn protect the cells from the accumulation of potentially toxic free purine nucleotides. The toxic effect of high intracellular dGTP and dATP is explained by their inhibitory effects on essential methylation of RNA, DNA, and proteins via S-adenosylhomocysteine hydrolase. A deficiency of PNP also results in accumulation of GTP in lymphocytes. It has been shown that GTP accumulation also has marked inhibitory effect on cell growth and is associated with ATP depletion (Mitchell and Sidi, 1985).

Recently we diagnosed a mild deficiency of zinc in six apparently normal human volunteers. The diagnosis was based on decreased cellular zinc levels in two out of three cell lines (<42 μg in granulocytes, <48 μg in lymphocytes, and <1.70 μg in platelets, per 10^{10} cells). In comparison with five zinc-sufficient subjects, six zinc-deficient subjects showed a decrease in the activity of NPase ($p = 0.01$), an increase in ADP ($p = 0.008$), a decreased ATP/ADP ratio ($p = 0.0001$), and an increase in both GTP ($p = 0.02$) and dGTP ($p = 0.04$) in the lymphocytes (Meftah and Prasad, 1989). Following zinc supplementation to zinc-deficient subjects, zinc levels in all cell lines and the activity of NPase in lymphocytes increased ($p < 0.01$), ADP decreased ($p = 0.01$), ATP/ADP increased ($p = 0.005$), GTP decreased ($p = 0.03$), and dGTP decreased ($p = 0.04$). These results (see Figs. 9-7 to 9-10 and Table 9-4) indicate that nucleotides accumulate in lymphocytes as a result of zinc deficiency and this may in part explain the abnormal functions of lymphocytes in such cases (Meftah and Prasad, 1989).

The exact role of increased GTP levels in inhibiting lymphocyte growth is not understood. It is very likely that several of these mechanisms are op-

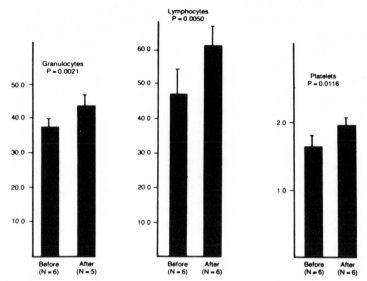

Figure 9-7. Zinc in granulocytes, lymphocytes and platelets before and after zinc supplementation. (Reprinted with permission from Meftah, S., and Prasad, A. S., 1989. Nucleotides in lymphocytes of human subjects with zinc deficiency, *J. Lab. Clin. Med.* 114:114.)

erative and that no single factor is solely responsible for the effects seen in lymphocytes as a result of zinc deficiency.

The ectoenzyme 5'-NT catalyzes the dephosphorylation of nucleotide 5'-monophosphates to yield the corresponding nucleoside. Lymphocytes are in general incapable of synthesizing purine *de novo;* hence, it has been postulated

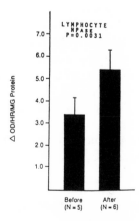

Figure 9-8. Changes in activity of nucleoside phosphorylase in lymphocytes before and after zinc supplementation in a group of subjects with zinc deficiency. (Reprinted with permission from Meftah, S., and Prasad, A. S., 1989. Nucleotides in lymphocytes of human subjects with zinc deficiency, *J. Lab. Clin. Med.* 114: 114.)

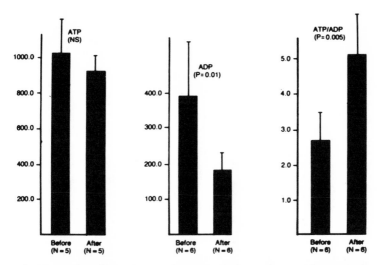

Figure 9-9. Changes in ATP, ADP, and ATP/ADP ratio in lymphocytes of a group of subjects with zinc deficiency before and after zinc supplementation. (Reprinted with permission from Meftah, S., and Prasad, A. S., 1989. Nucleotides in lymphocytes of human subjects with zinc deficiency, *J. Lab. Clin. Med.* 114:114.)

that 5'-NT provides necessary nucleic acid metabolites that can be transported across the cell membrane. A deficiency of this enzyme, which is known to be zinc dependent, may predispose to an immunodeficient state secondary to an arrest in lymphocyte differentiation and proliferation (Pilz *et al.*, 1982). Meftah *et al.* (1991) have shown that in zinc-deficient human subjects, lymphocyte 5'-NT activity is decreased. Thus, a decrease in lymphocyte 5'-NT activity may also contribute to lymphocyte function abnormality in zinc deficiency.

DNA synthesis and adenosine 5'-tetraphosphate 5'-adenosine ($A_{P4}A$) levels decreased in cells treated with EDTA (Grummt *et al.*, 1986). The inhibitory effect of EDTA could be reversed with micromolar amounts of zinc. Zinc in micromolar amounts inhibited $A_{P4}A$ hydrolase and stimulated amino acid-dependent $A_{P4}A$ synthesis, suggesting that Zn^{2+} may be modulating intracellular $A_{P4}A$ pools.

Previous studies indicated that the component inducing DNA synthesis is produced after mitogenic stimulation by the cell itself and that it is accumulated gradually during progression through G1 phase until a critical threshold concentration is reached which determines entry into S phase. A favorite candidate for such a signal molecule is $A_{P4}A$. $A_{P4}A$ has been demonstrated to bind as a specific ligand to calf thymus DNA polymerase alpha (deoxynucleosidetriphosphate:DNA deoxynucleotidyl transferase) holoen-

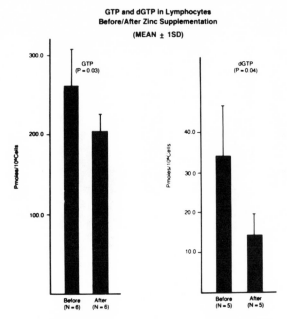

Figure 9-10. Changes in GTP and dGTP in lymphocytes of a group of subjects with zinc deficiency before and after zinc supplementation. (Reprinted with permission from Meftah, S., and Prasad, A. S., 1989. Nucleotides in lymphocytes of human subjects with zinc deficiency, *J. Lab. Clin. Med.* 114:114.)

zyme (M_r 404,000) and to affinity label a subunit of M_r 57,000 (Grummt *et al.*, 1979). In addition, the high-molecular-weight (660,000) and low-molecular-weight (14,500) forms of DNA polymerase alpha of human HeLa cells were shown to possess highly specific noncovalent, $A_{P4}A$ binding activity (Rappaport *et al.*, 1981). Evidence is now accumulating that $A_{P4}A$ could be involved in the primary reaction of mammalian DNA polymerase alpha holoenzyme.

The apparent interdependence of the availability of Zn^{2+} in the tissue culture medium, the intracellular $A_{P4}A$ and DNA synthesizing activity in mammalian cells raises the question whether mitogenic stimulation in growth arrest is correlated with Zn^{2+} uptake. It was observed that readdition of 10% serum to cells synchronized in the early G1 phase by serum deprivation resulted in an immediate increase in the ratio of Zn^{2+} uptake by about fivefold. On the other hand, serum withdrawal caused a fivefold decrease of uptake rates of ^{65}Zn. Thus, the intracellular Zn content could be controlled either by mitogenic stimulation or by growth arrest.

Table 9-4. Activity of Nucleoside Phosphorylase and Nucleotides in Lymphocytes of Human Subjects with Sufficient Zinc or with Zinc Deficiency

	Nucleoside phosphorylase	Nucleotides (pmol/10^6 cells)					
Subjects	OD/h per mg protein	ATP	ADP	ATP/ADP	GTP	dGTP
With sufficient zinc ($n = 5$)	4.8 ± 0.9[b]	675.7 ± 147.6	183.0 ± 56.2	3.78 ± 0.40	182.4 ± 17.30	9.9 ± 11.8
With zinc deficiency ($n = 6$)	3.40 ± 0.74	949.0 ± 184.1	379.3 ± 138.9	2.69 ± 0.73	262.03 ± 48.36	34.14 ± 12.84
p	<0.01	NS	<0.008	<0.001	0.02	0.04

[a] Adapted with permission from Meftah, S., and Prasad, A. S., 1989. Nucleotides in lymphocytes of human subjects with zinc deficiency, *J. Lab. Clin. Med.* 114:114.
[b] Values are means ± S.D.

These studies suggest that $A_{P_4}A$ acts as a chemical messenger of mitogenic induction similar to cAMP in hormone induction. The interaction between growth factors and their cognate receptors in cellular membranes could trigger signals which are transmitted into a cerebral intracellular response, i.e., 1000-fold expansion of the $A_{P_4}A$ pool. This increase in the $A_{P_4}A$ pool causes eventual intracellular response of mitogenic induction, i.e., onset of DNA replication at the G1/S-phase boundary of the cell cycle. The observation that mitogenic induction leading to increased Zn^{2+} uptake is essential for the increase in the intracellular $A_{P_4}A$ pool and DNA replication suggests that Zn^{2+} is a "second messenger" of mitogenic induction and $A_{P_4}A$ is a "third messenger."

T-helper-cell activity is depressed in zinc-deficient animals. T-helper cells produce IL-2, a lymphokine that in addition to triggering T-cell proliferation, augments NK activity. Our studies in zinc-deficient humans show that IL-2 production by peripheral blood mononuclear cells is markedly decreased (Prasad *et al.,* 1988). Thus, impaired T-helper activity in zinc deficiency could cause an *in vivo* deficiency of IL-2 which, in turn, could result in diminished *in vivo* activation of NK cells. Our recent studies (unpublished) indicate that the level of mRNA of IL-2 is decreased when Jurkat cells (malignant T-lymphoid cells) are grown in zinc-deficient medium. Whether or not zinc is involved in IL-2 gene expression needs to be established and further studies are required to show that this mechanism of zinc action is also applicable to human T cells.

Low NK activity has been detected in association with a variety of clinical conditions including malignancies, inflammatory bowel diseases, chronic renal failure, and systemic lupus erythematosus. Our results raise the possibility that zinc deficiency, which occurs frequently in these disorders and in SCA, may contribute significantly to the decreased NK activity in these conditions. Recent reports support this possibility. According to some investigators, NK cells play an important role in host defense against viral infections. The observation that neonates, who are uniquely susceptible to certain viral infections, have low NK activity is consistent with this.

Defective production or responsiveness to IL-2 may be another mechanism contributing to some of the immune abnormalities associated with zinc deficiency. IL-2 is a lymphokine produced primarily by $T4^+$ helper cells which plays a crucial role in T-cell proliferation, generation of cytotoxic T cells, and activation of NK lytic activity. The results of our studies suggest that IL-2 production in zinc-deficient subjects may be impaired (Prasad *et al.,* 1988).

As our study has provided evidence that zinc deficiency adversely affects serum thymulin activity, undoubtedly this may be an important mechanism whereby a deficiency of zinc adversely affects cell-mediated immune functions in humans.

Zinc is known to contribute to membrane stabilization, acting at the cytoskeletal level. An effect on membranes could explain the depression of phagocytosis, oxygen consumption, and bactericidal activity induced by zinc in phagocytic cells and the modification of Con A surface receptor availability on lymphoid cells.

It is clear from the above discussion that the effects of zinc deficiency on cell-mediated immune functions are multifactorial. It appears that a decreased serum thymulin activity as a result of zinc deficiency may play a crucial role in the overall picture accounting for T-cell disorders in zinc deficiency. Clearly, further investigations and controlled zinc supplementation trials must be carried out in order to understand the mechanism of immune dysfunction in zinc deficiency.

References

Allen, J. I., Perri, R. T., McClain, C. J., and Kay, N. E., 1983. Alterations in human natural killer cell activity and monocyte cytotoxicity induced by zinc deficiency, *J. Lab. Clin. Med.* 102: 577.

Antoniou, L. D., Shahloub, R. J., and Schecter, G. P., 1981. The effect of zinc on cellular immunity in chronic uremia, *Am. J. Clin. Nutr.* 32:1912.

Bach, J. F., 1983. Thymulin (FTS-Zn), *Clin. Immunol. Allergy* 3:133.

Bach, J. F., Dardenne, M., Pleau, J. M., and Rosa, J., 1977. Biochemical characterization of a serum thymic hormone, *Nature* 266:55.

Baer, M. T., King, J. C., Tamura, T., Mangen, S., Bradfield, R. B., Weston, W. L., and Daugherty, N. A., 1985. Nitrogen utilization, enzyme activity, glucose intolerance, and leukocyte chemotaxis in human experimental zinc depletion, *Am. J. Clin. Nutr.* 41:1220.

Ballester, O. F., and Prasad, A. S., 1983. Anergy, zinc deficiency and decreased nucleoside phosphorylase activity in patients with sickle cell anemia, *Ann. Intern. Med.* 98:180.

Ballester, O. F., Abdallah, J. M., and Prasad, A. S., 1986. Lymphocyte subpopulation abnormalities in sickle cell anemia, *Am. J. Hematol.* 21:23.

Barrett-Conner, E., 1971. Bacterial infection and sickle cell anemia, *Medicine* 59:97.

Beach, R. S., Gershwin, M. E., and Hurley, L. S., 1979. Altered thymic structure and mitogen responsiveness in postnatally zinc-deprived mice, *Dev. Comp. Immunol.* 3:725.

Beach, R. S., Gershwin, M. E., and Hurley, L. S., 1980a. Growth and development of postnatally zinc-deprived mice, *J. Nutr.* 110:201.

Beach, R. S., Gershwin, M. E., Makishima, R. K., and Hurley, L. S., 1980b. Impaired immunologic ontogeny in postnatal zinc deprivation, *J. Nutr.* 110:805.

Beach, R. S., Gershwin, M. E., and Hurley, L. S., 1981. Nutritional factors and autoimmunity. I. Immunopathology of zinc deprivation in New Zealand mice, *J. Immunol.* 126:1999.

Beach, R. S., Gershwin, M. E., and Hurley, L. S., 1982a. Gestational zinc deprivation in mice: Persistence of immunodeficiency for three generations, *Science* 218:469.

Beach, R. S., Gershwin, M. E., and Hurley, L. S., 1982b. The reversibility of developmental retardation following murine fetal zinc deprivation, *J. Nutr.* 112:1169.

Beach, R. S., Gershwin, M. E., and Hurley, L. S., 1982c. Nutritional factors and autoimmunity. II. Prolongation of survival in zinc-deprived NZB/W mice, *J. Immunol.* 128:308.

Beach, R. S., Gershwin, M. E., and Hurley, L. S., 1982d. Nutritional factors and autoimmunity. III. Zinc deprivation versus restricted food intake in MRL/1 mice—the distinction between interacting dietary influences, *J. Immunol.* 129:2686.

Bensman, A., Dardenne, M., Morgant, G., Vesmant, D., and Bach, J. F., 1985. Decrease of biological activity of serum thymic factor in children with nephrotic syndrome, *Int. J. Pediatr. Nephrol.* 5:201.

Blazsek, I., and Mathe, G., 1984. Zinc and immunity, *Biomed. Pharmacother.* 38:187.

Chandra, R. K., and Dayton, D. H., 1982. Trace element regulation of immunity and infection, *Nutr. Res.* 2:721.

Cunnigham-Rundles, C., Cunningham-Rundles, S., Iwata, T., Incefy, G., Garofalo, J. A., Menendez-Botet, C., Lewis, V., Twomey, J. J., and Good, R. A., 1981. Zinc deficiency, depressed thymic hormones, and T lymphocyte dysfunction in patients with hypogamma-globulinemia, *Clin. Immunol. Immunopathol.* 21:387.

Dardenne, M., Pleau, J. M., Nabarra, B., Lefrancier, P., Derrien, M., Choay, J., and Bach, J. F., 1982. Contribution of zinc and other metals to the biological activity of the serum thymic factor, *Proc. Natl. Acad. Sci. USA* 79:5370.

Dardenne, M., Savino, W., Wade, S., Kaiserlian, D., Lemonneir, D., and Bach, J. F., 1984. In vivo and in vitro studies of thymulin in marginally zinc deficient mice, *Eur. J. Immunol.* 14:454.

Douglas, R. G., Jr., Roberts, R. B., Romano, R., Metroka, C., Amberson, J., Soave, R., and Stover, D., 1984. Infectious complications in acquired immunodeficiency syndrome, in *Acquired Immune Deficiency Syndrome* (M. S. Gottlieb, Jr., and E. Groopman, eds.), Liss, New York, p. 321.

Duchateau, J., Delepesee, G., Virgens, R., and Collet, H. J., 1981. Beneficial effects of oral zinc supplementation on the immune response of old people, *Am. J. Med.* 70:1001.

Fernandes, G., Nair, M., Onoe, K., Tanaka, T., Floyd, R., and Good, R. A., 1979. Impairment of cell-mediated immunity functions by dietary zinc deficiency in mice, *Proc. Natl. Acad. Sci. USA* 76:457.

Fraker, P. J., 1983. Zinc deficiency: A common immunodeficiency state, *Surv. Immunol. Res.* 2: 155.

Fraker, P. J., DePasquale-Jardieu, P., Zwickl, C. M., and Luecke, R. W., 1978. Regeneration of T-cell helper function in zinc deficient adult mice, *Proc. Natl. Acad. Sci. USA* 75:5660.

Fraker, P. J., Haas, S., and Luecke, R. W., 1977. Effect of zinc deficiency on the immune response of the young adult A/J mouse, *J. Nutr.* 107:1889.

Fraker, P. J., Gershwin, M. E., Good, R. A., and Prasad, A. S., 1985. Interrelationships between zinc and immune function, *Fed. Proc.* 45:1474.

Frost, P., Rabbani, P., Smith, J., and Prasad, A. S., 1981. Cell mediated cytotoxicity and tumor growth in zinc deficient mice, *Proc. Soc. Exp. Biol. Med.* 167:333.

Gasintinel, L. N., Dardenne, M., Pleau, J. M., and Bach, J. F., 1984. Studies on the zinc binding to the serum thymic factor, *Biochim. Biophys. Acta* 797:147.

Giblett, E. R., Ammann, A. J., Wara, D. W., Sandman, R., and Diamond, L. K., 1975. Nucleoside phosphorylase deficiency in a child with severely defective T-cell immunity and normal B-cell immunity, *Lancet* 1:1010.

Golding, J. S., MacIver, J. E., and Went, L. H., 1959. The bone changes in sickle cell anemia and its genetic variants, *J. Bone Jt. Surg.* 41-B:711.

Golub, M. S., Gershwin, M. E., Hurley, L. S., Baly, D. L., and Hendrickx, A. G., 1984. Studies of marginal zinc deprivation in rhesus monkeys. I. Influences on pregnant dams, *Am. J. Clin. Nutr.* 39:265.

Good, R. A., and Fernandes, G., 1979. Nutrition, immunity, and cancer—A review. I: Influence of protein or protein-calorie malnutrition and zinc deficiency on immunity, *Clin. Bull.* 9:3.

Good, R. A., Fernandes, G., Garofalo, A., Cunningham-Rundles, C., Iwata, T., and West, A., 1982. Zinc and immunity, *Clinical, Biochemical, and Nutritional Aspects of Trace Elements* (A. S. Prasad, ed.), Liss, New York, p. 189.

Grummt, F., Waltl, G., Jantzen, H. M., Hamprecht, K., Huebscher, U., and Kuenzle, C. C., 1979. Diadenosine 5′,5″-D₁,D₄-tetraphosphate, a ligand of the 57-kilodalton subunit of DNA polymerase alpha, *Proc. Natl. Acad. Sci. USA* 76:6081.

Grummt, F., Weinman-Dorsch, C., Schneider-Schaulies, J., and Lux, A., 1986. Zinc as a second messenger of mitogenic induction, *Exp. Cell Res.* 163:191.

Haddad, J. D., John, J. F., and Pappas, A. A., 1984. Cytomegalovirus pneumonia in sickle cell disease, *Chest* 86:265.

Hardy, R. E., Cummings, C., Thomas, F., and Harrison, D., 1986. Cryptococcal pneumonia in a patient with sickle cell disease, *Chest* 89:892.

Haynes, D. C., Gershwin, M. E., Golub, M. S., Cheung, A. T. W., Hurley, L. S., and Hendrickx, A. G., 1985. Studies of marginal zinc deprivation in rhesus monkeys. VI. Influence on the immunohematology of infants in the first year, *Am. J. Clin. Nutr.* 42:252.

Laussac, J. P., Haran, R., Dardenne, M., Lefrancier, P., and Bach, J. F., 1985. Nuclear magnetic resonance study of the interaction of zinc with thymulin, *C.R. Acad. Sci. Ser. C* 301:471.

Ma, D. D., Wylwestrowicz, T., Janossy, G., and Hoffbrand, A. V., 1983. The role of purine metabolic enzymes and terminal deoxynucleotidyl transferase in intrathymic T cell differentiation, *Immunol. Today* 4:65.

Mann, J. R., Cotter, K. P., Walker, R. A., Bird, G. W. G., and Stuart, G., 1975. Anemia crisis in sickle cell disease, *J. Clin. Pathol.* 28:341.

Meftah, S., and Prasad, A. S., 1989. Nucleotides in lymphocytes of human subjects with zinc deficiency, *J. Lab. Clin. Med.* 114:114.

Meftah, S., Prasad, A. S., Lee, D.-Y., and Brewer, G. J., 1991. Ecto 5′-nucleotidase (5′NT) as a sensitive indicator of human zinc deficiency, *J. Lab. Clin. Med.* 118:309.

Mitchell, B. S., and Sidi, Y., 1985. Differential metabolism of guanine nucleosides by human lymphoid cell lines, *Proc. Soc. Exp. Biol. Med.* 179:427.

Nash, L., Iwata, T., Fernandes, G., Good, R. A., and Incefy, G., 1979. Effect of zinc deficiency on autologous rosette-forming cells, *Cell. Immunol.* 48:238.

Oleske, J. M., Westphal, M. L., Shore, S., Gorden, D., Bogden, J. D., and Nahimias, A., 1979. Zinc therapy of depressed cellular immunity in acrodermatitis enteropathica, *Am. J. Dis. Child.* 133:915.

Onwubalili, J. K., 1983. Sickle cell disease and infection, *Journal of Infection* 7(1):2.

Pilz, R. B., Willis, R. C., and Seegmiller, J. E., 1982. Regulation of human lymphoblast plasma membrane 5′nucleotidase, *J. Biol. Chem.* 257:13544.

Prasad, A. S., and Oberleas, D., 1971. Changes in activities of zinc dependent enzymes in zinc deficient tissues of rats, *J. Appl. Physiol.* 31:842.

Prasad, A. S., and Rabbani, P., 1981. Nucleoside phosphorylase in zinc deficiency, *Trans. Assoc. Am. Physicians* 94:314.

Prasad, A. S., Kaplan, J., and Abdallah, J., 1986. Zinc deficiency as a basis for immunodeficiencies in sickle cell anemia (SCA), in *Symposium: The Nature, Cellular and Biochemical Basis and Management of Immunodeficiencies* (R. A. Good and E. Lindenlaub, eds.), FKS Verlag, Stuttgart, p. 317.

Prasad, A. S., Meftah, S., Abdallah, J., Kaplan, J., Brewer, G. J., Bach, J. F., and Dardenne, M., 1988. Serum thymulin in human zinc deficiency, *J. Clin. Invest.* 82:1202.

Rabbani, P. I., Prasad, A. S., Tsai, R., Harland, B. F., and Fox, M. R. S., 1987. Dietary model of production of experimental zinc deficiency in man, *Am. J. Clin. Nutr.* 45:1514.

Rappaport, E., Zamecnik, P. C., and Baril, E. F., 1981. HeLa cell DNA polymerase a is tightly associated with tryptophanyl-tRNA synthetase and diadenosine 5',5''-P1,P4-tetraphosphate binding activities, *Proc. Natl. Acad. Sci. USA* 78:838.

Savino, W., Dardenne, M., Papiernik, M., and Bach, J. F., 1982. Thymic hormone-containing cells. Characterization and localization of serum thymic factor in young mouse thymus studied by monoclonal antibodies, *J. Exp. Med.* 156:628.

Sergeant, G. R., Topley, J. M., Mason, K., Sergeant, B. E., Pattison, S. E,. and Mohamed, R., 1981. Outbreak of aplastic crisis in sickle cell anemia associated with parvovirus like agent, *Lancet* 2:595.

Shulman, S. T., Bartlett, J., Clyde, W. A., and Ayoub, E. M., 1972. The unusual severity of mycoplasma pneumonia in children with sickle cell disease, *N. Engl. J. Med.* 287:164.

Tapazoglou, E., Prasad, A. S., Hill, G., Brewer, G. J., and Kaplan, J., 1985. Decreased natural killer cell activity in zinc deficient subjects with sickle cell disease, *J. Lab. Clin. Med.* 105: 19.

Zwickl, C. W., and Fraker, P. J., 1980. Restoration of the antibody mediated response of zinc/ caloric deficient neonatal mice, *Immunol. Commun.* 9:611.

Metabolism of Zinc 10

10.1 Dietary Sources of Zinc

The primary dietary sources of zinc are red meats, seafood, and cereals (Welsh and Marston, 1983). The zinc content of raw vegetables, legumes, cereals, and even meat is not retained after cooking and the availability of zinc in various foods may be adversely affected by ligands such as phytate. Because of the above, it is difficult to estimate the actual absorbed amounts of zinc solely on the basis of zinc as calculated from food tables.

10.2 Absorption

According to some investigators, zinc uptake by high-molecular-weight proteins in the intestinal mucosa is an active process requiring ATP (Menard and Cousins, 1983). Zinc uptake by low-molecular-weight proteins (such as metallothionein) may be passive. Uptake of zinc by the intestinal cells is bidirectional, inasmuch as zinc is taken up from the intestinal lumen and also from the blood supply. In rats, zinc absorption appears to occur throughout the small intestine.

One report showed that amino acids increased zinc absorption in the colon, a site not considered as important for zinc absorption by other investigators (Wapnir et al., 1983).

Studies by Menard and Cousins (1983) suggested that under conditions of normal dietary zinc uptake, the initial uptake of zinc to the brush border membrane of the mucosal cell appeared to depend on its net availability for uptake to, and transfer by, a membrane-associated carrier. The uptake increased significantly with zinc depletion. The net availability of zinc for uptake into the mucosal cell could be dependent on the relative affinity of zinc-binding ligands in the intestinal lumen and the membrane carrier.

We recently determined the intestinal site of zinc absorption in humans using the triple-lumen steady-state perfusion technique (Lee *et al.*, 1989). Seventeen healthy subjects participated in the study. During intestinal perfusion of a balanced electrolyte solution containing 0.1 mM zinc acetate, zinc absorption occurred throughout the entire small intestine (see Figs. 10-1 to 10-6). However, the jejunum had the highest rate of absorption (357 ± 14 nM/min per 40 cm) compared with the duodenum (230 ± 33 nM/min per 40 cm) and ileum (84 ± 10 nM/min per 40 cm). Over a range of zinc concentrations infused into the jejunum (0.1, 0.9, and 1.8 mM) there were linear increases in the rate of zinc absorption ($p < 0.05$). Intestinal absorption of zinc was significantly stimulated by the addition of glucose (20 mM). Zinc absorption increased from 459 ± 39 to 582 ± 45 nM/min per 40 cm ($p < 0.05$). Conversely, zinc (0.9 mM) also enhanced the absorption of glucose, which was increased from 293 ± 43 to 447 ± 27 mM/min per 40 cm ($p < 0.05$). The enhanced absorption of zinc in glucose was not accompanied by any increase in absorption of water and sodium. In contrast, increasing the concentration of zinc in the perfusate resulted in decreased absorption of sodium and water in a dose-related manner. In conclusion, our study demonstrated that zinc absorption is concentration dependent and occurs throughout the small intestine. The jejunum is the site of maximal absorption of zinc. The interaction between absorption of zinc and other solubles suggests that the transport process of zinc is carrier mediated.

The following factors are known to determine the net amount of zinc which may be absorbed by the intestinal cells: (1) ligands—zinc may be bound to one or more ligands, some of which will impede and others enhance absorption; (2) zinc status—zinc absorption is increased during zinc deficiency; (3) the active transport mechanism of zinc appears to be under metabolic control; and (4) endogenous zinc secretion—a significant amount of zinc is secreted into the intestinal lumen via the epithelial cells, bile, and pancreatic secretion.

10.2.1 Ligands

Most of the information concerning zinc absorption is limited to experimental animal model studies. During digestion, zinc is released from its dietary ligands (mostly proteins) and becomes associated with intestinal low-molecular-weight ligands which make zinc available to the intestinal microvilli. Some of these ligands such as histidine and other amino acids are of dietary origin but others such as metallothionein may be of endogenous origin. Factors affecting zinc absorption are listed in Table 10-1.

Figure 10-1. Rate of zinc absorption during infusion of 0.1 mM zinc acetate into jejunum over 60 min. Note that a steady state was achieved within the first 30 min of perfusion. Results from the last three periods were used for calculation of net solute movements in all subsequent studies. (Reprinted with permission from Lee, H. H., Prasad, A. S., Brewer, G. J., and Owyang, C., 1989. Zinc absorption in human small intestine, *Am. J. Physiol.* 256:G87.)

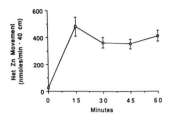

Putative ligands in milk include citric acid, picolinic acid, immunoglobulins, and lactoferrins. Whether picolinic acid or citric acid is the predominant low-molecular-weight zinc ligand in human milk has not been settled (Evans and Johnson, 1980; Hurley and Lonnerdal, 1982; Johnson and Evans, 1982). Although human milk contains less zinc than cow's milk, its availability is substantially better (Casey *et al.,* 1981). Some investigators have suggested that human milk contains a specific protein ligand for zinc, which is not present in large quantity in cow's milk (Eckhert, 1985; Eckhert *et al.,* 1977). The possibility also exists that the association of zinc with citric or picolinic acid is an artifact of isolation procedures. Although both citric acid and picolinic acid enhance zinc absorption, their physiological relevance is unclear.

Other compounds in milk which may or may not be ligands *per se* also facilitate zinc absorption. These include histidine, EDTA, D-penicillamine, essential fatty acids, and prostaglandins (Cunnane, 1988).

In one study, a human milk protein was isolated which enhanced zinc absorption (Eckhert, 1985). The consumption of cow's milk supplemented with this protein resulted in a significant elevation of human plasma zinc. During lactation, the zinc content of milk remained constant in spite of 60 mg zinc supplementation per day for 2 weeks, according to some investigators (Feeley *et al.,* 1983; Moore *et al.,* 1984). Prostaglandin E has been reported to increase zinc absorption, and in zinc-deficient rats a positive correlation between the zinc and prostaglandin E content of the gut has been observed

Figure 10-2. Rate of zinc, sodium, and water absorption during infusion of 0.1 mM zinc acetate in three segments (duodenum, jejunum, and ileum) of human small intestine. Values are means ± S.E. Significant difference from rates of absorption in duodenum. (Reprinted with permission from Lee, H. H., Prasad, A. S., Brewer, G. J., and Owyang, C., 1989. Zinc absorption in human small intestine, *Am. J. Physiol.* 256:G87.)

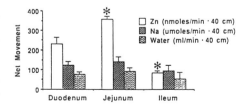

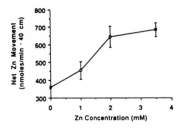

Figure 10-3. Relation between concentration of zinc acetate in perfusate and rate of zinc absorption in jejunum. Values are means ± S.E. (Reprinted with permission from Lee, H. H., Prasad, A. S., Brewer, G. J., and Owyang, C., 1989. Zinc absorption in human small intestine, *Am. J. Physiol.* 256:G87.)

(Song and Adham, 1985). Cottonseed oil improved the clinical condition of three cases of patients with acrodermatitis enteropathica prior to the advent of zinc therapy for this fatal condition (Cash and Berger, 1969). Since cottonseed oil does not contain appreciable amounts of zinc, citric acid, or picolinic acid, but does contain 54% linoleic acid, the observation in rats that essential fatty acids, excluding arachidonic acid, significantly increase zinc absorption may be pertinent to humans. Levels of linoleic acid and γ-linoleic acid are increased in the phospholipids of the intestinal mucosa of the zinc-deficient rats. It may be that although zinc is not bound to prostaglandins or essential fatty acids, they may facilitate its absorption by unknown mechanisms.

Calcium, copper, and iron compete with zinc for absorption. In animal studies, calcium has been observed to competitively inhibit absorption of zinc. In humans, dietary calcium alone does not inhibit zinc absorption. As a calcium phytate complex, inhibition of zinc absorption may occur in humans. Geophagia or pica (the practice of eating clay) may contribute to zinc deficiency because of its high content of calcium. In animal studies, competitive interaction of zinc and copper for absorption sites in the gut has been

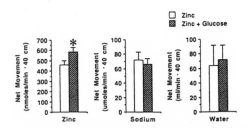

Figure 10-4. Effect of glucose on zinc absorption in jejunum. Significant enhancement of zinc absorption by addition of 20 mM glucose compared with control perfusate (0.9 mM zinc acetate). This was not accompanied by any significant effect on absorption of water and sodium. (Reprinted with permission from Lee, H. H., Prasad, A. S., Brewer, G. J., and Owyang, C., 1989. Zinc absorption in human small intestine, *Am. J. Physiol.* 256:G87.)

Figure 10-5. Effect of zinc on glucose absorption in jejunum. Significant enhancement of glucose absorption by addition of 0.9 mM zinc compared with control perfusate (20 mM glucose). (Reprinted with permission from Lee, H. H., Prasad, A. S., Brewer, G. J., and Owyang, C., 1989. Zinc absorption in human small intestine, *Am. J. Physiol.* 256:G87.)

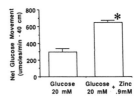

reported. These studies, however, used excessive levels of copper. It is therefore unlikely that in humans, under physiological conditions, copper exerts any inhibitory effect on zinc absorption. On the other hand, 50 mg of zinc supplementation taken three times a day by subjects with Wilson's disease results in a negative copper balance (Hill *et al.*, 1987).

In pregnant women, zinc absorption is decreased by oral iron supplements (Hambidge *et al.*, 1983; Meadows *et al.*, 1983). In a zinc tolerance test, a ratio of iron to zinc increasing from 0:1 to 3:1 given in aqueous solution (zinc kept constant at 25 mg), caused a progressive decrease in the plasma response to the ingested zinc in healthy nonpregnant women (Solomons and Jacob, 1981). The total number of ions as well as the iron-to-zinc ratio seem to be important because at an iron-to-zinc ratio of 2.5:1, absorption of zinc from water in fasting individuals was not different from controls (Sandstrom *et al.*, 1985).

In uremic subjects a 25% decrease versus 40% decrease in controls was observed in the peak plasma concentration when 300 mg of iron was given with 25 mg of zinc in a zinc tolerance test (Abu-Hamdan *et al.*, 1984). The apparently decreased zinc absorption was associated with lower fasting plasma zinc and was exacerbated by aluminum intake (Abu-Hamdan *et al.*, 1986). Because high aluminum intake is common in hemodialyzed patients, the above interactions are important for long-term management of these cases.

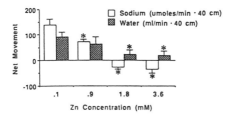

Figure 10-6. Effect of zinc on sodium and water absorption in jejunum. Increasing concentrations of zinc in perfusate resulted in decreased secretion of water and net secretion of sodium. Significant difference from rates of absorption when jejunum was perfused with 0.1 mM zinc acetate. (Reprinted with permission from Lee, H. H., Prasad, A. S., Brewer, G. J., and Owyang, C., 1989. Zinc absorption in human small intestine, *Am. J. Physiol.* 256:G87.)

Table 10-1. Factors Associated
with Decreased Zinc Absorption[a]

Dietary factors
  Calcium
  Copper
  Iron
  Fiber/phytate
  Alcohol
  Chelating agents
Absence of appropriate absorption ligands
  Acrodermatitis enteropathica
  Cystic fibrosis
  Pancreatic dysfunction
  Breast milk versus milk formulas
  Phenylketonuria
  Hypothyroidism
Gastrointestinal dysfunction
  Intestinal mucosal damage
  Malabsorption syndrome
  Gastrointestinal surgery

[a] Reprinted with permission from Cunnane, S. C., 1988. *Zinc: Clinical and Biochemical Significance,* CRC Press, Boca Raton, Fla., p. 34.

In contrast to adults, infants do not appear to be as prone to the inhibitory effects of iron on zinc metabolism. In healthy infants who are zinc sufficient, iron supplementation (up to 30 mg/day) does not appear to affect zinc nutriture (Yip *et al.,* 1985).

Phytate (phytic acid, myoinositol 1,2,3,4,5,6-hexakis dihydrogen phosphate) is a component of plant protein. Plant proteins, such as soy, increase the zinc requirement relative to proteins of animal origin in experimental animals. Excessive intake of phytate in unleavened bread has been implicated as an important factor responsible for zinc deficiency in Middle Eastern dwarfs (Prasad *et al.,* 1961). Negative zinc balance as a result of fiber supplementation has been demonstrated (Latta and Liebman, 1984). However, the effect has not been consistently observed in humans or in animals, which may reflect the different effects of soluble and insoluble fiber (Anderson *et al.,* 1981; King *et al.,* 1981; Swanson *et al.,* 1983). In vegetarians, phytate and fiber intake is significantly higher than in individuals eating an omnivorous diet and it is likely that their combined effect on zinc absorption is clinically more important.

Alcohol decreases zinc absorption in the rat (Antonson and Vanderhoff, 1983; Silverman *et al.,* 1979). In humans, the effect of alcohol on zinc absorption is not resolved. Sullivan *et al.* (1979), using a zinc tolerance test, showed that zinc absorption was decreased by concurrent alcohol intake. Recently, absorption of [65]Zn was shown to be lower in patients with alcoholic cirrhosis than in controls (Valberg *et al.,* 1985). In alcoholics consuming a standard meal, absorption of 50 mg but not 25 mg of zinc was decreased (Dinsmore *et al.,* 1985). Alcoholics also exhibit hyperzincuria and low levels of plasma zinc. Chronic use of alcohol appears to be a common cause of zinc deficiency in humans.

EDTA reduces zinc absorption in humans (Solomons *et al.,* 1979; Spencer *et al.,* 1966). Since EDTA is used to remove zinc from vegetables during processing, it appears to have two effects, one to decrease its content and the other to decrease its absorption. Thus, depending on the frequency and quantity of EDTA being used, one may observe a negative effect on zinc homeostasis in humans.

An intact pancreatic function appears to be necessary for normal zinc absorption. In hypothyroidism, a decreased bidirectional transport of zinc across the gut mucosa has been reported. This defect is correctable with thyroxine administration.

10.2.2 Zinc Status

Zinc absorption is increased in zinc-deficient rats and decreased in rats fed excess zinc (Cunnane, 1982; Evans *et al.,* 1975; Flanagan *et al.,* 1983). Zinc absorption is also increased in pregnant rats, and reverts to normal after parturition.

10.2.3 Intracellular Transport

The mechanism by which zinc is transferred to or across the mucosal surface of the microvilli is unknown. Once in the intestinal cell, zinc becomes associated with metallothionein and perhaps other proteins. This process appears to be dependent on protein synthesis, since cycloheximide and actinomycin D inhibit mucosal zinc uptake (Richards and Cousins, 1976). The precise role of intestinal metallothionein in zinc absorption is not well understood.

Both transferrin and albumin have been proposed as likely serum proteins involved in transport of zinc in blood subsequent to its absorption.

10.2.4 Endogenous Secretion

The secretion of endogenous zinc into the intestinal lumen in pancreatic secretions has been established in animal studies (Davies, 1980). Mateseche *et al.* (1980) have demonstrated that even under "normal" conditions, nearly as much zinc could be endogenously secreted from the pancreas as was absorbed. The amount of enteropancreatic circulation of zinc appears to depend on the origin of the dietary protein. The secretion of endogenous zinc is increased when the dietary protein is of plant origin rather than of animal origin.

Lee *et al.* (1990) have examined basal and cholecystokinin-stimulated pancreaticobiliary secretion of zinc in normal subjects on zinc-adequate and zinc-deficient diets, and whether basal and stimulated secretion of zinc was abnormal in Wilson's disease patients before and after zinc therapy. Gastroduodenal intubation was performed in six healthy subjects and five Wilson's disease patients. Following intravenous infusion of octapeptide of cholecystokinin (CCK-8) (40 ng/kg per h) the pancreaticobiliary secretion of zinc increased from a basal level of 283.1 ± 75.8 nM/min to a peak of 716.6 ± 175.3 nM/min in normal subjects. Normal subjects on a zinc-deficient diet had both lower basal (66.8 ± 15.8 nM/min) and stimulated (559.5 ± 31 nM/min) pancreaticobiliary secretion of zinc than while on a zinc-sufficient diet (Fig. 10-7). In contrast to the markedly reduced pancreaticobiliary secretion of copper, patients with Wilson's disease not treated with zinc had normal basal (226.6 ± 126 nM/min) and stimulated (728.7 ± 195.5 nM/min) zinc secretion. These studies indicate that a considerable amount of zinc is being secreted in pancreaticobiliary fluid in healthy subjects and there was no impairment of zinc secretion in Wilson's disease patients. Our data also indicate that pancreaticobiliary secretion of zinc is dependent on the zinc status of the

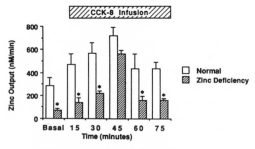

Figure 10-7. Pancreaticobiliary zinc secretion during basal and CCK-8 infusion (40 ng/kg per h) period in normal and zinc-deficient subjects. Asterisk denotes output significantly different from values obtained from normal subjects ($p < 0.05$). (Reprinted with permission from Lee, H. H., Hill, G. M., Sihka, V. K. N. M., Brewer, G. J., Prasad, A. S., and Owyang, C., 1990. Pancreaticobiliary secretion of zinc and copper in normal persons and patients with Wilson's disease, *J. Lab. Clin. Med.* 116:283.)

subjects, suggesting that endogenous secretion of zinc may play a significant role in the homeostasis of zinc.

10.3 Intake and Excretion of Zinc

Until recently, only meager information was available on the zinc requirement in man. Zinc-balance studies carried out nearly 30 years ago indicated that on a dietary zinc intake ranging from 4 to 6 mg/day, the zinc balance was negative in preschool children (Scoular, 1939). In children aged 8–12, the mean retention of zinc was 4.9 mg/day on a zinc intake ranging from 14 to 18 mg/day, indicating a high requirement of zinc during growth and development. In young adults aged 17 to 27, the retention of zinc was also very high, 5–8 mg/day, on a zinc intake ranging from 12 to 14 mg/day (Tribble and Scoular, 1954). In a study of preadolescent girls, about 30% of a dietary zinc intake averaging 7 mg/day was retained (Engel *et al.*, 1966). Variable results have been reported for adults. In early studies, the zinc balance was reported to be in equilibrium in three subjects on a zinc intake of 5 mg/day; however, the retention of zinc was similar in two others receiving twice this amount, while the zinc balance was more positive, ranging up to 2.6 mg/day on a zinc intake of 20 mg/day (McCance and Widdowson, 1942). In a long-term study of two adults receiving a self-selected diet, one subject was in a negative zinc balance, -4 mg/day, on a diet containing 11 mg zinc/day, while the other, receiving a dietary zinc intake of 18 mg/day, was in a positive zinc balance of 1 mg/day (Tipton *et al.*, 1969). In another study of young women, the zinc balance was in equilibrium on a dietary zinc intake of about 11 mg/day (White and Gynee, 1971), while in a study of young men receiving a synthetic diet having a zinc content of about 20 mg/day, the retention of zinc was high and ranged from 7 to 8 mg/day (Gormican and Catli, 1971).

The protein content of the diet appears to influence the absorption and retention of zinc. The zinc balance of preadolescent girls receiving a zinc intake of about 5 mg/day on a low-protein diet was in equilibrium, the zinc retention ranging from 0.5 to 0.8 mg/day (Price *et al.*, 1970), while on a high-protein diet that had only a slightly higher zinc content, about 6.7 mg/day, the retention of zinc was about 2 mg/day. The obligatory zinc retention during periods of growth and during stress was calculated by Sandstead (1973a,b). The zinc retention for growing children, age 11 years, and for young adults, age 17 years, was estimated to exceed 0.4 mg/day for pregnant women; the retention was 0.7 mg/day.

The zinc content of a diet depends on the dietary protein content. A diet containing about 1 g protein/kg body wt for a 70-kg man is expected to contain about 12.5 mg zinc. A diet that is adequate in calories but has a low protein content may contain less than half this amount of zinc, whereas the zinc content of a high-protein diet may be two to three times as high as that of the low-protein diet. The main source of the dietary zinc is meat. However, marked differences in the zinc content of different types of meat have been demonstrated, red meat having the highest zinc content (Rose and Willden, 1972). Fish also has a relatively high zinc content. Most other food items that constitute the daily diet have a zinc content of 1 mg or less per 100 g wet wt.

On normal zinc intake of about 12.5 mg/day, the urinary zinc excretion is low, ranging from 0.5 to 0.8 mg/day, while most of the zinc is excreted in the stool (Spencer et al., 1976). The fecal excretion of zinc is only partly the result of unabsorbed zinc. The remainder results from the secretion of zinc from the vacuolar space into the intestine. Studies of ^{65}Zn administered intravenously in man indicate that the intestinal secretion of ^{65}Zn is as high as 18% of the administered zinc tracer (Spencer et al., 1965a,b).

The zinc balances in adults were in equilibrium on a zinc intake of 12.5 mg/day in one study (Spencer et al., 1976). These balances were quite similar in consecutive study periods. It was also observed, however, that not all subjects are in zinc equilibrium on this intake, and that in some cases, a higher zinc intake of about 15 mg/day is needed to attain equilibrium. All zinc-balance studies in man reported thus far have to be considered as maximal balances, since the loss of zinc in sweat has not been considered and the balances are based only on the zinc intake and the excretion of zinc in the urine and stool. The loss of zinc in sweat in man was reported to be 1.15 mg/liter for whole sweat and 0.9 mg/liter for cell-free sweat (Prasad et al., 1963c).

Persons with a normal body weight who maintain themselves on a low-calorie intake for prolonged periods of time, e.g., patients with chronic alcoholism, show a very high retention of zinc when given a normal dietary intake. This high zinc retention was observed to continue for several months until the ideal body weight was attained.

When the dietary intake of zinc is completely eliminated, as in total starvation used for weight reduction in marked obesity, the rapid weight loss is associated with a considerable loss of zinc. This zinc loss occurs promptly with onset of the weight loss and results primarily from a marked increase of the urinary zinc excretion (Spencer et al., 1976).

The high excretion of zinc during starvation resulting in weight loss is most likely caused by the catabolic state and breakdown of muscle. Although muscle tissue has a very low concentration of zinc (Mansouri et al., 1970; Spencer et al., 1965a,b), the muscle mass is very large and contains a high percentage of the total body zinc (Schroeder et al., 1967). During weight loss, the loss of muscle tissue is considerable in addition to the loss of fat and water. In studies of zinc metabolism following injury and after surgical procedures, the increased excretion of zinc in urine was related to loss of muscle tissue (Fell *et al.*, 1973; Davies and Fell, 1974). This finding was supported by the observation that there was a good correlation between the urinary zinc/creatine ratio and the urinary excretion of nitrogen (Fell *et al.*, 1973).

Human metabolic balance as reported by Sandstead (1982) suggested that dietary protein intake influenced the dietary requirement for zinc. Dietary zinc requirements (mg/day) were calculated for men consuming mixed American diets containing four different levels of protein and phosphorus (Table 10-2). Zinc requirements were increased when dietary protein and phosphorus were high. The wide 95% confidence intervals suggested that some individuals maintained zinc homeostasis when their dietary intakes of zinc were substantially below levels that impaired zinc homeostasis of other individuals. According to this report, the mean requirement for zinc, exclusive of zinc loss in sweat, for Americans consuming 100 g of protein and 1500 mg of phosphorus per day, was 12.57 mg.

The mechanisms by which phosphorus and nitrogen (i.e., protein) intakes influence the requirement for zinc are not defined. Perhaps the increased dietary phosphate interferes with zinc absorption through formation of insoluble complexes with zinc and other substances such as fat in the intestinal lumen. The positive effect of dietary protein on the requirement for zinc may be related to the essential role of zinc in the utilization of amino acids for protein synthesis.

Table 10-2. Relationship of Dietary Zinc Requirements to Dietary Phosphorus and Protein Intakes of Men Fed Fixed American Diets[a,b,*]

Protein (g)	Zinc (mg)			
	40	60	80	100
Phosphorus (mg)				
1000	5.27	6.91	8.54	10.17
	(2.48–8.07)[c]	(4.11–9.70)	(5.74–11.33)	(7.38–12.96)
1500	9.11	10.27	11.42	12.57
	(6.32–11.91)	(7.47–13.06)	(8.62–14.21)	(9.78–15.36)
2000	12.95	13.63	14.30	14.97
	(10.16–15.75)	(10.83–16.42)	(11.50–17.09)	(12.18–17.76)
2500	16.79	16.99	17.18	17.37
	(14.00–19.59)	(14.19–19.78)	(14.38–19.97)	(14.58–20.16)

[a] Adapted with permission from Sandstead, H. H., 1982. Availability of zinc and its requirement in human subjects, in *Clinical, Biochemical, and Nutritional Aspects of Trace Elements* (A. S. Prasad, ed.), Liss, New York, p. 83.
[b] These data are exclusive of the zinc loss in sweat, which was 0.50 ± 0.38 mg for 88 24-h collections on 13 men under temperate conditions.
[c] Estimated 95% confidence interval.

Zinc content is a major determinant of the adequacy of foods as sources of zinc. The main sources of zinc are meat, poultry, fish, and dairy products followed by cereals and legumes. Because of their greater content and availability of zinc, foods such as meat, fish, and poultry are the best sources of this element.

10.4 Excretion of Zinc

Zinc excretion is primarily via the feces. Fecal excretion ranges from 5 to 10 mg/day and depends on the dietary zinc. In pathological conditions accompanied by diarrhea and malabsorption, the excessive fecal loss of zinc may rapidly result in negative zinc balance. Fecal zinc is comprised of unabsorbed dietary zinc and endogenous zinc loss from bile, pancreatic fluid, and intestinal mucosa.

Zinc is also excreted in the urine. Approximately 200 to 600 μg zinc is lost per day in the urine. Urinary zinc excretion appears to be sensitive to alterations in the zinc status (Prasad et al., 1977).

Zinc clearance studies in anesthetized dogs were performed during hydropenia, mannitol infusion, and infusion of mannitol plus $ZnSO_4$, $ZnCl_2$, or cysteine (Abu-Hamdan et al., 1981). Mannitol expansion caused no significant change in Zn clearance. $ZnSO_4$ infusion increased filtered Zn 13-fold without changing clearance. Zn excretion increased only 6-fold, indicating increased net Zn reabsorption. Cysteine infusion increased urinary Zn excretion 86-fold, indicating net tubular Zn secretion, some of which derived from nonplasma sources. Stop-flow studies localized Zn reabsorption to distal nephron during infusion of mannitol and mannitol plus $ZnSO_4$ or $ZnCl_2$. Net Zn secretion was shown to occur in the proximal tubule during cysteine infusion with reversal of the distal reabsorption pattern seen during $ZnSO_4$ and $ZnCl_2$ infusion. Despite increased urinary Zn excretion during $ZnSO_4$ infusion, calcium excretion was unaltered. During cysteine infusion, dissociation of tubular handing of Ca^{2+} and Zn occurred in both the proximal and distal tubule. These experiments demonstrate that the nephron under these experimental conditions is capable of both proximal secretion and distal reabsorption of Zn.

10.5 Requirements for Zinc

The U.S. RDA recommends that adults (both sexes) consume 15 mg of zinc daily (Table 10-3). Other government advisory bodies have set different recommendations for zinc intake. The Canadian Bureau of Nutritional Sci-

Table 10-3. Recommended Dietary Allowance for Zinc[a]

	Infants		Children	Males	Females	Pregnant	Lactating
Age (years)	0–0.5; 0.5–1.0		1–10	11–51+	11–51+		
Zinc (mg)	3	5	10	15	15	20	25

[a] Reprinted with permission from Sandstead, H. H., 1982. Availability of zinc and its requirement in human subjects, in *Clinical, Biochemical, and Nutritional Aspects of Trace Elements* (A. S. Prasad, ed.), Liss, New York, p. 83.

ences has recommended that a lower zinc intake (9 mg/day for adult males and 8 mg/day for females) is safe and adequate.

The U.S. RDA is viewed as controversial because in the United States, i.e., an ostensibly healthy population, only a small percentage of individuals consume the recommended intake of zinc (Sandstead, 1973a, 1973b; Sandstead, 1982). Thus, it has been argued that if the population is healthy, its zinc requirement cannot be as high as recommended by the U.S. RDA.

A comparison of the zinc content of some published diets with recommended daily allowances (USA and WHO) indicates that many individuals consume substantially less zinc than the U.S. RDA and that a significant proportion of the population may have marginal intakes by WHO standards (Table 10-4). This may therefore predispose to a mild or marginal deficiency of zinc. This is supported by clinical observations on low-income children who participated in a Head-Start program (Hambidge *et al.,* 1976), the zinc content of typical diets eaten by poor children in the southeastern United States (Price *et al.,* 1970; Sandstead, 1973a,b), assessments of zinc status of pregnant women with abnormalities of labor and delivery (Jameson, 1976), assessment of zinc nutriture of elderly persons (Greger, 1977; Prasad, 1988), and a recent report of zinc-related immunological abnormalities observed in so-called "healthy" adults in the United States (Prasad *et al.,* 1988; Meftah and Prasad, 1989).

In the United States, although a severe dietary deficiency of zinc in the absence of conditioning factors is unlikely, a mild and marginal deficiency of zinc is more likely to occur in subjects consuming a low-protein and high-phosphorus diet when cereals are the predominant food and animal products are infrequently consumed. Such foods are high in phytate and dietary fiber. The risk of zinc deficiency (moderate or severe) in such cases is substantially increased by conditioning factors such as blood loss by parasitic infestations, geophagia, and in excessive loss of zinc related to sweating, as reported from the Middle East (Prasad *et al.,* 1961, 1963a–c). Schoolboys from the same population without these conditioning factors were not overtly zinc deficient, but displayed zinc-responsive retardation of growth and maturation (Ronaghy

Table 10-4. Provisional Dietary Requirements for Zinc in Relation to Estimates of
Retention, Losses, and Availability[a]

Age	Peak daily retention (mg)	Urinary excretion (mg)	Sweat excretion (mg)	Total required (mg)	Milligrams necessary in daily diet if content of available zinc is		
					10%	20%	40%
Infants							
0–4 months	0.35	0.4	0.5	1.25	12.5	6.3	3.1
5–12 months	0.2	0.4	0.5	1.1	11.0	5.5	2.8
Males							
1–10 years	0.2	0.4	1.0	1.6	16.0	8.0	4.0
11–17 years	0.8	0.5	1.5	2.8	28.0	14.0	7.0
18 years plus	0.2	0.5	1.5	2.2	22.0	11.0	5.5
Females							
1–9 years	0.15	0.4	1.0	1.55	15.5	7.8	3.9
10–13 years	0.65	0.5	1.5	2.65	26.5	13.3	6.6
14–16 years	0.2	0.5	1.5	2.2	22.0	11.0	5.5
17 years plus	0.2	0.5	1.5	2.2	22.0	11.0	5.5
Pregnant women							
0–20 weeks	0.55	0.5	1.5	2.55	25.5	12.8	6.4
20–30 weeks	0.9	0.5	1.5	2.9	29.0	14.5	7.3
30–40 weeks	1.0	0.5	1.5	3.0	30.0	15.0	7.5
Lactating women	3.45	0.5	1.5	5.45	54.5	27.3	13.7

[a] Reprinted with permission from Sandstead, H. H., 1982. Availability of zinc and its requirement in human subjects, in *Clinical, Biochemical, and Nutritional Aspects of Trace Elements* (A. S. Prasad, ed.), Liss, New York, p. 83.

et al., 1974). Other studies have shown that such individuals eventually grow and achieve maturity without the benefit of zinc supplementation, but their final growth is below that of their genetic peers from higher economic strata (Prasad, 1966; Coble *et al.,* 1966).

Our own data indicate that endogenous gastrointestinal secretion of zinc in an adult male may be approximately 4.0 to 5.0 mg/day (unpublished observation). If one adds to this 0.5 mg of zinc excretion in the urine, 0.5 mg of dermal zinc loss, and additional loss in other secretions (e.g., seminal fluid), and presuming that only 40% of zinc is available for absorption on an average animal protein diet, 12 to 13 mg zinc intake per day would be needed to meet the daily requirements. In view of the above and a recent report that a marginal deficiency of zinc is not uncommon in the so-called "normal" population of the United States, the U.S. RDA of 15 mg/day zinc intake is not unreasonable for adults in Western countries.

It is extremely difficult to precisely determine the requirement of zinc in various populations, since many factors seem to affect the bioavailability of zinc and factors such as environmental and physiological demands in different age groups alter the requirements of zinc. Nonetheless, the U.S. guidelines, in my view, appear reasonable with respect to the RDA for zinc. Tables 10-5 and 10-6 give the zinc content of various foods and their contribution to dietary zinc intake.

10.6 Peripheral Utilization

Apart from studies of zinc uptake by the liver (Failla and Cousins, 1975; Pattison and Cousins, 1986a, 1986b), very little is known about the utilization

Table 10-5. Zinc Content of Common Household Portions of Select Foods[a]

Food	Portion	Zinc (mg)
Fish, light poultry meat, shellfish (except crab and oyster)	3 oz	<2.0
Poultry liver, dark chicken meat	3 oz	2.0/3.0
Pork, veal, crab, dark turkey meat, ground beef (77% lean)	3 oz	3.0/4.0
Beef liver, beef	3 oz	4.0/5.0
Oyster	3 oz	>5.0
Egg (whole)	1	0.5
Peanut butter	2 tbsp	0.9
Mature dried beans, lentils, chick-peas, split peas (boiled, drained)	$\frac{1}{2}$ cup	0.9/1.0
Cowpeas, black-eyed peas (boiled, drained)	$\frac{1}{2}$ cup	1.5
Milk: whole fluid	1 cup	0.9
Canned, evaporated	$\frac{1}{2}$ cup	1.0
Dried, nonfat, instant	$\frac{1}{3}$ cup	1.0
Ice cream	$1\frac{1}{2}$ cup	1.0
Cheddar cheese	3 slices ($1\frac{1}{2}$ oz)	1.6
Cooked oatmeal	1 cup	1.2
Cooked whole wheat cereal	1 cup	1.2
Wheat flakes	1 oz	0.6
Bran flakes (40%)	1 oz	1.0
Wheat germ (toasted)	1 tbsp	0.9
Corn flakes	1 oz	0.08
Cooked corn meal	1 cup	0.3
White wheat bread	1 slice	0.2
Whole wheat bread	1 slice	0.5
Cooked brown rice (hot)	1 cup	1.2
Cooked white rice (hot)	1 cup	0.8
Precooked white rice (hot)	1 cup	0.4

[a] Adapted with permission from Sandstead, H. H., 1982. Availability of zinc and its requirement in human subjects, in *Clinical, Biochemical, and Nutritional Aspects of Trace Elements* (A. S. Prasad, ed.), Liss, New York, p. 83.

Table 10-6. Contribution of Foods to Dietary Intake of Zinc[a]

Food group	mg/kg as purchased	mg consumed/ person per day	Percent of total food intake
Milk, cheese, ice cream	4.3	2.5	20
Meat, poultry, fish	18.6	5.5	43
Dry beans, peas, nuts	24.7	0.5	4
Eggs	12.7	0.6	5
Dark green and deep yellow vegetables	2.8	0.1	1
Citrus fruit, tomatoes	1.8	0.3	2
Potatoes	2.4	0.3	2
Other vegetables, fruit	2.0	0.7	6
Cereal, pasta	16.3	0.8	7
Flour, mixes	2.8	0.1	1
Bread	7.2	0.6	5
Other bakery products	6.0	0.4	3
Fats, oils	1.8	0.1	1
Sugar, sweets	0.6	0	0
Total food		12.5	100

[a] Reprinted with permission from Sandstead, H. H., 1982. Availability of zinc and its requirement in human subjects, in *Clinical, Biochemical, and Nutritional Aspects of Trace Elements* (A. S. Prasad, ed.), Liss, New York, p. 83.

of zinc by peripheral tissues. Studies of ^{65}Zn uptake by isolated rat hepatocytes in culture showed a rapid initial phase which was not carrier-mediated, and a slower second phase. Neither phase was affected by cyanide, prostaglandins, or sex steroids, suggesting that active transport was not involved (Stacey and Klassen, 1981; Failla and Cousins, 1975). High (65,000)- and low (metallothionein, about 6000)-molecular-weight proteins were involved in zinc uptake by hepatocytes. When excess zinc was available, the lower weight proteins were favored.

According to a recent study, kinetic analysis of zinc uptake by isolated rat liver parenchymal cells defined two intracellular pools (Pattison and Cousins, 1986a, 1986b). In one pool zinc was bound relatively weakly and equilibrated rapidly with the medium at 37°C (labile pool). In the other pool zinc appeared to be bound tightly and equilibrated slowly with the medium at 37°C. Zinc uptake was temperature dependent and was inhibited by both N-ethylmaleimide and iodoacetamide, suggesting that sulfhydryls may be involved in one or more steps in the translocation/binding process. Metabolic inhibitors (e.g., azide, cyanide, and oligomycin) inhibited uptake of zinc. The factors that augmented the uptake/exchange of zinc, namely glucocorticoids, glucagon, epinephrine, and dibutyryl cyclic AMP, were also those that stimulated metallothionein gene expression in hepatocytes. Changes in zinc flux into intracellular pools were directly related to the metallothionein content of hepatocytes. Characterization of the labile zinc pool suggested that it served as the initial intermediate in zinc metabolism by hepatocytes as well as more general aspects of liver functions related to zinc.

It has been suggested that the peripheral utilization of zinc is abnormal in patients with myotonic dystrophy, thereby causing localized zinc depletion in cardiac and skeletal muscle. ^{65}Zn kinetic analysis in humans showed that muscle, erythrocytes, and active secretion by the gut were secondary sites governing the handling of excess zinc (Horrobin and Morgan, 1980). Muscle catabolism is also an important mechanism by which zinc can be released to maintain fetal growth in zinc-deficient pregnant rats (Masters *et al.,* 1986).

In pregnancy, there is a reduced maternal fecal zinc excretion, suggesting that peripheral utilization of zinc by the fetoplacental unit may be increased.

10.7 Blood Transport—Zinc in Plasma and Red Cells

Zinc in serum is 16% higher than in plasma (Foley *et al.,* 1968). The higher content of zinc in serum has been attributed to the liberation of zinc from the platelets during the process of clotting and to invisible hemolysis of red cells, which occurs regularly. Values for the plasma zinc in normal subjects obtained by different investigators using various techniques are with few exceptions in reasonably good agreement. Better methods of avoiding contamination and more precise analytical tools now provide accurate data for plasma zinc. Using our techniques the plasma zinc concentration (mean ± S.D.) in normal subjects is 110.7 ± 14.8 µg/dl (Whitehouse *et al.,* 1982).

Plasma zinc levels in the newborn are in the same range as adults. The levels fall below those of the adult within the first week of life and continue to decline until 3 months. At 4 months of age, the plasma zinc level attains the normal adult level. Some investigators have reported decreasing plasma

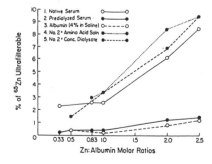

Figure 10-8. Percentage of ultrafilterable ^{65}Zn plotted against zinc/albumin molar ratios. (Reprinted with permission from Prasad, A. S., and Oberleas, D., 1970. Binding of zinc to amino acids and serum proteins in vitro, *J. Lab. Clin. Med.* 76:416.)

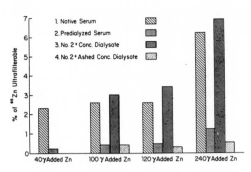

Figure 10-9. Percentage of ultrafilterable ^{65}Zn in native serum, predialyzed serum, predialyzed serum reconstituted with concentrated dialysate, and predialyzed serum reconstituted with ashed concentrated dialysate at different levels of Zn protein molar ratios (0.33, 0.83, 1.0, and 2.0). (Reprinted with permission from Prasad, A. S., and Oberleas, D., 1970. Binding of zinc to amino acids and serum proteins in vitro, *J. Lab. Clin. Med.* 76:416.)

zinc values with increasing age over 60 years (Lindeman *et al.*, 1972; Wilden and Robinson, 1975; Hallbook and Hedelin, 1977). Whether or not this decrease implies an increased requirement for zinc in the elderly subjects remains to be established.

The binding of zinc to amino acids and serum protein was studied *in vitro* by Prasad and Oberleas (1970). Following incubation of ^{65}Zn with pooled native human serum *in vitro,* ultrafilterable zinc was determined to be 2–8% of the total serum zinc, when the zinc/albumin molar ratio was varied from 0.33 to 2.5 (see Figs. 10-8 to 10-13). Under similar conditions, 0.2–1.2% of zinc was ultrafilterable when predialyzed serum was used. At physiological concentrations, addition of amino acids to predialyzed serum increased ultrafilterable ^{65}Zn severalfold. Histidine, glutamine, threonine, cystine, and lysine showed the most marked effect in this regard. It was suggested that the amino-acid-bound fraction of zinc may have an important role in biological transport of this element. By means of starch-block electrophoresis of pre-

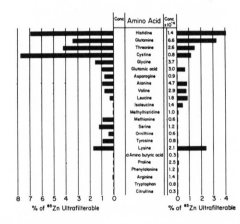

Figure 10-10. Effect of addition of single amino acids to predialyzed serum on the percentage of ultrafilterable ^{65}Zn at a Zn/albumin ratio of 2.0. The concentration of amino acids is expressed at 10^{-4} M. (Reprinted with permission from Prasad, A. S., and Oberleas, D., 1970. Binding of zinc to amino acids and serum proteins in vitro, *J. Lab. Clin. Med.* 76:416.)

Figure 10-11. Effects of additions of physiological concentrations of amino acids in various combinations to samples of predialyzed serum compared with the sum of their individual effects. Zn/albumin molar ratio, 2.0. (Σ) Sum of individual effects; (Cys) cysteine; (Cit) citrulline; (His) histidine; (Gln) glutamine; (Thr) threonine; (Lys) lysine; (Gly) glycine; (Glu) glumatic acid; (Ala) alanine; (Val) valine; (Leu) leucine; (Ile) isoleucine; (Me His) 1-methyl histidine; (Asn) asparagine; (Met) methionine; (Ser) serine; (Orn) ornithine; (Tyr) tyrosine; (αABA) α-aminobutyric acid; (Pro) proline; (Phe) phenylalanine; (Arg) arginine; (Trp) tryptophan. (Re-

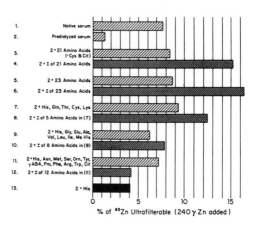

printed with permission from Prasad, A. S., and Oberleas, D., 1970. Binding of zinc to amino acids and serum proteins in vitro, *J. Lab. Clin. Med.* 76:416.)

dialyzed serum, the stable zinc content was determined to be highest in the albumin fraction, although smaller concentrations of zinc were found in α-, β-, and γ-globulins as well. The results obtained using ^{65}Zn-incubated predialyzed serum, however, indicated a difference in the behavior of exogenous zinc compared with the endogenous zinc bound to various serum proteins. *In vitro* studies employing predialyzed albumin, haptoglobin, ceruloplasmin, α_2-macroglobulin, transferrin, and IgG, incubated with ^{65}Zn, revealed that zinc was bound to all of these proteins, and that the binding of zinc to IgG was electrostatic in nature. Whereas amino acids competed effectively with

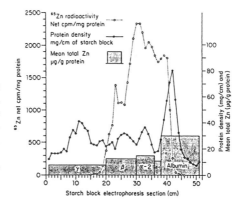

Figure 10-12. Distribution of endogenous stable zinc and exogenous ^{65}Zn on starch block, following electrophoresis of predialyzed normal human serum incubated with ^{65}Zn (without carrier zinc). (Reprinted with permission from Prasad, A. S., and Oberleas, D., 1970. Binding of zinc to amino acids and serum proteins in vitro, *J. Lab. Clin. Med.* 76:416.)

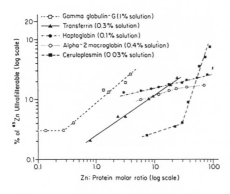

Figure 10-13. Changes in percentages of ultrafilterable ^{65}Zn at different Zn/protein molar ratios for various predialyzed proteins. Data are presented on a log–log scale. (Reprinted with permission from Prasad, A. S., and Oberleas, D., 1970. Binding of zinc to amino acids and serum proteins in vitro, *J. Lab. Clin. Med.* 76:416.)

albumin, haptoglobin, transferrin, and IgG for binding of zinc, a similar phenomenon was not observed with respect to ceruloplasmin and α_2-macroglobulin, suggesting that the latter two proteins exhibited a specific binding property of zinc.

The uptake of transferrin-bound zinc by human lymphocytes is stimulated *in vitro* by several agents such as prostaglandin E_1, epinephrine, glucagon, histamine, and serotonin (Phillips, 1978). Prostaglandin $F_{2\alpha}$, however, is inhibitory. The mechanism is unknown.

10.8 Homeostasis of Zinc

There is no known capacity in mammals to store zinc with the possible exception of zinc stored in the bone. Homeostasis of zinc thus depends on

Table 10-7. Body Fluid, Hair, and Blood Cell Levels of Zinc in Humans

	Zn (mean ± S.D.)	Reference
Plasma	110.7 ± 14.8 μg/dl	Whitehouse *et al.* (1982)
Erythrocytes	40.67 ± 3.60 mg/g Hb	Whitehouse *et al.* (1982)
Lymphocytes	50.9 ± 5.9 μg/10^{10} cells	Wang *et al.* (1989)
Granulocytes	45.9 ± 6.4 μg/10^{10} cells	Wang *et al.* (1989)
Platelets	3.3 ± 0.3 μg/10^{10} cells	Wang *et al.* (1989)
Hair	193 ± 18 μg/g	Prasad (1976)
Urine	643 ± 198 μg/dl	Prasad (1966)
Milk	1.2 ± 3.6 μg/ml	Fransson and Lonnerdal (1983)
Bile	171 ± 106 μg/ml	Strain *et al.* (1974)
Whole sweat	115 ± 30 μg/dl	Prasad *et al.* (1963c)
Cell-free sweat	93 ± 26 μg/dl	Prasad *et al.* (1963c)
Seminal plasma	692 ± 24 μg/ml	Schoenfeld *et al.* (1979)
Cerebrospinal fluid	0.03 ± 0.02 μg/ml	Agarwal and Henkin (1982)

Table 10-8. Zinc Concentration in Human and Animal Tissues (mg/kg Dry wt)

Tissue	Human	Rat		Calf		Pig	
		Normal	Zinc-deficient	Normal	Zinc-deficient	Normal	Zinc-deficient
Liver	141–245[a]	101 ± 13	89 ± 12	101	84	150.8 ± 12	96.1 ± 8
Kidney	184–230	91 ± 3	80 ± 3	73	76	97.8 ± 3.0	90.8 ± 4.0
Lung	67–86	81 ± 3	77 ± 9	81	72		
Muscle	197–226	45 ± 5	31 ± 6	86	78		
Pancreas	115–135					139.5 ± 4.0	88.3 ± 4.0
Heart	100	73 ± 16	67 ± 9				
Bone	218	168 ± 8	69 ± 6	78	63	95 ± 1.8	47 ± 1.6
Prostate							
Normal	520						
Hyperplasia	2330						
Cancer	285						
Eye							
Retina	571						
Choroid	562						
Ciliary body	288						
Testis		176 ± 12	132 ± 16	79	70	54 ± 2.0	59 ± 2.0
Esophagus		108 ± 17	88 ± 10			88.1 ± 3.0	97.6 ± 5.0

[a] Values are means ± S.D.

the balance between absorption and excretion. Zinc excretion is decreased when animals or humans are zinc deficient (Prasad *et al.*, 1976; Coppen and Davies, 1987; Wada *et al.*, 1985). Turnover of zinc is decreased and tissue zinc levels may be increased as a result of hypothyroidism (Yadav *et al.*, 1980).

Zinc excretion in both urine and feces may be extremely low in zinc-deficient rats. In addition, zinc redistribution from bone into muscle may occur (Guigliano and Millward, 1984; Jackson *et al.*, 1982). In pregnant rats, muscle catabolism in the dam releases zinc which may be utilized by the fetus (Masters *et al.*, 1986). Conversely, in zinc-supplemented animals, endogenous secretion of zinc is increased which helps to maintain zinc homeostasis.

Body fluid, hair, and blood cell levels of zinc in humans are shown in Table 10-7. Zinc levels in lymphocytes, granulocytes, and platelets appear to be sensitive indicators of zinc status in humans. Table 10-8 shows the zinc concentrations of various human tissue.

References

Abu-Hamdan, D. K., Migdal, S. D., Whitehouse, R., Rabbani, P., Prasad, A. S., and McDonald, F. D., 1981. Renal handling of zinc: Effect of cysteine infusion, *Am. J. Physiol.* 241:F487.

Abu-Hamdan, D. K., Mahajan, S. K., Migdal, S. D., Prasad, A. S., and McDonald, F. D., 1984. Zinc absorption in uremia: Effects of phosphate binders and iron supplements, *J. Am. Coll. Nutr.* 3:283.

Abu-Hamdan, D. K., Mahajan, S. K., Migdal, S. D., Prasad, A. S., and McDonald, F. D., 1986. Zinc tolerance test in uremia. Effect of ferrous sulfate and aluminum hydroxide, *Ann. Intern. Med.* 204:50.

Agarwal, R. P., and Henkin, R. I., 1982. Zinc and copper in human cerebrospinal fluid, *Biol. Trace Elem. Res.* 4:117.

Anderson, B. M., Gibson, R. S., and Sabry, J. H., 1981. The iron and zinc status of long-term vegetarian women, *Am. J. Clin. Nutr.* 34:1042.

Antonson, D. L., and Vanderhoff, A., 1983. Effect of chronic ethanol ingestion on zinc absorption in rat small intestine, *Dig. Dis. Sci.* 28:604.

Casey, C. E., Walravens, P. A., and Hambidge, K. M., 1981. Availability of zinc in loading tests with human milk, cow's milk and infant formulae, *Pediatrics* 68:394.

Cash, R., and Berger, C. K., 1969. Acrodermatitis enteropathica: Defective metabolism of unsaturated fatty acids, *J. Pediatr.* 74:717.

Coble, Y. D., Schulert, A. R., and Farid, Z., 1966. Growth and sexual development of male subjects in an Egyptian oasis, *Am. J. Clin. Nutr.* 18:421.

Coppen, D. E., and Davies, N. T., 1987. Studies on the effect of dietary zinc dose on zinc[65] absorption in vivo and the effects of zinc status on zinc[65] absorption and body loss in young rats, *Br. J. Nutr.* 57:35.

Cunnane, S. C., 1982. Maternal essential fatty acid supplementation enhances zinc absorption in neonatal rats: Relevance to the defect in zinc in acrodermatitis enteropathica, *Pediatr. Res.* 16:599.

Cunnane, S. C., 1988. *Zinc: Clinical and Biochemical Significance*, CRC Press, Boca Raton, Fla. p. 69.

Davies, J. W. L., and Fell, G. S., 1974. Tissue catabolism in patients with burns, *Clin. Chim. Acta* 51:83.

Davies, N. T., 1980. Studies on the absorption of zinc by rat intestine, *Br. J. Nutr.* 43:189.

Dinsmore, W. W., Callender, M. E., McMaster, D., and Love, A. H. G., 1985. The absorption of zinc from a standardized meal in alcoholics and in normal volunteers, *Am. J. Clin. Nutr.* 42:688.

Eckhert, C. D., 1985. Isolation of a protein from human milk that enhances zinc absorption in humans, *Biochem. Biophys. Res. Commun.* 130:264.

Eckhert, C. D., Sloan, M. V., Duncan, J. R., and Hurley, L. S., 1977. Zinc binding: A difference between human and bovine milk, *Science* 195:789.

Engel, R. W., Miller, R. E., and Price, N. O., 1966. Metabolic patterns in preadolescent children. XIII. Zinc balance, in *Zinc Metabolism* (A. S. Prasad, ed.), Thomas, Springfield, Ill., p. 326.

Evans, G. W., and Johnson, E. C., 1980. Zinc concentration of liver and kidneys from rat pups nursing dams fed supplemental zinc dipicolinate or zinc acetate, *J. Nutr.* 110:2121.

Evans, G. W., Grace, C. I., and Votava, H. J., 1975. A proposed mechanism for zinc absorption in the rat, *Am. J. Physiol.* 228:501.

Failla, M. L., and Cousins, R. J., 1975. Zinc accumulation and metabolism in primary cultures of adult rat liver cells, *Biochim. Biophys. Acta* 543:293.

Feeley, R. M., Eitenmiller, R. R., Jones, J. B., and Barnhart, H., 1983. Copper, iron and zinc contents of human milk at early stages of lactation, *Am. J. Clin. Nutr.* 37:443.

Fell, G. S., Fleck, A., Cuthbertson, D. P., Queen, K., Morrison, C., Bessent, R. G., and Husain, S. L., 1973. Urinary zinc level as an indication of muscle catabolism, *Lancet* 1:290.

Flanagan, P. R., Haist, J., and Valberg, L. S., 1983. Zinc absorption, intraluminal zinc and intestinal metallothionein in zinc deficient and zinc replete rodents, *J. Nutr.* 113:962.

Foley, B., Johnson, S. A., Hackley, B., Smith, J. C., Jr., and Halsted, J. A., 1968. Zinc content of human platelets, *Proc. Soc. Exp. Biol. Med.* 128:265.

Fransson, G. B., and Lonnerdal, B., 1983. Distribution of trace elements and minerals in human cow's milk, *J. Pediatr.* 17:912.

Gormican, A., and Catli, E., 1971. Mineral balance in young men fed a fortified milk-base formula, *Nutr. Metab.* 13:364.

Greger, J. L., 1977. Dietary intake and nutritional status in regard to zinc of institutionalized aged, *J. Gerontol.* 32:549.

Guigliano, R., and Millward, D. J., 1984. Growth and zinc homeostasis in the severely zinc deficient rat, *Br. J. Nutr.* 52:545.

Hallbook, T., and Hedelin, H., 1977. Zinc metabolism and surgical trauma, *Br. J. Surg.* 64:271.

Hambidge, K. M., Walravens, P. A., White, S., Anthony, M. L., and Roth, M. L., 1976. Zinc nutrition of preschool children in the Denver Head Start Program, *Am. J. Clin. Nutr.* 29: 734.

Hambidge, K. M., Krebs, N. F., Jacobs, M. A., Favier, A., Guyette, L., and Ikle, D. N., 1983. Zinc nutritional status during pregnancy: A longitudinal study, *Am. J. Clin. Nutr.* 37:429.

Hill, G. M., Brewer, G. J., Prasad, A. S., Hydrick, C. R., and Hartmann, D. E., 1987. Treatment of Wilson's disease with zinc. I. Oral zinc therapy regimens, *Hepatology* 7:522.

Horrobin, D. F., and Morgan, R. D., 1980. Myotonic dystrophy: A disease caused by functional zinc deficiency due to an abnormal zinc-binding ligand, *Med. Hypoth.* 6:375.

Hurley, L. S., and Lonnerdal, B., 1982. Zinc binding in human milk: Citrate versus picolinate, *Nutr. Rev.* 40:65.

Jackson, M. J., Jones, D. A., and Edwards, R. H. T., 1982. Tissue zinc level as an index of body zinc status, *Clin. Physiol.* 2:333.

Jameson, S., 1976. Effects of zinc deficiency in human reproduction, *Acta Med. Scand.* 197(Suppl.):3.

Johnson, W. T., and Evans, G. W., 1982. Tissue uptake of zinc in rats following the administration of zinc dipicolinate or zinc histidinate, *J. Nutr.* 112:914.

King, J. C., Stein, T., and Doyle, M., 1981. Effect of vegetarianism on the zinc status of pregnant women, *Am. J. Clin. Nutr.* 34:1049.

Latta, D., and Liebman, M., 1984. Iron and zinc status of vegetarian and non-vegetarian males, *Nutr. Rep. Int.* 30:141.

Lee, H. H., Prasad, A. S., Brewer, G. J., and Owyang, C., 1989. Zinc absorption in human small intestine, *Am. J. Physiol.* 256:G87.

Lee, H. H., Hill, G. M., Sikha, V. K. N. M., Brewer, G. J., Prasad, A. S., and Owyang, C., 1990. Pancreaticobiliary secretion of zinc and copper in normal persons and patients with Wilson's disease, *J. Lab. Clin. Med.* 116:283.

Lindeman, R. D., Bottomley, R. G., Cornelison, R. L., Jr., and Jacobs, L. A., 1972. Influence of acute tissue injury on zinc metabolism in man, *J. Lab. Clin. Med.* 79:452.

McCance, R. A., and Widdowson, E. M., 1942. The absorption and excretion of zinc, *Biochem. J.* 36:692.

Mansouri, K., Halsted, J., and Gombos, E., 1970. Zinc, copper, magnesium and calcium in dialyzed and non dialyzed uremic patients, *Arch. Intern. Med.* 125:88.

Masters, D. G., Keen, C. L., Lonnerdal, B., and Hurley, L. S., 1986. Release of zinc from maternal tissues during zinc deficiency or simultaneous zinc and calcium deficiency in the pregnant rat, *J. Nutr.* 116:2148.

Mateseche, J. W., Phillips, S. F., Malagelada, J. R., and McCall, J. T., 1980. Recovery of dietary iron and zinc from the proximal intestine of healthy man. Studies of different meals and supplements, *Am. J. Clin. Nutr.* 33:1946.

Meadows, N. J., Grainger, S. L., Ruse, W., Keeling, P. W. N., and Thompson, R. P. H., 1983. Oral iron and the bioavailability of zinc, *Br. Med. J.* 287:1013.

Meftah, S., and Prasad, A. S., 1989. Nucleotides in lymphocytes of human subjects with zinc deficiency, *J. Lab. Clin. Med.* 114:114.

Menard, M. P., and Cousins, R. J., 1983. Zinc transport by brush border membrane vesicles for rat intestine, *J. Nutr.* 113:1434.

Moore, M. E. C., Moran, J. R., and Greene, H. L., 1984. Zinc supplementation in lactating women—Evidence for mammary control of zinc secretion, *J. Pediatr.* 105:660.

Pattison, S. E., and Cousins, R. J., 1986a. Zinc uptake and metabolism by hepatocytes, *Fed. Proc.* 45:2805.

Pattison, S. E., and Cousins, R. J., 1986b. Kinetics of zinc uptake and exchange by primary cultures of rat hepatocytes, *Am. J. Physiol.* 250:E677.

Phillips, J. L., 1978. Uptake of transferrin-bound zinc by human lymphocytes, *Cell. Immunol.* 35:318.

Prasad, A. S., 1966. Metabolism of zinc and its deficiency in human subjects, in *Zinc Metabolism* (A. S. Prasad, ed.), Thomas, Springfield, Ill. p. 250.

Prasad, A. S., 1976. Zinc, in *Trace Elements and Iron in Human Metabolism* (A. S. Prasad, ed.), Plenum Press, New York, p. 251.

Prasad, A. S., 1988. Clinical spectrum and diagnostic aspects of human zinc deficiency, in *Essential and Toxic Trace Elements in Human Health and Disease* (A. S. Prasad, ed.), Liss, New York, p. 3.

Prasad, A. S., and Oberleas, D., 1970. Binding of zinc to amino acids and serum proteins in vitro, *J. Lab. Clin. Med.* 76:416.

Prasad, A. S., Halsted, J. A., and Nadimi, M., 1961. Syndrome of iron deficiency, anemia, hepatosplenomegaly, dwarfism, hypogonadism and geophagia, *Am. J. Med.* 31:532.

Prasad, A. S., Miale, A., Jr., Farid, Z., Sandstead, H. H., and Schulert, A. R., 1963a. Zinc metabolism in patients with syndrome of iron deficiency anemia, hepatosplenomegaly, dwarfism, and hypogonadism, *J. Lab. Clin. Med.* 61:537.

Prasad, A. S., Schulert, A. R., Miale, A. J., Farid, Z., and Sandstead, H. H., 1963b. Zinc and iron deficiencies in male subjects with dwarfism but without ancylostomiasis, schistosomiasis, or severe anemia, *Am. J. Clin. Nutr.* 12:437.

Prasad, A. S., Schulert, A. R., Sandstead, H. H., Miale, A., Jr., and Farid, Z., 1963c. Zinc, iron, and nitrogen content of sweat in normal and deficient subjects, *J. Lab. Clin. Med.* 62:84.

Prasad, A. S., Rabbani, P., Abbasi, A., Bowersox, E., and Spivey-Fox, M. R. S., 1977. Experimental zinc deficiency in humans, *Ann. Intern. Med.* 89:483.

Prasad, A. S., Meftah, S., Abdallah, J., Kaplan, J., Brewer, G. J., Bach, J. F., and Dardenne, M., 1988. Serum thymulin human zinc deficiency, *J. Clin. Invest.* 82:1202.

Price, N. O., Bunce, G. E., and Engel, R. W., 1970. Copper, manganese and zinc balance in preadolescent girls, *Am. J. Clin. Nutr.* 23:258.

Richards, M. P., and Cousins, R. J., 1976. Zinc-binding protein: Relationship to short term changes in zinc metabolism, *Proc. Soc. Exp. Biol. Med.* 153:52.

Ronaghy, H. A., Reinhold, J. G., Mahloudji, M., Ghavami, P., Fox, M. R. S., and Halsted, J. A., 1974. Zinc supplementation of malnourished schoolboys in Iran: Increased growth and other effects, *Am. J. Clin. Nutr.* 27:112.

Rose, G. A., and Willden, E. G., 1972. Whole blood, red cell, and plasma total and ultrafilterable zinc levels in normal subjects and in patients with chronic renal failure with and without hemodialysis, *Br. J. Urol.* 44:281.

Sandstead, H. H., 1973a. Zinc nutrition in the United States, *Am. J. Clin. Nutr.* 26:1251.

Sandstead, H. H., 1973b. Zinc nutrition in the United States, *Am. Med. Assoc.* 240:2188.

Sandstead, H. H., 1982. Availability of zinc and its requirement in human subjects, in *Clinical, Biochemical, and Nutritional Aspects of Trace Elements* (A. S. Prasad, ed.), Liss, New York, p. 83.

Sandstrom, B. M., Davidson, L., and Cederblad, A., 1985. Oral, iron, dietary ligands and zinc absorption, *J. Nutr.* 115:411.

Schoenfeld, C., Amelar, R. D., Dubin, L., and Numeroff, M., 1979. Prolactin, fructose and zinc levels found in human seminal plasma, *Fertil. Steril.* 32:206.

Schroeder, H. A., Nason, A. P., Tipton, I. H., and Balassa, J. J., 1967. Essential trace metals in man. Zinc: Relation to environmental cadmium, *J. Chronic Dis.* 20:179.

Scoular, F. L., 1939. A quantitative study, by means of spectrographic analysis of zinc in nutrition, *J. Nutr.* 17:103.

Silverman, B., Kwaitkowski, D., Pinto, J., and Rivlin, R., 1979. Disturbances in zinc binding to jejunal proteins induced by ethanol ingestion in rats, *Clin. Res.* 27:555A.

Solomons, N. W., and Jacob, R. A., 1981. Studies on the bioavailability of zinc in man. IV. Effect of heme and non-heme iron on the absorption of zinc, *Am. J. Clin. Nutr.* 34:475.

Solomons, N. W., Jacob, R. A., Pineda, O., and Viteri, F. E., 1979. Studies on the bioavailability of zinc in man. II. Absorption of zinc from organic and inorganic sources, *J. Lab. Clin. Med.* 94:335.

Song, M. K., and Adham, N. F., 1985. Relationship between zinc and prostaglandin metabolism in plasma and small intestine of rats, *Am. J. Clin. Nutr.* 41:1201.

Spencer, H., Rosoff, B., Feldstein, A., Cohn, S. H., and Gusmano, E., 1965a. Metabolism of zinc[65] in man, *Radiat. Res.* 24:432.

Spencer, H., Vankinscott, V., Lewin, I., and Samachson, J., 1965b. Zinc[65] metabolism during low and high calcium intake in man, *J. Nutr.* 86:169.

Spencer, H., Rosoff, B., Lewin, I., and Samachson, J., 1966. Studies of zinc[65] metabolism in man, in *Zinc Metabolism* (A. S. Prasad, ed.), Thomas, Springfield, Ill., p. 339.

Spencer, H., Osis, D., Kramer, L., and Norris, C., 1976. Intake, excretion and retention of zinc in man, in *Trace Elements in Human Health and Disease* (A. S. Prasad, ed.), Academic Press, New York, p. 345.

Stacey, N. H., and Klassen, C. D., 1981. Zinc uptake by isolated rat hepatocytes, *Biochim. Biophys. Acta* 640:693.

Strain, W. H., Macon, W. L., Pories, W. J., Perim, C., Adams, F. D., and Hill, O. A., 1974. Excretion of trace elements in bile, in *Trace Element Metabolism in Animals,* Volume 2 (W. H. Hoekstra, J. W. Suttie, H. E. Ganther, and W. Mertz, eds.), University Park Press, Baltimore, p. 644.

Sullivan, J. F., Jetton, M. M., and Burch, R. E., 1979a. A zinc tolerance test, *J. Lab. Clin. Med.* 93:485.

Sullivan, J. F., Williams, R. V., and Burch, R. E., 1979b. Metabolism of zinc and selenium in cirrhotic patients during 6 weeks of zinc ingestion, *Alcoholism* 3:235.

Swanson, C. A., Turnlund, J. R., and King, J. C., 1983. Effect of dietary sources and pregnancy on zinc utilization in adult women fed controlled diets, *J. Nutr.* 113:2557.

Tipton, I. H., Stewart, P. L., and Dickson, J., 1969. Patterns of elemental excretion in long term balance studies, *Health Phys.* 16:455.

Tribble, H. M., and Scoular, F. I., 1954. Zinc metabolism of young college women on self-selected diets, *J. Nutr.* 52:209.

Valberg, L. S., Flanagan, P. R., Brennan, J., and Chamberlain, M. J., 1985. Does the oral zinc tolerance test measure zinc absorption? *Am. J. Clin. Nutr.* 41:37.

Wada, L., Turnlund, J. R., and King, J. C., 1985. Zinc utilization in young men fed adequate and low zinc intakes, *J. Nutr.* 115:1345.

Wang, H., Prasad, A. S., and DuMouchelle, E. A., 1989. Zinc in platelets, lymphocytes, and granulocytes by flameless atomic absorption spectrophotometry, *J. Micronutr. Anal.* 5:181.

Wapnir, R. A., Khani, D. E., Bayne, M. A., and Lifshitz, F., 1983. Absorption of zinc by the rat ileum: Effects of histidine and other low-molecular ligands, *J. Nutr.* 113:1346.

Welsh, S. O., and Marston, R. M., 1983. Trends in levels of zinc in the U.S. food supply, 1909–1981, in *Bioavailability of Zinc,* ACS Symp. Ser. 210 (G. E. Inglett, ed.), American Chemical Society, Washington, D.C., p. 15.

White, H. S., and Gynee, T. M., 1971. Utilization of inorganic elements by young men eating iron-fortified foods, *J. Am. Diet. Assoc.* 59:27.

Whitehouse, R. C., Prasad, A. S., Rabbani, P. I., and Cossack, Z. T., 1982. Zinc in plasma, neutrophils, lymphocytes, and erythrocytes as determined by flameless atomic absorption spectrophotometry, *Clin. Chem.* 28. Washington, D.C., 475.

Wilden, E. G., and Robinson, M. R. G., 1975. Plasma zinc levels in prostatic disease, *Br. J. Urol.* 47:295.

Yadav, H. S., Nagpal, K. K., Sharma, B. N., and Chaundhuri, B. N., 1980. Influence of thyroxine and temperature on zinc metabolism, *Indian J. Exp. Biol.* 18:993.

Yip, R., Reeves, J. D., Lonnerdal, B., Keen, C. L., and Dallman, P. R., 1985. Does iron supplementation compromise zinc nutrition in healthy infants? *Am. J. Clin. Nutr.* 42:683.

Clinical Spectrum of Human Zinc Deficiency

11

During the past two decades, a spectrum of clinical deficiency of zinc in human subjects has been recognized. At one end, the manifestations of zinc deficiency may be severe, and at the other end, zinc deficiency may be mild or marginal.

11.1 Severe Manifestations of Zinc Deficiency

A severe deficiency of zinc has been reported to occur in patients with acrodermatitis enteropathica (AE), following total parenteral nutrition (TPN) without zinc, following excessive use of alcohol, and following penicillamine therapy.

11.1.1 Acrodermatitis Enteropathica

AE is a lethal, autosomal recessive trait that usually occurs in infants of Italian, Armenian, or Iranian lineage. The disease is not present at birth but usually develops in the early months of life soon after weaning from breast feeding.

The dermatologic manifestations of severe zinc deficiency in patients with AE include bullous pustular dermatitis of the extremities and the oral, anal, and genital areas (around the orifices), combined with paronychia and generalized alopecia. Ophthalmic signs may include blepharitis, conjunctivitis, photophobia, and corneal opacities.

Neuropsychiatric signs include irritability, emotional disorders, tremors, and occasional cerebellar ataxia. Patients with AE generally exhibit weight loss, growth retardation, and males exhibit hypogonadism.

A high incidence of congenital malformation of fetuses and infants born of pregnant women with AE has been reported (Hambidge *et al.*, 1978). Similar malformations have been reported in the offspring of maternal rats that are zinc deficient (Hurley, 1976). Thus, the human fetus appears to be susceptible to the teratogenic effects of maternal zinc deficiency.

Patients with AE have an increased susceptibility to infections. In AE, thymic hypoplasia, absence of germinal centers in lymph nodes, and plasmacytosis in the spleen are found consistently (Good et al., 1982). Lack of delayed hypersensitivity to bacterial and fungal antigens and occasional lack of IgA have been reported. Decreased peripheral T cell number and a depression of T-cell-mitogen-induced blast transformation have been observed. All of these T-cell-mediated functional abnormalities are completely corrected with zinc supplementation. Abnormal chemotaxis correctable with zinc therapy has also been reported in AE patients (Hambidge et al., 1978). In general, the clinical course is downhill with failure to thrive and complicated by intercurrent bacterial, fungal, and other opportunistic infections.

Gastrointestinal disturbances are usually severe, including chronic diarrhea, malabsorption, steatorrhea, and lactose intolerance.

The disease, if unrecognized and untreated, is fatal. Zinc supplementation results in complete recovery (Barnes and Moynahan, 1973). The genetic basis of zinc malabsorption is unknown.

11.1.2 Total Parenteral Nutrition

Zinc deficiency following TPN (without zinc) was first recognized by Kay and Tasman-Jones (1975) and by Okada et al. (1976) and Arakawa et al. (1976) in adults and children, respectively. The syndrome as it occurs during TPN consists of the development of a rash around the nasolabial folds and the mouth (Figs. 11-1 to 11-4). The rash may also occur over the extensor surfaces and, indeed, may spread over the entire body. The rash may vary from papular, scaly lesions to weeping open erosions. The lesions become infected with both bacterial and fungal organisms.

Patients on TPN with diarrhea may lose 6 to 12 mg of zinc daily. This excessive loss of zinc may result in a severe deficiency of zinc. In such cases, not only dermatologic manifestations but also alopecia, neuropsychiatric manifestations, weight loss, and intercurrent infections, particularly involving opportunistic infections, are observed. Carbohydrate utilization is impaired, and there is a negative nitrogen balance (Jeejeebohy, 1982). If zinc deficiency in such cases is not recognized and treated, the condition may become fatal.

11.1.3 Excess Alcohol

Previous observations have indicated that a certain percentage of alcoholic subjects excreted increased amounts of zinc, even though clinical or laboratory

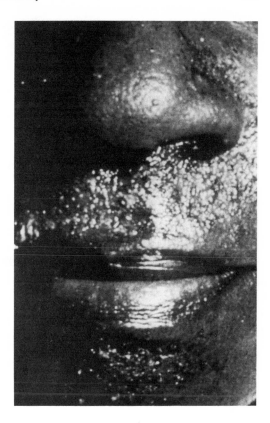

Figure 11-1. A patient with acquired acrodermatitis enteropathica following total parenteral nutrition before zinc supplementation. (Nutrition Today, Inc, Syllabus Alpha Series Teaching Aid, No. 41, 1982.)

evidence of chronic liver disease was absent. An absolute increase in renal clearance of zinc in 33% of the alcoholics demonstrable at both normal and high serum zinc concentration has been observed (Gudbjarnason and Prasad, 1969; Allan *et al.,* 1975).

Excessive ingestion of alcohol may lead to a severe deficiency of zinc as reported by Weismann *et al.* (1976). In this case, widespread eczema craquelé, hair loss, steatorrhea, dysproteinemia with edema, and mental disturbances related to zinc deficiency were observed. Therapy with zinc reversed these manifestations. A similar clinical syndrome has been seen among Ugandan blacks addicted to banana gin.

Alcohol ingestion causes excessive loss of zinc in the urine (hyperzincuria) and this may lead to zinc deficiency in alcoholic mothers. Infants and children born to alcoholic mothers exhibit characteristic clinical features. Fetal alcohol syndrome is characterized by prenatal and postnatal growth retardation, mental deficiency, small head size, and minor anomalies of the face, eyes, heart, joints, and external genitalia.

Figure 11-2. The patient shown in Fig. 11-1 after zinc supplementation. (Nutrition Today, Inc, Syllabus Alpha Series Teaching Aid, No. 41, 1982.)

The pathogenesis of the teratogenic effects of alcohol remains unknown. In rats, a deficiency of zinc during crucial stages of gestation is known to be teratogenic. Many enzymes required for DNA synthesis are zinc dependent, and an adverse effect of zinc deficiency on the activity of deoxythymidine kinase has been reported. Because alcohol ingestion may adversely affect zinc balance, we may speculate that the teratogenic effect of alcohol is mediated through zinc enzymes involved in DNA synthesis. Further studies must be carried out in order to understand fully the role of zinc in the pathogenic mechanism of fetal alcohol syndrome. In animal studies, alcohol has been shown to increase both urinary and fecal losses of zinc and thus contribute to a negative zinc balance.

11.1.4 Penicillamine Therapy

A patient with Wilson's disease was treated with penicillamine, off and on, for almost 5 years when he showed variable alopecia with "dead" hair, blepharitis, and rhinitis (Klingberg *et al.,* 1976). The dose of penicillamine was 1 to 2 g/day. Later he developed areas of parakeratosis over his knees, elbows, in the axillae, groin, and around the anus and on the buttocks. Other symptoms included a sudden reduction of visual acuity. Eye examination

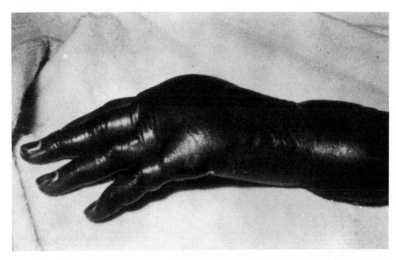

Figure 11-3. Another patient with acquired acrodermatitis enteropathica following total parenteral nutrition before zinc supplementation. (Nutrition Today, Inc, Syllabus Alpha Series Teaching Aid, No. 41, 1982.)

revealed centrocecal scotoma, "stippled" appearance of the cornea, and punctate keratitis.

In July, 1970, when the patient was 18, 6 years after the diagnosis of Wilson's disease and initiation of penicillamine therapy, his father, a farmer, questioned whether the skin and hair involvement might be similar to swine parakeratosis caused by zinc deficiency. Histologically, parakeratosis of the skin was confirmed (Figs. 11-5 to 11-9). Zinc deficiency was substantiated by analysis of plasma and red cells for zinc and urinary excretion of zinc. The deficiency of zinc presumably occurred as a result of the chelating effect of penicillamine, which depleted the body of its zinc content. After the diagnosis of zinc deficiency was made, penicillamine therapy was discontinued and zinc sulfate, 200 mg three times a day, was begun orally. Improvement was seen in his skin and hair in the first 2 weeks. Because of gastrointestinal irritation, zinc sulfate was discontinued and zinc as acetate in similar doses was started. The patient made a remarkable recovery (Figs. 11-10 and 11-11). He gained back all hair on the scalp. Skin became normal. He also grew hair in the axillae, pubic area, on his chest, and a full-fledged beard and moustache developed, requiring shaving.

The authors speculated that keratitis and centrocecal scotoma were also caused by zinc deficiency. The corneal lesions cleared following zinc therapy, but the centrocecal scotoma was irreversible. The actual relationship of zinc deficiency to such eye lesions is thus not well established. It is clear from the

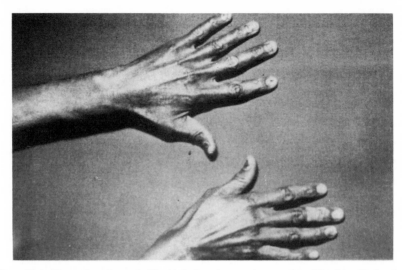

Figure 11-4. The patient shown in Fig. 11-3 after zinc supplementation. (Nutrition Today, Inc, Syllabus Alpha Series Teaching Aid, No. 41, 1982.)

case study that one must consider this complication when using penicillamine for therapeutic purposes.

In summary, the manifestations of severe zinc deficiency in humans include bullous pustular dermatitis, alopecia, diarrhea, emotional disorder, weight loss, intercurrent infections relating to cell-mediated immune dysfunctions, hypogonadism in males, neurosensory disorders, and problems with healing of ulcers. If this condition is unrecognized and untreated, it becomes fatal.

11.2 Moderate Deficiency of Zinc

A moderate level of zinc deficiency has been reported in a variety of conditions. These include nutritional relating to dietary factors, malabsorption syndrome, alcoholic liver disease, chronic renal disease, sickle-cell disease, and chronically debilitated conditions.

11.2.1 Nutritional

Growth retardation, hypogonadism in males, poor appetite, mental lethargy, rough skin, and intercurrent infections were the classical clinical features

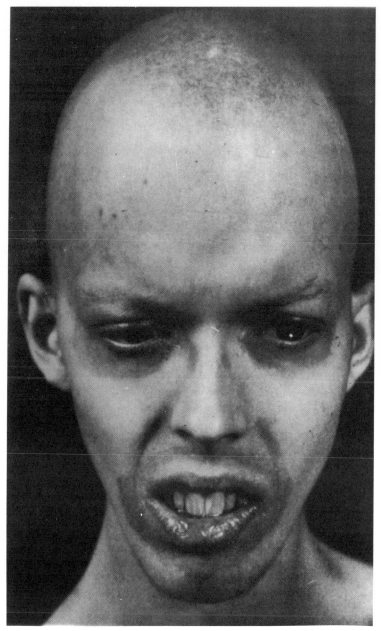

Figure 11-5. Alopecia. Orbital and perioral acanthosis. [Reprinted with permission from Klingberg, W. G., Prasad, A. S., and Oberleas, D., 1976. Zinc deficiency following penicillamine therapy, in *Trace Elements in Human Health and Disease,* Vol. I (A. S. Prasad, ed.), Academic Press, New York, p. 53.]

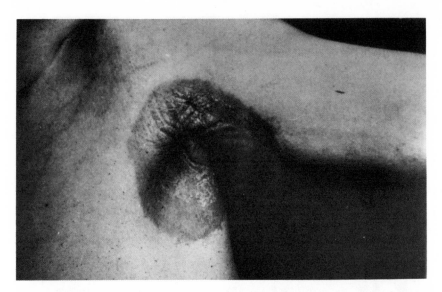

Figure 11-6. Parakeratosis of axilla. [Reprinted with permission from Klingberg, W. G., Prasad, A. S., and Oberleas, D., 1976. Zinc deficiency following penicillamine therapy, in *Trace Elements in Human Health and Disease,* Vol. I (A. S. Prasad, ed.), Academic Press, New York, p. 54.]

of chronically zinc-deficient subjects from the Middle East as reported by Prasad and co-workers in the early 1960s (Prasad *et al.,* 1961, 1963a,b; Prasad, 1966). The basis for zinc deficiency was nutritional inasmuch as zinc was poorly available from their diet because of the high content of phytate and phosphate. All of the above-mentioned features were corrected by zinc supplementation (Sandstead *et al.,* 1967). Liver and spleen were found to be enlarged in the zinc-deficient dwarfs, and improved following zinc supplementation. The mechanism of spleen and liver enlargement in this syndrome is not well understood. Later, zinc deficiency in children in the United States was described by Hambidge *et al.* (1972).

As previously mentioned, a nutritional deficiency of zinc in humans is fairly prevalent throughout the world, particularly in areas where cereal proteins are a primary feature of the local diet. Just as in Iran, geophagia is a common problem in Turkey and a majority of adolescents with geophagia exhibit both iron and zinc deficiencies (Cavdar *et al.,* 1980a). Growth retardation and hypogonadism have been related to zinc deficiency in such cases, and zinc supplementation has resulted in complete correction of these problems. The poor dietary habits of Turkish villagers appear to be the major factor responsible for these deficiencies. The village diet consists of wheat bread rich in phytate; the availability of iron and zinc is thus less than optimal.

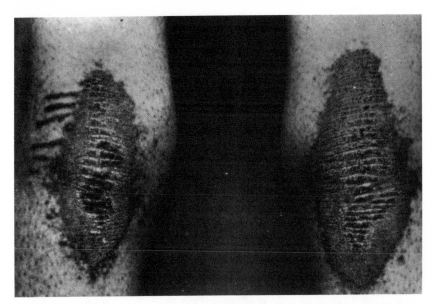

Figure 11-7. Parakeratosis of knees. [Reprinted with permission from Klingberg, W. G., Prasad, A. S., and Oberleas, D., 1976. Zinc deficiency following penicillamine therapy, in *Trace Elements in Human Health and Disease,* Vol. I (A. S. Prasad, ed.), Academic Press, New York, p. 54.]

The authors have concluded that zinc deficiency is one of the major nutritional problems in Turkey.

In another study, Cavdar *et al.* (1980b) found the zinc level to be decreased in almost 30% of pregnant women in Turkey, all of whom were of low socioeconomic status. Their diet consisted mainly of cereals. In view of the serious teratogenic effects of maternal zinc deficiency in experimental animals as well as epidemiological evidence that maternal zinc deficiency may be responsible for severe congenital malformation of the central nervous system in humans, more studies and correction of this nutritional problem are urgently needed.

Zinc deficiency has been found in aboriginal communities in Australia's tropical northwest (Holt *et al.,* 1980). The incidence of hypozincemia was 24.4%, and was most prevalent in children during the important pre- and postadolescent growth periods (31–67%). The diet consumed in these communities is predominantly white flour and refined sugar. Geophagia is common. Clinical features of zinc deficiency such as growth retardation, hypogonadism, and intercurrent infections are also prevalent.

Although we had observed in the Middle East that intercurrent infections were serious problems in the zinc-deficient dwarfs, the nature of the infections

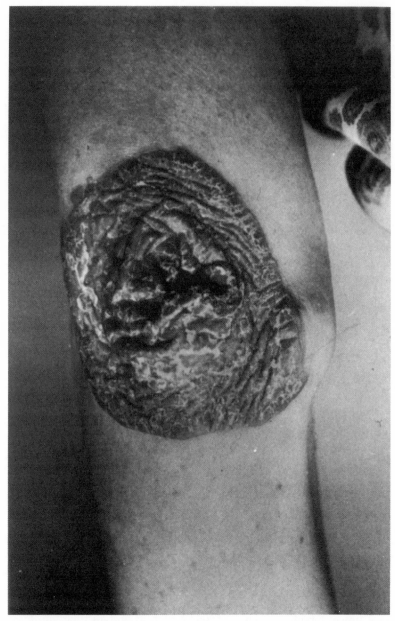

Figure 11-8. Parakeratosis of elbow. [Reprinted with permission from Klingberg, W. G., Prasad, A. S., and Oberleas, D., 1976. Zinc deficiency following penicillamine therapy, in *Trace Elements in Human Health and Disease,* Vol. I (A. S. Prasad, ed.), Academic Press, New York, p. 55.]

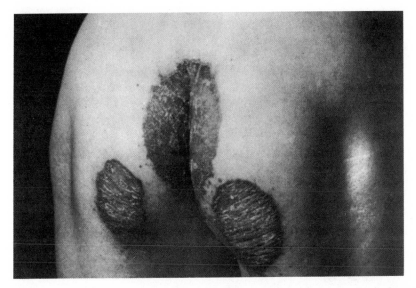

Figure 11-9. Parakeratosis of buttocks. [Reprinted with permission from Klingberg, W. G., Prasad, A. S., and Oberleas, D., 1976. Zinc deficiency following penicillamine therapy, in *Trace Elements in Human Health and Disease,* Vol. I (A. S. Prasad, ed.), Academic Press, New York, p. 56.]

and their immunological basis were not clear (Prasad *et al.,* 1961, 1963a,b). Only recently has it been known that zinc is essential for cell-mediated immune functions and that indeed an abnormal function of thymus-dependent lymphocytes (T cells) may be the key mechanism involved in infections seen in zinc-deficient human subjects.

11.2.2 Gastrointestinal Disorders and Liver Disease

A moderate level of zinc deficiency has been observed in many gastrointestinal disorders. These include malabsorption syndrome, Crohn's disease, regional ileitis, and steatorrhea. Because this topic has been covered elsewhere, no discussion will be provided here.

A low serum and hepatic zinc and, paradoxically, hyperzincuria were demonstrated in patients with cirrhosis of the liver by Vallee *et al.* (1956). These observations have since been confirmed by several groups of investigators. It has been suggested that zinc deficiency in the alcoholic cirrhotic patient may be a conditioned deficiency, somehow related to alcohol ingestion.

Patek and Haig (1939) reported that some patients with cirrhosis of the liver had night blindness which did not respond to vitamin A therapy. Mor-

Figure 11-10. Photograph taken following zinc therapy. [Reprinted with permission from Kling-berg, W. G., Prasad, A. S., and Oberleas, D., 1976. Zinc deficiency following penicillamine therapy, in *Trace Elements in Human Health and Disease,* Vol. I (A. S. Prasad, ed.), Academic Press, New York, p. 60.]

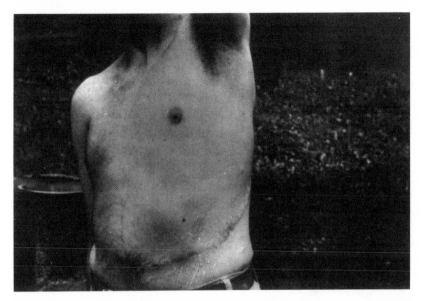

Figure 11-11. Photograph of axilla following zinc therapy. Notice disappearance of skin lesion. [Reprinted with permission from Klingberg, W. G., Prasad, A. S., and Oberleas, D., 1976. Zinc deficiency following penicillamine therapy, in *Trace Elements in Human Health and Disease,* Vol. I (A. S. Prasad, ed.), Academic Press, New York, p. 61.]

rison *et al.* (1978) observed a similar phenomenon but treated their subjects with zinc orally and observed an improvement in dark adaptation. Night vision is dependent on the interaction of the photochemically active vitamin A aldehyde (retinaldehyde) which must be supplied continuously to the rods to form visual pigment. Illumination of the retina results in the release of retinaldehyde, which is reduced to retinol. Retinaldehyde is regenerated from retinol by retinol alcohol dehydrogenase, a zinc metalloenzyme. Zinc-deficient rats have decreased activity of this retinal enzyme.

Many patients with cirrhosis are known to exhibit markedly enhanced sensitivity to drugs. Hepatic coma may be precipitated by administration of methionine to cirrhosis patients with an Eck fistula. Similarly, elevated blood ammonia seems to be intimately related to the development of hepatic coma. It is known that zinc-deficient rats have a defect in the metabolism of sulfur-containing amino acids. Zinc deficiency may also affect urea synthesis and, thus, abnormalities related to metabolism of amino acids and ammonia may act in concert to produce the picture of hepatic coma. We have reported an elevated level of plasma ammonia in human subjects as a result of dietary zinc restriction (Prasad *et al.,* 1978). Rabbani and Prasad (1978) observed a decrease in hepatic ornithine carbamoyl transferase (OCT) activity and an increase in plasma ammonia levels in zinc-deficient rats. An increased activity of the purine nucleotide enzyme AMP deaminase as a result of zinc deficiency, has also been observed; it is possible that several factors may account for increased plasma ammonia levels in zinc deficiency associated with cirrhosis of the liver. Recent papers suggest that zinc therapy may be beneficial to subjects with hepatic encephalopathy (Reding *et al.*, 1984; Coulnaud, 1985). Clearly more studies are needed in this important area.

It is likely that some of the clinical features of cirrhosis of the liver, such as loss of body hair, testicular hypofunction, poor appetite, mental lethargy, difficulty in healing, abnormal cell-mediated immune functions, and night blindness, may indeed by related to the secondary zinc-deficient state in this disease. Careful clinical trials with zinc supplementation must be carried out in order to determine whether zinc is beneficial to patients with chronic liver disease.

11.2.3 Renal Disease

Mahajan *et al.* (1979) documented that patients with chronic renal failure had low concentrations of zinc in plasma, leukocytes, and hair as well as increased plasma ammonia levels and increased activity of plasma ribonuclease. Patients with uremia, irrespective of whether or not they were on dialysis, had a mean plasma zinc level significantly less than that in controls. Patients undergoing maintenance hemodialysis and peritoneal dialysis had plasma zinc levels similar to those of patients not on dialysis. The concentration of zinc in hair and leukocytes was also significantly decreased in all groups with chronic renal failure compared with controls.

The mean erythrocyte zinc content was significantly higher in each group of uremic patients than in controls. There was no significant difference in erythrocyte zinc between dialyzed and nondialyzed uremic patients.

The mean plasma ammonia and ribonuclease activity were significantly elevated in all uremic patients. Ribonuclease activity was in most patients greater than five times the control value. Hematological studies, including serum iron, folic acid, and vitamin B_{12}, were within normal ranges in all uremic patients. There was no evidence of hemolysis or macrocytosis as judged by the peripheral smear and red blood cell indices. Neither the prescribed amount of protein in the diets nor the albumin levels correlated significantly with plasma zinc concentrations in dialyzed and nondialyzed uremic patients.

Our observation of low plasma zinc levels in uremic patients is similar to previously published reports (Mansouri *et al.*, 1970; Rose and Willden, 1972; Condon and Freeman, 1970). However, the plasma zinc levels in patients undergoing hemodialysis have been reported to be low (Condon and Freeman, 1970; Halsted and Smith, 1970), normal (Mansouri et al., 1970), or high (Rose and Willden, 1972; Mahler et al., 1971). The normalization of plasma zinc in some patients has been attributed to active uptake of zinc from the zinc plaster used in the construction of coils, or to a high dietary protein intake. Our finding of similar plasma zinc levels in uremic patients regardless of dialysis suggests that neither the liberalization of a prescribed amount of protein in the diet nor dialysis plays a significant role in the correction of plasma zinc. Plasma zinc levels, although very constant in individuals in good health, fluctuate markedly in situations of stress, including alcoholism, acute and chronic infections, myocardial infarction, etc. (Prasad, 1976), and the stress of renal failure may also influence plasma zinc levels in uremic patients. It is unknown at present whether a low concentration of plasma zinc is indicative of zinc deficiency in uremia. It seems likely that a low concentration of zinc reflects an impaired zinc intake in many, whereas in others it may reflect a shift of zinc from the plasma into another body pool.

Leukocytes are rich in zinc, but very limited data are available in the literature especially in regard to use of leukocyte zinc as an indicator in the diagnosis of zinc deficiency. Our observation of a decreased concentration of zinc in leukocytes from dialyzed and nondialyzed uremic patients suggests that these patients were zinc deficient. Michael et al. (1978), however, reported no difference in the zinc content of leukocytes between normal controls and uremic subjects. Their data are difficult to interpret: their results were expressed as milligrams of zinc per kilogram of dry solids (leukocytes), whereas we expressed our results in terms of micrograms of zinc per 10^{10} cell. Furthermore, their method of isolation and preparation of leukocytes is different, and therefore the results of the two studies are not comparable.

Blomfield et al. (1969) were the first to report high erythrocyte zinc levels in hemodialysis patients and attributed that to active uptake of zinc by erythrocytes during dialysis. Subsequent reports, however, did not show any increase in erythrocyte zinc following 12 h of dialysis with a Kill type of dialyzer (Mansouri et al., 1970; Rose and Willden, 1972). The erythrocyte zinc concentration in nondialyzed uremic patients has been reported to be normal (Mansouri et al., 1970) or high (Rose and Willden, 1972). Our observation of high erythrocyte zinc in all uremic patients with or without dialysis suggests that high erythrocyte zinc in the presence of low plasma zinc may be related to factor(s) other than dialysis. Most erythrocyte zinc exists as a part of the carbonic anhydrase enzyme. Only a small portion of erythrocyte zinc equilibrates freely with the plasma pool. In the presence of low plasma and leukocyte zinc concentrations, high erythrocyte zinc levels in uremia might represent an abnormal shift of plasma zinc into erythrocytes or ineffective erythropoiesis associated with a decreased rate of cell division and maturation of normoblasts.

A similar plasma erythrocyte zinc distribution has been reported in other states of ineffective erythropoiesis, such as is seen in folic acid and vitamin B_{12} deficiency (Prasad, 1966, 1976). Our patients, however, had normal levels of serum folate and vitamin B_{12} and there was no evidence of megaloblastosis. One of the major factors underlying the anemia of chronic renal failure is diminished erythrocyte production secondary to decreased erythropoietin levels. It is therefore possible that the protein-bound zinc pool in the precursors of erythrocytes remains high inasmuch as the normoblasts do not divide and mature normally.

The concentration of zinc in hair appears to reflect the nutritional status of zinc provided that the hair has been growing at a reasonable rate. The turnover rate of zinc in hair is slow. Thus, low values would indicate chronic zinc deficiency. Decreased hair zinc concentration in all patients with uremia suggests that long-standing zinc deficiency persists despite initiation of regular dialysis treatment.

These biochemical changes in chronic renal failure suggest that zinc deficiency is a complicating feature of uremia. It was further concluded that chronic dialysis therapy did not correct this deficiency in such subjects.

Diminished taste acuity may account for persistence of protein and calorie malnutrition observed in a majority of hemodialysis patients in spite of liberalization of the prescribed amount of dietary protein. Twenty-two patients undergoing thrice-weekly hemodialysis for more than 6 months were tested for taste acuity and plasma zinc concentration, after which a double-blind study was instituted using a zinc supplement (50 mg of elemental zinc as zinc acetate per day) or a placebo. The threshold of taste detection and recognition for salt (NaCl), sweet (sucrose), and bitter (urea), but not for sour (HCl) improved significantly in all patients on zinc supplementation. There was no improvement in those taking the placebo. During the study period, the mean plasma zinc level increased from 75 ± 8 to 97 ± 10 $\mu g/dl$ ($p < 0.001$) in patients receiving zinc acetate. There was no significant change in plasma zinc level in the placebo group (75 ± 15 to 80 ± 15 $\mu g/dl$). The results of this study showed that uremic hypogeusia improved in association with zinc supplementation and elevation of plasma zinc concentration (Mahajan *et al.*, 1980).

Impotence is common in uremic males and is not improved by hemodialysis (HD). The etiology of impotence in uremia is unknown, but the presence of gonadal dysfunction has been implicated. Because zinc deficiency in uremics has been documented, a double-blind clinical trial was carried out using zinc acetate (ten HD patients) to determine the effect of zinc on uremic gonadal dysfunction (Mahajan *et al.*, 1982). Serum testosterone, sperm counts, serum luteinizing hormone (LH), serum follicle-stimulating hormone (FSH), and plasma zinc were measured; sexual function history was obtained in all patients before, during, and after a 6-month study period. Before therapy, 15 of the 20 patients had total or partial impotence. All had low plasma zinc, serum testosterone, and sperm counts but high serum LH and FSH levels. Following therapy, the zinc-treated group demonstrated significant increases in plasma zinc, serum testosterone, and sperm counts and a fall in serum LH and FSH, and also reversal of impotence. In contrast, none of the patients receiving placebo showed any improvement in either gonadal dysfunction or impotence. The results of this study thus suggested that zinc deficiency and gonadal dysfunction were reversible causes of sexual dysfunction in uremia.

Moderate zinc deficiency in hemodialysis patients resulted in granulocyte zinc depletion, decreased chemotaxis, and decreased chemokinetic activity (Briggs et al., 1982). The effect of zinc supplementation on lymphocyte function was also studied in chronically uremic patients (Antoniou et al., 1981). A skin test for delayed hypersensitivity to mumps antigen was carried out in 25 apparently well-nourished men with a prior history of mumps infection who were receiving regular hemodialysis because of end-stage renal diseases. Nine patients were given zinc in dialysis baths for the treatment of hypogonadism. Only 1 patient in the zinc-treated group was anergic to mumps. In contrast, anergy to mumps and other antigens was observed in 11 of 16 patients not treated with supplemental zinc. Of 4 anergic patients who were subsequently treated with zinc, the skin sensitivity test was restored to normal in 3. Thus, zinc deficiency may be a major cause of impaired cellular immunity in patients with chronic renal failure.

Hyperprolactinemia in patients with chronic renal failure persists despite adequate dialysis; the cause is unknown (Nagel et al., 1973). Studies in animals have suggested that zinc inhibits prolactin secretion by the pituitary (LaBelle et al., 1973; Login et al., 1983; Judd et al., 1984). The possibility of an interaction between zinc and prolactin in zinc-treated and untreated hemodialysis patients was explored recently in our laboratory (Mahajan et al., 1985). Zinc and prolactin levels were measured in 32 male hemodialysis patients; 12 were receiving 50 mg zinc/day as zinc acetate, and 20 were not (Figs. 11-12 and 11-13). Zinc-treated subjects had significantly higher plasma zinc levels and lower serum prolactin levels than untreated patients ($p < 0.001$). Plasma zinc and serum prolactin were inversely related in zinc-treated and untreated patients ($r = -0.79$, $p < 0.001$). The mechanism by which zinc inhibits pituitary prolactin secretion remains to be elucidated.

The cause of zinc deficiency in uremia is unknown. The deficiency may be related to low dietary intake of zinc as a result of protein restriction or poor appetite. In addition, decreased gastrointestinal absorption of dietary zinc, as well as decreased bioavailability of zinc as a result of drug interactions may be present. Ferrous sulfate and phosphate binders are routinely prescribed

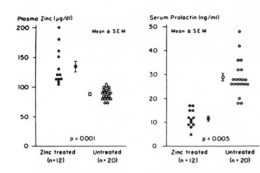

Figure 11-12. Plasma zinc and serum prolactin levels in zinc-treated and untreated hemodialysis patients. (Reprinted with permission from Mahajan, S. K., Flamenbaum, W., Hamburger, R. J., Prasad, A. S., and McDonald, F. D., 1985. Effect of zinc supplementation on hyperprolactinemia in uremic men, *Lancet* 2:750.)

for patients with renal failure. Various iron salts have been shown to interfere with zinc absorption in humans (Solomons *et al.*, 1983).

A simple oral zinc tolerance test has been used as an indirect indicator of zinc absorption for the gastrointestinal tract, and recently Valberg *et al.* (1985) have validated this test by comparing it with the direct measurements of Zn absorption and retention over a 1-week period.

We have recently utilized an oral zinc tolerance test to assess indirectly the gastrointestinal absorption of zinc and determine the effect of concurrent administration of ferrous sulfate or aluminum hydroxide on the plasma zinc appearance curve in patients with chronic renal disease. Our results confirmed the presence of diminished zinc absorption in patients with renal failure and showed that ferrous sulfate and aluminum hydroxide, which worsen this deficit, also impair zinc absorption in normal subjects (Abu-Hamdan *et al.*, 1986).

11.3 Zinc Deficiency in Sickle-Cell Disease

Our studies in the past have suggested the occurrence of zinc deficiency in adult sickle-cell anemia (SCA) patients (Prasad *et al.*, 1975, 1976, 1979,

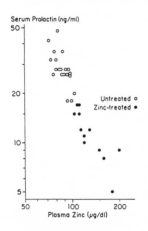

Figure 11-13. Relation between plasma zinc and serum prolactin levels in zinc-treated and untreated hemodialysis patients. (Reprinted with permission from Mahajan, S. K., Flamenbaum, W., Hamburger, R. J., Prasad, A. S., and McDonald, F. D., 1985. Effect of zinc supplementation on hyperprolactinemia in uremic men, *Lancet* 2:751.)

1981; Abbasi *et al.*, 1976; Warth *et al.*, 1981; Ballester and Prasad, 1983). Growth retardation, hypogonadism in males, hyperammonemia, abnormal dark adaptation, and cell-mediated immune disorder in SCA have been related to a deficiency of zinc. The biochemical evidence of zinc deficiency in SCA included a decreased level of zinc in plasma, erythrocytes, and hair, hyperzincuria, decreased activities of certain zinc-dependent enzymes such as carbonic anhydrase in the erythrocytes, alkaline phosphatase in the neutrophils, deoxythymidine kinase activity in newly synthesizing skin connective tissue and collagen, and hyperammonemia. Because zinc is known to be an inhibitor of ribonuclease, an increased activity of this enzyme in the plasma of SCA subjects was regarded as evidence of zinc deficiency (Prasad *et al.*, 1975, 1976). A limited trial with zinc supplementation in SCA subjects resulted in significant improvement in secondary sexual characteristics, normalization of plasma ammonia level, and reversal of dark adaptation abnormality (Tables 11-1, 11-2, and Fig. 11-14) (Prasad *et al.*, 1979, 1981; Warth *et al.*, 1981). As a result of zinc supplementation, the zinc level in plasma, erythrocytes, and neutrophils increased, and an expected response to supplementation was observed in the activities of the zinc-dependent enzymes (Prasad *et al.*, 1979, 1981; Warth *et al.*, 1981).

We have reported a beneficial effect of zinc on longitudinal growth and body weight in 14- to 18-year-old patients with SCA (Prasad and Cossack, 1984). In the first experiment, ten growth-retarded male SCA subjects between the ages of 14 and 17 were subdivided randomly into two groups. Five subjects

Table 11-1. Mean 30-min Threshold, $t_{1/2}$, and Neutrophil and Plasma Zinc Values
for Sickle-Cell Anemia Patients with Normal and Abnormal
Dark Adaptation and Controls[a]

	Normal dark adaptation ($N = 7$)	Abnormal dark adaptation ($N = 6$)	Controls	p^b
Mean 30-min threshold				
(log relative threshold)	7.41 ± 0.13^c	8.35 ± 0.45	7.59 ± 0.15	<0.01
$t_{1/2}$ (s)^d	639.14 ± 96.16	1078.17 ± 797.45	632.50 ± 58.85	NS
Neutrophil zinc				
($\mu g/10^{10}$ cells)	86.71 ± 19.53	37.33 ± 11.88	102.00 ± 13.02	<0.01
Plasma zinc ($\mu g/100$ μl)	105.86 ± 13.31	93.83 ± 8.26	112.90 ± 13.60	>0.05, <0.1

[a] Adapted with permission from Warth, J. A., Prasad, A. S., Zwas, F., and Frank R. N., 1981. Abnormal dark adaptation in sickle cell anemia, *J. Lab. Clin. Med.* 98:189.
[b] Normal versus abnormal dark adaptation.
[c] Values are means ± S.D.
[d] $t_{1/2}$ = time for rod threshold to fall to the midpoint of its descent.

Table 11-2. Mean 30-min Threshold and Zinc Levels in Plasma and Neutrophils in
Four Patients with Abnormal Dark Adaptation before and after Treatment
(Zinc or Placebo)[a]

Patient no.	Therapy	Mean 30-min threshold[b]		Plasma Zn (μ/100 μl)		Neutrophil Zn (μg/10^{10} cells)	
		Before	After	Before	After	Before	After
1	Zn	8.0	7.7	88	97	28	132
2	Zn	7.7	7.6	110	224	48	87
3	Zn	8.4	7.6	94	108	41	102
4	Placebo	8.0	8.6	88	92	38	23

[a] Adapted with permission from Warth, J. A., Prasad, A. S., Zwas, F., and Frank, R. N., 1981. Abnormal dark
adaptation in sickle cell anemia, *J. Lab. Clin. Med.* 98:189.
[b] Log relative threshold at 30 min.

received placebo twice a day, and the other five received 15 mg of zinc supplementation as acetate twice a day for 1 year (see Figs. 15 and 16).

Height and body weight were carefully recorded initially and at 3-month
intervals throughout the study by a single observer between 9 and 10 a.m.
Bone age was determined radiographically twice initially, and at the end of
the 1-year treatment period.

Zinc was determined in the plasma and erythrocytes once a month. Two
zinc-dependent enzymes, neutrophil alkaline phosphatase and erythrocyte

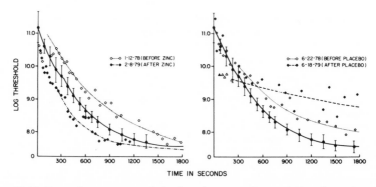

Figure 11-14. Pre- and posttreatment dark-adaptation curves for two sickle-cell anemia patients.
Log threshold (log relative threshold) versus time. A normal curve ± S.D. (circles) is shown with
each. Left, Patient 2. Right, Patient 4. (Reprinted with permission from Warth, J. A., Prasad,
A. S., Zwas, F., and Frank, R. N., 1981. Abnormal dark adaptation in sickle cell anemia, *J. Lab.
Clin. Med.* 98:192.)

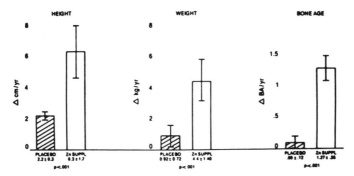

Figure 11-15. Figure showing that the increases in height, body weight, and bone age in sickle-cell disease patients were significantly greater in the zinc-treated group. Increased activities of neutrophil alkaline phosphatase and erythrocyte nucleoside phosphorylase and higher levels of serum testosterone in zinc-supplemented subjects were observed. [Prasad, A. S., 1988. Clinical spectrum and diagnostic aspects of human zinc deficiency, in *Essential and Toxic Trace Elements in Human Health and Disease* (A. S. Prasad, ed.), Liss, New York, p. 22.]

nucleoside phosphorylase, were also assayed simultaneously once a month (Prasad and Rabbani, 1981). Alkaline phosphatase activity in the neutrophils was measured quantitatively by an established technique. Nucleoside phosphorylase was assayed in the erythrocytes by the technique of Kalckar (Prasad and Rabbani, 1981). The data obtained during the last 3 months of each treatment phase were averaged and analyzed for statistical purposes. Basal serum testosterone was measured in duplicate every 6 months.

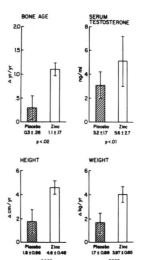

Figure 11-16. Figure showing the changes in height, body weight, bone age, and serum testosterone during placebo and zinc-treated periods in experiment II. It is clear that the gains in height, body weight, and bone age were significantly greater when subjects were switched to zinc treatment. Serum testosterone also increased significantly following zinc supplementation. [Prasad, A. S., 1988. Clinical spectrum and diagnostic aspects of human zinc deficiency, in *Essential and Toxic Trace Elements in Human Health and Disease* (A. S. Prasad, ed.), Liss, New York, p. 23.]

In the second experiment, six growth-retarded male subjects between the ages of 14 and 18 were supplemented with zinc (15 mg twice a day as acetate) for 1 year. Prior to supplementation with zinc, they received a placebo for 1 year. Thus, the subjects on zinc supplementation served as their own controls. The clinical and biochemical parameters followed were the same as in the first experiment. Additionally we measured zinc in the neutrophils in this study by methods published earlier (Whitehouse et al., 1982).

In experiment I, the mean plasma and erythrocyte zinc in the placebo and zinc-treated groups at the end of 1 year were as follows: plasma zinc, 90 ± 0.8 versus 124 ± 9 $\mu g/dl$, $p < 0.001$; erythrocyte zinc, 29.0 ± 2.0 us 36.4 ± 5.7 $\mu g/g$ Hb, $p < 0.02$.

In experiment II, zinc levels in the plasma, erythrocytes, and neutrophils during placebo and zinc supplementation respectively were as follows: plasma, 89.4 ± 5.2 versus 121.7 ± 11.5 $\mu g/dl$, $p < 0.005$; erythrocytes, 31.0 ± 5.4 versus 35.4 ± 3.7 $\mu g/Hb$, $p < 0.02$; neutrophils, 82.2 ± 10.6 versus 115.0 ± 5.8 $\mu g/10^{10}$ cells, $p < 0.001$. These values were obtained by an older cell separation technique. No attempt was made to remove platelets which contaminated the granulocyte pool. The activities of neutrophil alkaline phosphatase and erythrocyte nucleoside phosphorylase in placebo and zinc-treated periods were as follows: 5.47 ± 1.9 versus 10.4 ± 3.5 sigma units/mg protein, $p < 0.005$; and 8.3 ± 1.07 versus 11.8 ± 1.1 OD/h per mg Hb, $p < 0.008$.

Growth retardation in SCA is a well-known clinical entity (Daeschner et al., 1981; Olambiwonnu et al., 1975). The rate of growth is decreased and many fail to attain normal stature. Our studies in two sets of experiments establish a beneficial role of zinc on growth and development of adult SCA subjects.

For the supplementation study we selected only male subjects, since we were interested in observing the effect of zinc supplementation on both growth and testicular development. Although ovaries do not appear to be sensitive target tissues for zinc, female dwarfs responsive to zinc therapy have been reported from Iran and Turkey (Prasad, 1978).

From a nutritional standpoint, the nonhematological complications of zinc deficiency in subjects with SCA—e.g., poor growth and development, hypogonadism in males, abnormal dark adaptation, chronic leg ulcers which do not heal promptly, and anergy related to T-cell dysfunction (Prasad et al., 1975, 1976, 1979, 1981; Abbasi et al., 1976; Warth et al., 1981; Ballester and Prasad, 1983)—are considerable and basically preventable with oral zinc administration. We recommend that patients with SCA be screened for zinc deficiency and, if deficient, supplemented with zinc.

Cell-mediated immunity was evaluated in 26 adult patients with SCA by using skin tests for delayed-type hypersensitivity (DTH) reactions (Ballester and Prasad, 1983). Patients with impaired DTH reactions had lower zinc levels in plasma, erythrocytes, and neutrophils than did control subjects or patients with normal DTH reactions.

The activity of nucleoside phosphorylase, an enzyme essential for T-lymphocyte function, was significantly lower in zinc-deficient SCA subjects with anergy. Three anergic patients received oral zinc supplementation (45 mg/day as zinc acetate) and were reevaluated 6 months later. All three subjects showed a correction of neutrophil zinc levels and significantly improved nucleoside phosphorylase activity. On repeated skin tests, one patient had a positive response to two skin antigens, and the other two patients had positive responses to one test each. These new findings show that zinc deficiency in SCA patients is associated with impaired DTH which is correctable by zinc supplementation.

Natural killer (NK) cell activity was also studied in adults with SCA and in two normal volunteers rendered zinc deficient by dietary restriction of zinc alone (adequate intake of all other nutrients, including calories and proteins, was ensured throughout the study) (Tapazoglou et al., 1985). NK cell activity was significantly lower in zinc-deficient patients with SCA compared with control subjects. In the two volunteers, NK cell activity declined during zinc restriction and returned to near normal levels with zinc repletion. These results suggest that zinc deficiency is associated with a lowering of NK cell activity as has been observed in animal model systems.

In summary, the manifestations of a moderate deficiency of zinc include growth retardation and male hypogonadism in the adolescents, rough skin, poor appetite, mental lethargy, delayed wound healing, cell-mediated immune dysfunctions, and abnormal neurosensory changes.

11.4 Mild Deficiency of Zinc

Although the clinical, biochemical, and diagnostic aspects of severe and moderate levels of zinc deficiency in humans are fairly well defined, the recognition of mild levels of zinc deficiency has been difficult. Zinc assay in plasma, urine, and hair have been proposed as potential indicators of body zinc status (Prasad, 1982a–c). Currently, plasma zinc appears to be the most widely used parameter for assessment of human zinc status, and it is known to be decreased in cases of severe and moderate deficiency of zinc. However, it is also known that several physiologic and pathologic conditions may affect zinc levels in the plasma and urine; thus, these parameters may not be good indicators of low body zinc status. Zinc in hair and erythrocytes do not reflect active or recent status of body zinc, inasmuch as these tissues are slowly turning over. Furthermore, in cases of mild deficiency of zinc in humans, the plasma levels of zinc may remain normal and clinically there may not be any overt evidence of zinc deficiency, thus creating a difficult diagnostic problem.

We have, therefore, utilized assay of zinc in more rapidly turning over blood cells such as lymphocytes, granulocytes, and platelets as indicators of zinc status in human subjects. With the use of these data we have defined mild deficiency of zinc in humans in three groups of subjects.

The first group of subjects were volunteers in whom we induced a mild state of zinc deficiency by dietary means. Two adult male volunteers, ages 23 and 25, were hospitalized at the Clinical Research Center of the University of Michigan Medical School Hospital. A semipurified diet which supplied approximately 3.0 mg of zinc on a daily basis was used in order to produce zinc deficiency. The details of the methodology have been published elsewhere (Rabbani et al., 1987).

The volunteers were ambulatory and were encouraged to do daily moderate exercise throughout the study period. Prior to the study, a thorough history, physical examination, and routine laboratory tests including CBC, liver function tests, SMA-12, and serum electrolytes were performed and found to be normal. Zinc levels in lymphocytes, granulocytes, and platelets were in the normal range.

They were given a hospital diet containing animal protein daily for 4 weeks. This diet averaged 12 mg of zinc daily, consistent with the recommended dietary allowance of the National Research Council, National Academy of Sciences. Following this, they received 3.0 mg of zinc a day while consuming a soy protein-based experimental diet. This regime was continued for 28 weeks, at the end of which two cookies containing 27 mg of zinc supplement were added to the experimental diet. The supplementation was continued for 12 weeks.

Throughout the study the levels of all nutrients including protein, amino acids, vitamins, and minerals (both macro and micro elements) were kept constant, meeting the standards set by the RDA except for zinc which was varied as outlined above. By this technique we were able to induce a specific deficiency of zinc in human volunteers.

The serum level of biologically active thymulin was evaluated by a rosette assay described in detail elsewhere (Dardenne and Bach, 1975), and shown by us and several other investigators to be strictly thymus-specific (Fabris and Mocchegiani, 1985; Iwata et al., 1979). The assay analyzes the conversion of relatively azathioprine (Az)-resistant spleen of adult thymectomized mice to theta-positive rosette-forming cells that are more sensitive to Az (Dardenne and Bach, 1975).

The peripheral blood cells (lymphocytes, granulocytes, and platelets) for zinc assay were isolated by a modification of a previously published method (Whitehouse et al., 1982). Special care was taken to remove red cells from the granulocytes, platelets from the granulocytes and lymphocytes, and trapped plasma from the platelets. Extreme care was exercised to avoid exogenous zinc contamination throughout the assay procedure. Zinc was assayed in the samples by means of an atomic absorption spectrophotometer equipped with a furnace and auto sampling system Instrumentation Laboratory 555 AA spectrophotometer with a 655 furnace and 254 Fastac Auto Sampler (Instrumentation Laboratory, Inc., Lexington, Mass.).

Normal activity of circulating thymulin was observed in the two healthy young volunteers before initiation of the zinc-deprived diet. Thymulin activity began to decrease 3 months after beginning the zinc deprivation and was undetectable after 6 months. *In vivo* zinc supplementation induced within 1 month a rapid return to normal levels of thymulin activity. The final level of activity was even higher than those observed before zinc restriction (Prasad *et al.*, 1988).

The zinc in platelets decreased within 1 month. However, the zinc concentration of lymphocytes and granulocytes decreased only after 2 months of zinc-restricted dietary intake (Prasad *et al.*, 1988). The maximum decline of zinc in the cells was observed at the end of 6 months of restricted zinc intake. Following zinc supplementation, the cellular zinc levels returned to the original baseline levels within 3 months.

Although the body of an adult 70-kg male contains about 2300 mg zinc, only 10% exchanges with an isotopic dose within 1 week (Foster *et al.*, 1979). Approximately 28% of zinc resides in the bone, 62% in the muscles, 1.8% in the liver, and 0.1% in the plasma pool. In an adult animal model, zinc concentrations of muscle and bone do not change as a result of mild or marginal zinc deficiency (Underwood, 1977). It appears that in cases of mild or marginal deficiency of zinc in humans, one cannot expect a uniform distribution of the deficit over the entire body pool, but that most likely those compartments with high turnover rates (liver and peripheral blood cells such as lymphocytes, neutrophils, and platelets) would suffer a disproportionate deficit.

In our experimental human model studies, we created a negative zinc balance of approximately 1 mg/day (Prasad *et al.*, 1978). Although in 6 months this amounts to 180 mg of total negative zinc balance, this is still a small fraction of the total body zinc. However, if one were to consider only that 200 to 400 mg of zinc which is represented by liver zinc and mobile exchangeable pool, a negative balance of 180 mg may be a considerable fraction of the exchangeable pool.

Our present studies clearly show that at the mild level of zinc restriction used in our experimental diet, a measurable effect on zinc level in cells such as lymphocytes, neutrophils, and platelets was observed thus providing a basis for utilizing cellular zinc levels as indicators of human body zinc status. The levels of cellular zinc in this study are lower in the control subjects than previously reported values (Whitehouse *et al.*, 1982). This is because in previous studies platelets as contaminants in lymphocytes and granulocytes were not removed adequately and gave those preparations higher values.

In additional experiments, two out of four subjects showed abnormal dark adaptation and decreased lean body mass as a result of zinc restriction which were corrected on supplementation with zinc. In a taste test, all four subjects showed hypogeusia as a result of zinc restriction. Zinc supplementation corrected this clinical abnormality.

We assayed NK cell activity in four subjects during baseline, zinc restriction, and following supplementation with zinc in the experimental human model. NK cell activity declined during zinc restriction and returned to near-normal levels with zinc repletion (Tapazoglou *et al.*, 1985). IL-2 activity was assayed in two subjects during baseline and following zinc restriction. A significant decrease in the activity of IL-2 was observed as a result of mild zinc restriction (Prasad *et al.*, 1988).

We have previously reported the effects of mild zinc deficiency induced by dietary means on gonadal functions in male volunteers (Abbasi *et al.*, 1980). These subjects had normal serum androgens, FSH, LH, and sperm count prior to zinc restriction.

The sperm count declined slightly during zinc restriction and continued to decline in the early phase of zinc repletion. Oligospermia (total sperm count per ejaculate <40 million) was observed in four out of five subjects as a result of dietary zinc restriction (see Tables 11-3 to 11-5, Figs. 11-17 to 11-19). The sperm concentration (mean ± S.E.) during the baseline period was 305.9 ± 76.4 million/ml and total sperm count per ejaculate was 1114.7 ± 411.6 million. Following a zinc-restricted period of several weeks, sperm concentration for the entire group was 59.2 ± 17.4 million/ml, and the total sperm count per ejaculate was 53.6 ± 12.5 million during the early phase of zinc repletion; the difference between these two values and those of the stabilization period were statistically significant. During the late phase of zinc repletion, the sperm concentration increased to 207 ± 47.9 million/ml, and the total sperm count per ejaculate increased to 416.3 ± 102.9 million. These values are considered to be within the normal range. There was a significant correlation between the sperm concentration and the total sperm count ($r = 0.90$, $p < 0.001$). Changes in sperm motility and morphology throughout the study period were unremarkable.

The baseline serum testosterone decreased significantly during the early phase of zinc repletion and returned to normal levels during the late phase of zinc repletion. There was a slight decline in the maximal rise of serum testosterone after GnRH stimulation during zinc restriction and a more significant decline during the early phase of zinc repletion with recovery to normal level during the late phase of zinc repletion. The changes of serum dihydro-

Table 11-3. Experimental Dietary Design[a]

Subjects	Stabilization		Zinc restriction semipurified	Zinc repletion	
	Diet:Hospital	Semipurified		Semipurified	Hospital
Subject 1					
Duration, weeks	2	6	24	12	8
Daily intake of zinc, mg	10	2.7 + 10	2.7	2.7 + 30	10 + 30
Subjects 2 and 3					
Duration, weeks	3	5	40	8	8
Daily intake of zinc, mg	10	3.5 + 30	3.5	3.5 + 30	10 + 30
Subject 4					
Duration, weeks	4	—	40	12	8
Daily intake of zinc, mg	10		5	5 + 10	10 + 10[b]
Subject 5					
Duration, weeks	4	—	32	—	
Daily intake of zinc, mg	10		5		10[c]

[a] Adapted with permission from Abbasi, A. A., Prasad, A. S., and Rabbani, P., 1979. Experimental zinc deficiency in man: Effect of spermatogenesis, *Trans. Assoc. Am. Physicians* 17:292.
[b] Zinc, 10 mg was mixed with the diet.
[c] Home diet equivalent to hospital diet.

testosterone were similar to those of serum testosterone, but statistically not significant.

Although the mean maximal rise in LH after GnRH stimulation was highest during the early phase of zinc repletion, the values were not statistically significant relative to the other periods. Baseline mean serum FSH was highest during the zinc-restricted period, but again the values were not statistically significant compared with the other periods.

In our study, mild zinc deficiency as induced by dietary zinc restriction has a definite effect on gonadal function. The sperm count was susceptible to dietary restriction of zinc. Although there was only a slight decrease in the sperm count during zinc restriction, oligospermia became significant during the early phase of zinc repletion.

The nature of the delayed effect of zinc deficiency on sperm count is not well understood. However, an explanatory hypothesis can be drawn on the basis of other studies. The developmental progression of spermatogenesis,

Table 11-4. Erythrocyte and Plasma Zinc in Experimental Subjects[a]

	Erythrocyte zinc (μg/g Hb)			Plasma zinc (μg/dl)		
Subject	Stabilization	Restriction	Repletion	Stabilization	Restriction	Repletion
1	34.4 ± 2.8^b	28.8 ± 2.2	37.3 ± 2.9	115 ± 9	81 ± 8	110 ± 20
	(4)	(7)	(4)	(2)	(4)	(4)
		$p < 0.05^c$	$p < 0.01^d$		$p < 0.02^c$	$p < 0.01^d$
2	49.2 ± 5.9	43.8 ± 2.4	46.2 ± 4	85 ± 4	80 ± 2.7	99.5 ± 11
	(4)	(4)	(4)	(2)	(5)	(4)
		NS^c	NS^d		NS^c	$p < 0.01^d$
3	56.8 ± 2.7	49.9 ± 2.1	60.4 ± 1.9	106 ± 1.4	92 ± 4	117 ± 2.3
	(4)	(4)	(4)	(2)	(4)	(3)
		$p < 0.01^c$	$p < 0.01^d$		$p < 0.02^c$	$p < 0.001^d$
4	39.6 ± 3.7	30.3 ± 1.3	31.4 ± 2.7	105 ± 2.8	81.4 ± 8.6	81.7 ± 5.2
	(4)	(4)	(4)	(4)	(5)	(5)
		$p < 0.001^c$	NS^d		$p < 0.001^c$	NS^d
5	41.9 ± 1.8	36.4 ± 2.1	39.8 ± 1.5	120.7 ± 8.9	93.2 ± 10.4	85.7 ± 8.1
	(4)	(3)	(2)	(4)	(4)	(2)
		$p < 0.01^c$	NS^d		$p < 0.01^c$	NS^d

[a] Adapted with permission from Abbasi, A. A., Prasad, A. S., and Rabbani, P., 1979. Experimental zinc deficiency in man: Effect of spermatogenesis, *Trans. Assoc. Am. Physicians* 17:292.
[b] Values are means $\pm$ S.D. Number of determinations is given in parentheses.
[c] Comparison between stabilization and zinc restriction.
[d] Comparison between zinc restriction and zinc repletion.

from the origin of spermatozoa in the germinal epithelium to mature spermatozoa, is a prolonged process. The duration of human spermatogenesis is 74 ± 4.5 days (Heller and Clermont, 1964). Therefore, an insult affecting the germinal cells may not be evident until several months later. The effect of certain therapeutic drugs on spermatogenesis is an example. In patients treated with cyclophosphamide, oligospermia or azoospermia occurred several months after the treatment was started (Kumar *et al.*, 1972; Qureshi *et al.*, 1972; Fairley *et al.*, 1972; Buchanan *et al.*, 1975). Similarly, spermatogenesis returned to normal 15–49 months after cyclophosphamide therapy was stopped. It is therefore not surprising for an insidious insult to germinal cells, such as zinc deficiency, to be manifested after such a long period of time. If the temporal relationship between the onset of injury to germinal epithelium and the sperm count is true, one would expect an equal period of time to be required for the recovery of spermatogenesis once the ensuing disorder has been corrected. In our subjects, the recovery of oligospermia occurred in similar fashion. The sperm count returned to the baseline level after several months of zinc supplementation.

The long duration of human spermatogenesis, however, provides only a partial explanation for the effects of zinc on gonadal function, inasmuch as

Table 11-5. Sperm Count and Serum Testosterone in Experimental Subjects[a]

	Sperm count (million/ml)			Serum testosterone after LRH stimulation (ng/ml)		
Subject	Stabilization	Post-zinc restriction	Repletion	Stabilization	Post-zinc restriction	Repletion
1	187 ± 43^b (7)	26 (1) $p < 0.02^c$	113 ± 35 (2) NS^d	5.21 ± 1.06 (13)	4.70 ± 0.39 (4) NS^c	7.65 ± 1.10 (4) $p < 0.005^d$
2	628 ± 161 (5)	27 ± 21 (4) $p < 0.001^c$	267 ± 173 (3) $p < 0.05^d$	8.75 ± 2.21 (12)	5.21 ± 1.03 (8) $p < 0.001^c$	6.98 ± 1.32 (4) $p < 0.05^d$
3	234 ± 104 (2)	42 ± 15 (6) $p < 0.005^c$	179 ± 73 (4) $p < 0.005^d$	7.50 ± 0.76 (3)	2.03 ± 0.05 (4) $p < 0.01^c$	2.93 ± 1.0 (8) NS^d
4	219 ± 103 (7)	23 ± 16 (3) $p < 0.02^c$	86 ± 19 (3) $p < 0.02^d$	5.0 ± 1.02 (9)	3.28 ± 0.91 (8) $p < 0.005^c$	3.18 ± 1.02 (4) NS^d
5	180 ± 40 (5)	45 ± 4 (2) $p < 0.01^c$	109 ± 35 (2) NS^d	7.83 ± 1.02 (4)	2.66 ± 0.99 (8) $p < 0.001^c$	4.43 ± 1.47 (4) $p < 0.005^d$

[a] Adapted with permission from Abbasi, A. A., Prasad, A. S., and Rabbani, P., 1979. Experimental zinc deficiency in man: Effect of spermatogenesis, *Trans. Assoc. Am. Physicians* 17:292.
[b] Values are means ± S.D. Number of determinations is given in parentheses.
[c] Comparison between stabilization and zinc restriction.
[d] Comparison between zinc restriction and zinc repletion.

we also observed a similar delayed effect of dietary zinc restriction on plasma testosterone levels in our subjects. Another contributory factor for the lack of synchronism in testicular hypofunction with the periods of zinc restriction and zinc repletion is perhaps the design of our study. The slow induction of zinc deficiency by dietary means, with its nutritional and metabolic conse-

Figure 11-17. Decrease in sperm count following zinc restriction and its recovery after zinc repletion in five subjects. (Reprinted with permission from Abbasi, A., Prasad, A. S., and Rabbani, P., 1979. Experimental zinc deficiency in man: Effect of spermatogenesis, *Trans. Assoc. Am. Physicians* 92: 292.)

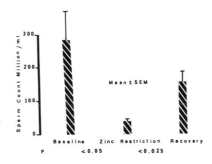

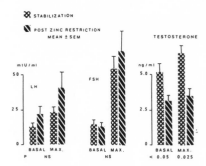

Figure 11-18. Changes in basal and maximal increments after LRH stimulation of serum LH, FSH, and testosterone, following zinc restriction in five subjects. LH, luteinizing hormone; FSH, follicle-stimulating hormone; MAX., maximal increment after LH releasing hormone stimulation; NS, not significant. (Reprinted with permission from Abbasi, A., Prasad, A. S., and Rabbani, P., 1979. Experimental zinc deficiency in man: Effect of spermatogenesis, *Trans. Assoc. Am. Physicians* 92:292.)

quences, required a relatively long period of time to be reflected clinically on the testicular function. Body stores of zinc were very slowly depleted and later were slowly replenished with zinc supplementation. The dosage of supplemental zinc in these subjects was within a physiologic rather than a therapeutic range. It is evident from our data that replenishment of body zinc was not accomplished until the late phase of zinc repletion which extended up to 12 months, and this correlated well with the observed effects on testicular function in our study.

One surprising finding in our studies in the human experimental model was that the plasma ammonia level increased during the period of zinc re-

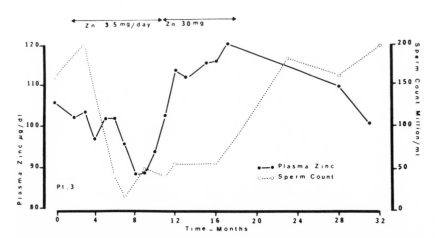

Figure 11-19. Relationship of sperm count values to plasma zinc concentrations during zinc restriction and zinc repletion periods in one subject. (Reprinted with permission from Abbasi, A., Prasad, A. S., and Rabbani, P., 1979. Experimental zinc deficiency in man: Effect of spermatogenesis, *Trans. Assoc. Am. Physicians* 92:292.)

striction (Prasad *et al.*, 1978). This was corrected following supplementation with zinc. We have reported similar findings in zinc-deficient rats (Rabbani and Prasad, 1978) and related this observation to abnormalities induced in the activity of ornithine transcarbamoylase activity in the liver, an enzyme known to be important in ammonia utilization and urea synthesis.

Thus, in our studies examining a mild deficiency of zinc in males induced by dietary means, decreased serum testosterone level, oligospermia, decreased NK cell activity, decreased production of IL-2 by T-helper cells, decreased thymulin activity, hyperammonemia, hypogeusia, decreased dark adaptation, and decreased lean body mass were observed. Thus, a mild deficiency of zinc in humans affects clinical, biochemical, and immunological functions adversely.

The second group of subjects in whom a mild deficiency of zinc was documented consisted of six volunteers between the ages of 25 and 58 who were laboratory personnel and medical students in apparently good health. Five were males and one was female. One subject was on chlorothiazide diuretic for treatment of mild hypertension. They were diagnosed to have mild deficiency of zinc based on their cellular zinc levels. Zinc levels in plasma and erythrocytes were normal. A mild state of zinc deficiency was defined by a decreased level (1 S.D. or more below the normal mean) in any two cell types (lymphocytes < 48, granulocytes < 42, and platelets < 1.6 μg/10^{10} cells).

Two types of studies were carried out in this group of subjects. Serum thymulin activity and levels of nucleotides in lymphocytes before and after zinc supplementation (50 mg zinc as acetate/day for 3 months) were measured.

Abnormally low levels of active thymulin were found in the six zinc-deficient subjects relative to age-matched healthy controls. Zinc supplementation restored thymulin activity to normal levels in the sera (Prasad *et al.*, 1988).

We also assayed the activities of nucleoside phosphorylase and ecto 5′-nucleotidase in the lymphocytes before and after zinc supplementation. Both of these were decreased during zinc deficiency but were corrected following zinc supplementation. (Meftah and Prasad, 1989; Meftah *et al.*, 1991). Our results suggested that the assays of 5′-nucleotidase and serum thymulin activity were sensitive indicators of zinc status in humans.

The third group of subjects in whom we have recognized a mild deficiency of zinc recently are elderly subjects between the ages of 65 and 85 (Prasad *et al.*, 1993). Deterioration of T-cell immune function is known to be associated with aging (Sandstead *et al.*, 1982) and the possibility that poor zinc nutriture may contribute to this phenomenon exists, inasmuch as zinc is required for cell-mediated immune functions (Fraker *et al.*, 1985).

In one study 15 anergic institutionalized apparently healthy persons (mean age 81) were given 100 mg zinc daily for 1 month, while 15 control anergic subjects (mean age 79.6) were given placebo (Duchateau *et al.*, 1981). The group receiving zinc had an increased percentage of circulating T lymphocytes, an increased frequency and magnitude of delayed hypersensitivity skin reactions to purified antigens, and a greater IgG antibody response to tetanus toxoid. Although the zinc status of the elderly subjects was not evaluated and the relation of the changes observed to their prestudy zinc nutriture is unclear, it seems evident that certain aspects of cell-mediated immune functions were influenced beneficially in elderly subjects by zinc supplementation. Whether or not the benefits were the result of a therapeutic effect of zinc or a simple correction of nutritional deficiency of zinc was not clear from the report. Other suggestive but by no means definitive clinical evidence of zinc deficiency in elderly subjects include decreased taste acuity, problems with healing of ulcers, and decreased serum testosterone levels in males (Sandstead *et al.*, 1982).

Plasma zinc has been assayed in the elderly; however, a decrease in the plasma level of zinc has not been observed in a consistent fashion (Sandstead *et al.*, 1982; Lindeman *et al.*, 1971). Because plasma zinc is not regarded as a sensitive indicator of zinc deficiency if the latter is mild, we assayed zinc in the granulocytes, lymphocytes, and platelets by recently established improved techniques in 118 healthy elderly subjects (ages 65 to 85) and were able to document for the first time a mild level of zinc deficiency in 36 subjects (Prasad *et al.*, 1993). The estimated mean daily zinc intake was 9.06 mg (69% the U.S. RDA).

Zinc levels in the plasma and erythrocytes were normal in the elderly subjects. Zinc levels in the lymphocytes and granulocytes were significantly decreased in the elderly subjects compared with the younger age controls ($p < 0.01$). Plasma copper was increased and IL-1 production and serum thymulin activity were significantly decreased. Reduced response to a skin test antigen panel and decreased taste acuity were observed in the elderly.

Thirteen elderly zinc-deficient subjects were supplemented with 30 mg zinc orally daily. Zinc supplementation corrected zinc deficiency and normalized plasma copper levels. Serum thymulin activity, IL-1 production, and lymphocyte 5'-NT increased significantly, and improvement in response to skin test antigen and taste acuity were observed following zinc supplementation.

Our studies thus suggest that a mild deficiency of zinc may have induced anergy, decreased IL-1 production by the peripheral blood mononuclear cells, decreased serum thymulin activity, decreased lymphocyte 5'-NT activity, and induced hypogeusia in the elderly subjects. Our studies also demonstrate that measurement of zinc in cells such as lymphocytes and granulocytes is useful in diagnosing mild zinc deficiency in humans.

In a recent study, Newsome *et al.* (1988) have reported a beneficial effect of zinc supplementation on macular degeneration associated with age. The elderly subjects received zinc sulfate 100 mg orally twice a day. Unfortunately, their zinc status was not defined in this study.

Burnet (1982) has suggested a role of zinc deficiency in senile dementia. He suggested that age-associated dementias might represent localized functional deficiency of zinc. This hypothesis was based on the broader concept of "error catastrophes" as suggested by Orgel (1963) with the additional suggestion that lack of zinc could increase inefficiency and the occurrence of informational errors during replication of DNA or its activation for protein synthesis in somatic cells.

According to Orgel (1963), random errors in protein synthesis occur at any age, but initially this phenomenon is probably of low frequency. The gradual accumulation of errors in various enzymes involved in nuclei and protein synthesis becomes amplified resulting in a decrease in cell functions or cell death with increasing age.

In Burnet's model, the progressive weakening of the thymus-dependent immune system responsible for immunological surveillance may be responsible for increased somatic mutations leading to cancer and autoimmune diseases in the elderly (Burnet, 1982). Walford (1980) also suggested that a somatic mutation in immunocytes leading to antigenic modification may lead to increased incidence of autoimmune disorders in the aging population.

It has been observed that antioxidants extend the life span of some rodents, thus providing support for the involvement of free radicals in the theory of aging. Because zinc is known to be involved in DNA synthesis, in free radical reactions, and in cell-mediated immunity, one must consider that an intracellular deficiency of zinc may play a role in the aging process (Prasad, 1984).

Several studies suggest that dietary zinc intake declines with advancing age, and for many older people falls below the RDA of 15 mg/day of the National Research Council (Sandstead *et al.,* 1982; Garry *et al.,* 1982; Abdulla *et al.,* 1977). Factors accounting for this observation include preference for cereal protein, decreased energy intake, and economic restraint, all of which have been noted in elderly subjects by some investigators (Sandstead *et al.,* 1982). Additional factors accounting for zinc deficiency in the elderly may be related to decreased gastrointestinal absorption of zinc as has been reported by some investigators and use of drugs such as diuretics which may increase excretion of zinc (Sandstead *et al.,* 1982; Lindeman *et al.,* 1971).

Whether or not aging *per se* affects zinc metabolism is unknown. Clearly more studies are required in this field. A well-controlled zinc supplementation study in the elderly must be carried out in order to determine whether some or all of the zinc-related clinical, biochemical, and immunological problems as discussed above are correctable. This may have great impact on the nutritional management of the growing population of elderly in this country and elsewhere.

In summary, the clinical manifestations of a mild level of zinc deficiency in humans include decreased serum testosterone level and oligospermia in males, decreased lean body mass, hyperammonemia, neurosensory changes, anergy, decreased serum thymulin activity, decreased IL-2 and IL-1 production by peripheral blood mononuclear cells, and decreased activity of 5'-NT in lymphocytes.

11.5 Miscellaneous Clinical Conditions

Several clinical conditions have been reported in the literature implicating a role of zinc in their pathogeneses (Goldberg and Sheehy, 1982; Kumar and Anday, 1984; Dura-Trave et al., 1984; Flynn et al., 1981; Taper et al., 1985; Kutti and Kutti, 1984; Shambaugh, 1985; Simmer and Thompson, 1985).

Convulsions occurring around the fifth day of life in otherwise normal babies have aroused considerable interest in recent years, partly because the etiology is unknown (Pryor et al., 1981). A 3-year prospective study of the cause of fifth day fits was carried out by Goldberg and Sheehy (1982). They found a highly significant relationship between low cerebrospinal fluid zinc levels and fifth day fits. No attempt was made to reverse the syndrome by supplementing zinc, inasmuch as the convulsions were short-lived. This interesting clinical observation needs further confirmation.

Flynn et al. (1981) studied the zinc status of 25 alcoholic and 25 nonalcoholic pregnant women and compared this with fetal outcome. The alcoholic women had significantly lower plasma zinc levels than the nonalcoholic women. Fetal cord plasma zinc levels were also lower in the offspring of alcoholic women. Infants from alcoholic mothers had more defects than infants from nonalcoholic mothers, and the defects of the nonalcoholic group more closely resembled those described by Jones and Smith (1973). Fetal dysmorphogenesis involved craniofacial, cutaneous, cardiovascular, and neurological organs in the alcoholic group. In this study, fetal dysmorphogenesis appeared to be related to zinc status. A mild deficiency of zinc in experimental animals is known to produce teratogenic effects (Hurley, 1976). Further controlled studies must be done in order to document the effects of zinc supplementation in prevention of this serious complication of alcoholic mothers.

An additional positive balance of approximately 375 mg is required during normal pregnancy. In one study, mild deficiency of zinc in pregnant women was reported to be associated with increased maternal morbidity, abnormal taste sensation, prolonged gestation, inefficient labor, atonic bleeding, and increased risks to the fetus (Jameson, 1980). Zinc supplementation to the deficient mothers resulted in reduced frequency of the above complications.

Dura-Trave *et al.* (1984) have reported poor uterine contractility in hypozincemic pregnant women, an observation which supports the previous report of increased maternal morbidity in zinc deficient mothers (Jameson, 1980). Simmer and Thompson (1985) have implicated zinc deficiency as a factor responsible for intrauterine growth retardation. These observations are provocative and need further confirmation.

Kumar and Anday (1984) have described edema, hypoproteinemia, and zinc deficiency in low-birth-weight infants. Edema and hypoproteinemia were observed between 5 and 9 weeks of age. Nine of the infants had diarrhea, liver disease, or urinary protein loss. Treatment with oral zinc supplements led to rapid resolution of the edema, with increases in values for serum proteins and for alkaline phosphatase activity in the serum. The authors suggest that zinc deficiency should be included in the list of causes of generalized edema in the low-birth-weight infants.

Anorexia nervosa is a poorly understood disorder of unclear etiology and is associated with high morbidity and mortality (Herzog and Copeland, 1985). The treatment is unsatisfactory. Kutti and Kutti (1984) have published encouraging results of zinc supplementation in patients with anorexia nervosa. It is important that a double-blind clinical trial of zinc supplementation in such patients be done in order to determine its therapeutic efficacy.

Although zinc is known to be present in high concentrations in the iris, the choroid, and the retina, it has not been appreciated that the inner ear (cochlea and vestibule) is also very rich in zinc (Shambaugh, 1985). Shambaugh has suggested that some cases of sensorineural hearing loss and imbalance may be responsive to zinc supplementation. Future studies are needed in this area.

References

Abbasi, A. A., Prasad, A. S., Ortega, J., Congco, E., and Oberleas, D., 1976. Gonadal function abnormalities in sickle cell anemia: Studies in adult male patients; *Ann. Intern. Med.* 85: 601.

Abbasi, A. A., Prasad, A. S., Rabbani, P., and DuMouchelle, E., 1980. Experimental zinc deficiency in man: Effect on testicular function, *J. Lab. Clin. Med.* 96:544.

Abdulla, M., Jagerstand, M., Norden, A., Qvist, I., and Svennson, S., 1977. Dietary intake of electrolytes and trace elements in the elderly, *Nutr. Metab. Suppl.* 1:21.

Abu-Hamdan, D. K., Mahajan, S. K., Migdal, S. D., Prasad, A. S., and McDonald, F. D., 1986. Zinc tolerance test in uremia: Effect of ferrous sulfate and aluminum hydroxide, *Ann. Intern. Med.* 104:50.

Allan, J. F., Fell, G. S., and Russell, R. I., 1975. Urinary zinc in hepatic cirrhosis, *Scott. Med. J.* 109:109.

Antoniou, L. D., Shalhoub, R. J., and Schechter, G. P., 1981. The effect of zinc on cellular immunity in chronic uremia, *Am. J. Clin. Nutr.* 34:1912.

Arakawa, T., Tamura, T., and Igarashi, Y., 1976. Zinc deficiency in two infants during parenteral alimentation for diarrhea, *Am. J. Clin. Nutr.* 29:197.

Ballester, O. F., and Prasad, A. S., 1983. Anergy, zinc deficiency, and decreased nucleoside phosphorylase activity in patients with sickle cell anemia, *Ann. Intern. Med.* 98:180.

Barnes, P. M., and Moynahan, E. J., 1973. Zinc deficiency in acrodermatitis enteropathica: Multiple dietary intolerance treated with synthetic diet, *Proc. R. Soc. Med.* 66:327.

Blomfield, J., McPherson, J., and George, C. R. P., 1969. Active uptake of copper and zinc during hemodialysis, *Br. Med. J.* 2:141.

Briggs, W. A., Pedersen, M., Mahajan, S., Sillix, D., Prasad, A. S., and McDonald, F., 1982. Lymphocyte and granulocyte function in zinc treated and zinc deficient hemodialysis patients, *Kidney Int.* 21:827.

Buchanan, J. D., Fairley, K. F., and Barnie, J. U., 1975. Return of spermatogenesis after stopping cyclophosphamide therapy, *Lancet* 2:156.

Burnet, F. M., 1982. New horizons in the role of zinc in cellular function, in *Clinical Applications of Recent Advances in Zinc Metabolism* (A. S. Prasad, I. E. Dreosti, and B. S. Hetzel, eds.), Liss, New York, p. 181.

Cavdar, A. O., Arcasoy, A., Cin, S., and Gumus, H., 1980a. Zinc deficiency in geophagia in Turkish children and response to treatment with zinc sulphate, *Hematologie* 65:403.

Cavdar, A. O., Babacan, E., Arcasoy, A., and Ertein, U., 1980b. Effect of nutrition on serum zinc concentration during pregnancy in Turkish women, *Am. J. Clin. Nutr.* 33:542.

Condon, C. H., and Freeman, R. M., 1970. Zinc metabolism in renal failure, *Ann. Intern. Med.* 73:531.

Coulnaud, C. L., 1985. Trailement de K'encephalopathie portosystemique aigue par le zinc, *Chirrurgie* III:575.

Daeschner, C. W., III, Matustik, M. C., Carpentieri, U., and Haggard, M. E., 1981. Zinc and growth in patients with sickle cell disease, *J. Pediatr.* 98:778.

Dardenne, M., and Bach, J. F., 1975. The sheep cell rosette assay for the evaluation of thymic hormones, in Biological activity of thymic Hormones, (D. V. Bekkum, ed.), Kooyker Scientific Publications, Rotterdam, p. 235.

Duchateau, J., Delepesse, G., Vrigens, R., and Collet, H., 1981. Beneficial effects of oral zinc supplementation on the immune response of old people, *Am. J. Med.* 60:1001.

Dura-Trave, T., Puig-Abuli, M., Monreal, I., and Villa-Eluzaga, I., 1984. Relation between maternal plasmatic zinc levels and uterine contractility, *Gynecol. Obstet. Invest.* 17:247.

Fairley, K. F., Barnie, J. U., and Johnson, W., 1972. Sterility and testicular atrophy related to cyclophosphamide therapy, *Lancet* 1:568.

Fabris, N., and Mocchegiani, E., 1985. Endocrine control of thymic serum factor production in young adult and old mice, *Cell. Immunol.* 91:325.

Flynn, A., Martier, S. S., Sokol, R. F., Miller, S. I., Golden, N. L., and DelVillano, B. C., 1981. Zinc status of pregnant alcoholic women: A determinant of fetal outcome, *Lancet* 1:572.

Foster, D. M., Aamodt, R. L., Henkin, R. I., and Berman, M., 1979. Zinc metabolism in humans: A kinetic model, *Am. J. Physiol.* 237:R340.

Fraker, P. J., Gershwin, M. E., Good, R. A., and Prasad, A. S., 1985. Interrelationships between zinc and immune function, *Fed. Proc.* 45:1474.

Garry, P. J., Gordon, J. S., Hunt, W. C., Hooper, E. M., and Loednar, A. G., 1982. Nutritional status in a healthy elderly population. Dietary and supplemental intakes, *Am. J. Clin. Nutr.* 36:319.

Ghavami-Maibodi, S. Z., Collipp, P. J., Castro-Magna, M., Stewart, C., and Chen, S. Y., 1983. Effect of oral zinc supplements on growth, hormonal levels, and zinc in healthy short children, *Ann. Nutr. Metab.* 27:214.

Goldberg, H. J., and Sheehy, E. M., 1982. Fifth day fits: An acute zinc deficiency syndrome? *Arch. Dis. Child.* 57:633.

Good, R. A., Fernandes, G., Garofalo, J. A., Cunningham-Rundles, C., Iwata, T., and West, A., 1982. Zinc and immunity, in *Clinical, Biochemical, and Nutritional Aspects of Trace Elements* (A. S. Prasad, ed.), Liss, New York, p. 189.

Gudbjarnason, S., and Prasad, A. S., 1969. Cardiac metabolism in experimental alcoholism, in *Biochemical and Clinical Aspects of Alcohol Metabolism* (V. Sardesai, ed.), Thomas, Springfield, Ill., p. 266.

Halsted, J. A., and Smith, J. C., 1970. Plasma zinc in health and disease, *Lancet* 1:322.

Halsted, J. A., Ronaghy, H. A., Abadi, P., Haghshenass, M., Amirkakimi, G. H., Barakat, R. M., and Reinhold, J. G., 1972. Zinc deficiency in man: The Shiraz experiment, *Am. J. Med.* 53:277.

Hambidge, K. M., Hambidge, C., Jacobs, M., and Baum, J. D., 1972. Low levels of zinc in hair, anorexia, poor growth, and hypogeusia in children, *Pediatr. Res.* 6:868.

Hambidge, K. M., Neldner, K. H., Walravens, P. A., Weston, W. L., Silverman, A., Sabol, J. L., and Brown, R. M., 1978. Zinc and acrodermatitis enteropathica, in *Zinc and Copper in Clinical Medicine* (K. M. Hambidge and B. L. Nichols, eds.), Spectrum Publications, New York, p. 81.

Heller, C. H., and Clermont, Y., 1964. Kinetics of the germinal epithelium in man, *Recent Prog. Horm. Res.* 20:545.

Herzog, D. B., and Copeland, P. M., 1985. Eating disorders, *N. Engl. J. Med.* 313:295.

Holt, A. B., Spargo, R. M., Iveson, J. B., Faulkner, G. S., and Cheek, D. B., 1980. Serum and plasma zinc, copper, and iron concentrations in aboriginal communities of northwestern Australia, *Am. J. Clin. Nutr.* 33:119.

Hurley, L. S., 1976. Perinatal effects of trace element deficiencies, in *Trace Elements in Human Health and Disease,* Volume 1 (A. S. Prasad, ed.), Academic Press, New York, p. 301.

Iwata, T., Incefy, G., Fernandez, T. G., Menendez-Botet, C. J., Pih, I., and Good, R. A., 1979. Circulating thymic hormone levels in zinc deficiency, *Cell. Immunol.* 47:101.

Jameson, S., 1980. Zinc and pregnancy, in *Zinc in the Environment, Part II* (J. O. Nriagu, ed.), Wiley, New York, p. 183.

Jeejeebhoy, K., 1982. Trace element requirements during total parenteral nutrition, in *Clinical, Biochemical, and Nutritional Aspects of Trace Elements* (A. S. Prasad, ed.), Liss, New York, p. 469.

Jones, K. L., and Smith, D. W., 1973. Recognition of the fetal alcohol syndrome in early infancy, *Lancet* 2:999.

Judd, A. M., MacLeod, R. M., and Login, I. S., 1984. Zinc acutely selectively and reversibly inhibits pituitary prolactin secretion, *Brain Res.* 294:190.

Kay, R. G., and Tasman-Jones, C., 1975. Zinc deficiency and intravenous feeding, *Lancet* 2:605.

Klingberg, W. G., Prasad, A. S., and Oberleas, D., 1976. Zinc deficiency following penicillamine therapy, in *Trace Elements in Human Health and Disease,* Volume I (A. S. Prasad, ed.), Academic Press, New York, p. 51.

Kumar, R., Biggard, J. D., McEnvoy, J., and McGrown, M. G., 1972. Cyclophosphamide and reproductive function, *Lancet* 1:1212.

Kumar, S. P., and Anday, E. K., 1984. Edema, hypoproteinemia, and zinc deficiency in low-birth-weight infants, *Pediatrics* 73:327.

Kutti, S. S., and Kutti, J., 1984. Zinc and anorexia nervosa, *Ann. Intern. Med.* 100:317.

LaBelle, F., Dular, R., Vivian, S., and Queen, G., 1973. Pituitary hormone releasing or inhibiting activity of metal ions present in hypothalamic extracts, *Biochem. Biophys. Res. Commun.* 52:786.

Lindeman, R. D., Clark, M. L., and Colmore, J. P., 1971. Influence of age and sex on plasma and red cell zinc concentrations, *J. Gerontol.* 26:358.

Login, I. S., Thorner, M. O., and MacLeod, R. M., 1983. Zinc may have physiological role in regulating pituitary prolactin secretion, *Neuroendocrinology* 37:317.

Mahajan, S. K., Prasad, A. S., Rabbani, P., Briggs, W. A., and McDonald, F. D., 1979. Zinc metabolism in uremia, *J. Lab. Clin. Med.* 94:693.

Mahajan, S. K., Prasad, A. S., Lambjun, J., Abbasi, A. A., Rabbani, P., Briggs, W. A., and McDonald, F. D., 1980. Improvement of uremic hypogeusia by zinc: A double-blind study, *Am. J. Clin. Nutr.* 33:1517.

Mahajan, S. K., Prasad, A. S., Briggs, W. A., and McDonald, F. D., 1982. Correction of taste abnormalities and sexual dysfunction by zinc (Zn) in uremia: A double-blind study, *Ann. Intern. Med.* 97:357.

Mahajan, S. K., Flamenbaum, W., Hamburger, R. J., Prasad, A. S., and McDonald, F. D., 1985. Effect of zinc supplementation on hyperprolactinaemia in uremic men, *Lancet* 2:750.

Mahler, D. J., Walsh, J. R., and Haynie, G. D., 1971. Magnesium, zinc, and copper in dialysis patients, *Am. J. Clin. Pathol.* 56:17.

Mansouri, K., Halsted, J., and Gambos, E. A., 1970. Zinc, copper, magnesium and calcium in dialyzed and non-dialyzed uremic patients, *Arch. Intern. Med.* 125:88.

Meftah, S., and Prasad, A. S., 1989. Nucleotides in lymphocytes of human subjects with zinc deficiency, *J. Lab. Clin. Med.* 114:114.

Meftah, S., Prasad, A. S., Lee, D.-Y., and Brewer, G. J., 1991. Ecto 5′nucleotidase (5′NT) as a sensitive indicator of human zinc deficiency, *J. Lab. Clin. Med.* 118:309.

Michael, J., Hilton, P. J., and Jones, N. F., 1978. Zinc and the sodium pump in uremia, *Am. J. Clin. Nutr.* 31:1945.

Morrison, S. A., Russell, R. M., Carney, E. A., and Oaks, E. V., 1978. Zinc deficiency: A cause of abnormal dark adaptation in cirrhotics, *Am. J. Clin. Nutr.* 31:276.

Nagel, R. C., Freinkel, N., Bell, R. H., Friesen, H., Wilber, J. F., and Metzger, B. E., 1973. Gynecomastia, prolactin and other peptide hormones in patients undergoing chronic hemodialysis, *J. Clin. Endocrinol. Metab.* 36:428.

Newsome, D. A., Swartz, M., Leone, N. C., Elston, R. C., and Miller, E., 1988. Oral zinc in macular degeneration, *Arch. Ophthalmol.* 106:192.

Okada, A., Takagi, Y., Itakura, T., Satani, M., Manabe, H., Iida, Y., Tanigaki, T., Iwasaki, M., and Kasahara, N., 1976. Skin lesions during intravenous hyperalimentation. Zinc deficiency, *Surgery* 80:629.

Olambiwonnu, N. O., Penny, R., and Frasier, S. D., 1975. Sexual maturation in subjects with sickle cell anemia: Studies of serum gonadotropin concentration, height, weight, and skeletal age, *J. Pediatr.* 87:459.

Orgel, L. E., 1963. The maintenance of the accuracy of protein synthesis and its relevance to aging, *Proc. Natl. Acad. Sci. USA* 49:517.

Patek, A. J., and Haig, C., 1939. The occurrence of abnormal dark adaptation and its relation to vitamin A metabolism in patients with cirrhosis of the liver, *J. Clin. Invest.* 18:609.

Prasad, A. S., 1966. Metabolism of zinc and its deficiency in human subjects, in *Zinc Metabolism* (A. S. Prasad, ed.), Thomas, Springfield, Ill., p. 250.

Prasad, A. S., 1976. Deficiency of zinc in man and its toxicity, in *Trace Elements in Human Health and Disease* (A. S. Prasad, ed.), Academic Press, New York, p. 1.

Prasad, A. S., 1978. Zinc, in *Trace Elements and Iron in Human Metabolism* (A. S. Prasad, ed.), Plenum Press, New York, p. 251.

Prasad, A. S., 1982a. Clinical and biochemical spectrum of zinc deficiency in human subjects, in *Clinical, Biochemical, and Nutritional Aspects of Trace Elements* (A. S. Prasad, ed.), Liss, New York, p. 3.

Prasad, A. S., 1982b. History of zinc in human nutrition, in *Clinical Applications of Recent Advances in Zinc Metabolism* (A. S. Prasad, I. E. Dreosti, and B. S. Hetzel, eds.), Liss, New York, p. 1.

Prasad, A. S., 1982c. Recent developments in the diagnosis of zinc deficiency in man, in *Clinical Applications of Recent Advances in Zinc Metabolism* (A. S. Prasad, I. E. Dreosti, and B. S. Hetzel, eds.), Liss, New York, p. 141.

Prasad, A. S., 1984. Discovery and importance of zinc in human nutrition, *Fed. Proc.* 43:2829.

Prasad, A. S., and Cossack, Z. T., 1984. Zinc supplementation and growth in sickle cell disease, *Ann. Intern. Med.* 100:367.

Prasad, A. S., and Rabbani, P., 1981. Nucleoside phosphorylase in zinc deficiency, *Trans. Assoc. Am. Physicians* 94:314.

Prasad, A. S., Halsted, J. A., and Nadimi, M., 1961. Syndrome of iron deficiency anemia, hepatosplenomegaly, hypogonadism, dwarfism and geophagia, *Am. J. Med.* 31:532.

Prasad, A. S., Miale, A., Farid, Z., Sandstead, H. H., and Darby, W. J., 1963a. Biochemical studies on dwarfism, hypogonadism, and anemia, *Arch. Intern. Med.* 111:407.

Prasad, A. S., Miale, A., Farid, Z., Schulert, A., and Sandstead, H. H., 1963b. Zinc metabolism in patients with the syndrome of iron deficiency anemia, hypogonadism, and dwarfism, *J. Lab. Clin. Med.* 61:537.

Prasad, A. S., Schoomaker, E. B., Ortega, J., Brewer, G. J., Oberleas, D., and Oelshlegel, F. J., 1975. Zinc deficiency in sickle cell disease, *Clin. Chem.* 21: Washington, D.C., 582.

Prasad, A. S., Ortega, J., Brewer, G. J., and Oberleas, D., 1976. Trace elements in sickle cell disease, *J. Am. Med. Assoc.* 235:2396.

Prasad, A. S., Rabbani, P., Abbasi, A., Bowersox, E., and Fox, M. R. S., 1978. Experimental zinc deficiency in humans, *Ann. Intern. Med.* 89:483.

Prasad, A. S., Rabbani, P., and Warth, J. A., 1979. Effect of zinc on hyperammonemia in sickle cell anemia subjects, *Am. J. Hematol.* 7:323.

Prasad, A. S., Abbasi, A. A., Rabbani, P., and DuMouchelle, E., 1981. Effect of zinc supplementation on serum testosterone level in adult male sickle cell anemia subjects, *Am. J. Hematol.* 10:119.

Prasad, A. S., Meftah, S., Abdallah, J., Kaplan, J., Brewer, G. J., Bach, J. F., and Dardenne, M., 1988. Serum thymulin in human zinc deficiency, *J. Clin. Invest.* 82:1202.

Prasad, A. S., Fitzgerald, J. T., Hess, J. W., Kaplan, J., Pelen, F., and Dardenne, M., 1993. Zinc deficiency in the elderly, *Nutrition* 9:218.

Pryor, D. S., Don, N., and Macourt, D. C., 1981. Fifth day fits: A syndrome of neonatal convulsions, *Arch. Dis. Child.* 56:753.

Qureshi, M. S., Goldsmith, H. J., Pennington, J. H., and Cox, P. E., 1972. Cyclophosphamide therapy and sterility, *Lancet* 2:1290.

Rabbani, P., and Prasad, A. S., 1978. Plasma ammonia and liver ornithine transcarbamoylase activity in zinc deficient rat, *Am. J. Physiol.* 235:E203.

Rabbani, P., Prasad, A. S., Tsai, R., Harland, B. F., and Fox, M. R. S., 1987. Dietary model for production of experimental zinc deficiency in man, *Am. J. Clin. Nutr.* 45:1514.

Reding, P., Duchateau, J., and Bataille, C., 1984. Oral zinc supplementation improves hepatic encephalopathy, *Lancet* 2:493.

Rose, G. A., and Willden, E. G., 1972. Whole blood, red cell, and plasma total and ultrafilterable zinc levels in normal subjects and in patients with chronic renal failure with and without hemodialysis, *Br. J. Urol.* 44:281.

Sandstead, H. H., Prasad, A. S., Schulert, A. R., Farid, Z., Miale, A., Bassily, S., and Darby, W. J., 1967. Human zinc deficiency, endocrine manifestations and response to treatment, *Am. J. Clin. Nutr.* 20:422.

Sandstead, H. H., Henriksen, L. K., Greger, J. L., and Prasad, A. S., 1982. Zinc nutriture in the elderly in relation to taste acuity, immune response, and wound healing, *Am. J. Clin. Nutr.* 36:1046.

Shambaugh, G. E., Jr., 1985. Zinc and presbycusis, *Am. J. Otolaryngology* 6:116.

Simmer, K., and Thompson, R. P. H., 1985. Maternal zinc and intrauterine growth retardation, *Clin. Sci.* 68:395.

Solomons, N. W., Pineda, O., Viteri, F., and Sandstead, H. H., 1983. Studies in the bioavailability of zinc in humans: Mechanism of the intestinal interactions of non-heme iron and zinc, *J. Nutr.* 113:337.

Tapazoglou, E., Prasad, A. S., Hill, G., Brewer, G. J., and Kaplan, J., 1985. Decreased natural killer cell activity in patients with zinc deficiency and sickle cell disease, *J. Lab. Clin. Med.* 105:19.

Taper, L. J., Oliva, J. T., and Ritchey, S. J., 1985. Zinc and copper retention during pregnancy: The adequacy of prenatal diets with and without dietary supplementation, *Am. J. Clin. Nutr.* 41:1184.

Underwood, E. J., 1977. Zinc, in *Trace Elements in Human and Animal Nutrition* (E. Underwood, ed.), Academic Press, New York, p. 196.

Valberg, L. S., Flannagan, P. R., Brennan, J., and Chamberlain, M. J., 1985. Does the oral zinc tolerance test measure zinc absorption? *Am. J. Clin. Nutr.* 41:37.

Vallee, B. L., Wacker, W. E. C., Bartholomay, A. F., and Robin, E. D., 1956. Zinc metabolism in hepatic dysfunction. I. Serum zinc concentrations in Laennec's cirrhosis and their validation by sequential analysis, *N. Engl. J. Med.* 255:403.

Walford, R. L., 1980. Immunology and aging, *Am. J. Clin. Pathol.* 74:247.

Warth, J. A., Prasad, A. S., Zwas, F., and Frank, R. N., 1981. Abnormal dark adaptation in sickle cell anemia, *J. Lab. Clin. Med.* 98:189.

Weismann, K., Roed-Petersen, J., Hjorth, N., and Kopp, H., 1976. Chronic zinc deficiency syndrome in a beer drinker with a Billroth II resection, *Int. J. Dermatol.* 15:757.

Whitehouse, R. C., Prasad, A. S., Rabbani, P., and Cossack, Z. T., 1982. Zinc in plasma, neutrophils, lymphocytes, and erythrocytes as determined by flameless atomic absorption spectrophotometry, *Clin. Chem.* 28: Washington, D.C., 475.

Interactions of Zinc with Other Micronutrients

12

12.1 Introduction

During the past two decades, it has become clear that the addition of a trace element to the animal diet alters the metabolism of other elements. In some instances the interaction between the added element and the responding element is complementary. For example, the dietary level of iron needed to maintain a given concentration of hemoglobin is dependent on the dietary copper level (Hill and Matrone, 1961). A majority of other interactions, however, are of antagonistic nature. For example, high levels of zinc added to diets are known to precipitate copper deficiency in animals and humans (Hill and Matrone, 1970; Prasad *et al.*, 1978a). Several years ago, Hill and Matrone (1961) proposed that those elements whose electronic structure of the valence shell of the ions was the same, would act antagonistically to each other in biological systems (Hill, 1976).

This theory was based on the assumption that the biological interaction of elements must be dependent on the physicochemical properties of their ions and that ions whose valence shell electronic structures were similar would be antagonistic to each other biologically. In some cases ions may exhibit antagonistic relationships based on chemical interactions, and they may not show any similarity in electronic structures.

The electronic structure of the cuprous ion is d^{10} while that of the cupric ion is d^8p^1. Zn^{2+}, Cd^{2+}, and Hg^{2+} have the same structure of the valence shell as the cuprous ion, whereas Ag^{2+} has the same structure as the cupric ion. Biological antagonism between copper and zinc is well documented in the literature (Hill, 1976, 1988). A zinc level of 200 ppm is known to worsen the anemia of copper-deficient but not copper-supplemented chicks. Mortality was increased in the presence of 100 ppm zinc in the copper-deficient chicks and as little as 50 ppm zinc decreased growth of the copper-deficient chicks but not the copper-supplemented chicks.

A significant antagonistic interaction between cadmium and zinc as far as growth was concerned and between copper and cadmium as far as mortality was concerned, were documented in the chicks (Hill, 1976). Ag^{2+} also may act as an antagonist of copper. It has been shown that Ag^{2+} reduced growth, increased mortality, reduced hemoglobin levels, and reduced aortic elastin in the absence of copper in the diet of chicks, but these effects were prevented when copper was added to the diet.

12.2 Interactions of Zinc and Iron in Humans

There has been a great deal of concern that an antagonistic effect of iron on zinc absorption may have a detrimental effect on zinc nutrition, particularly in population groups that are routinely supplemented with iron (Flanagan and Valberg, 1988; Solomons, 1988). Both iron and zinc appear in the first transition series of the periodic table where they share an identical outer electronic configuration with manganese, cobalt, and nickel. Zinc and iron are similar in that both are essential for normal growth and development and are similar in amounts normally ingested, and absorbed iron represents only a fraction of the total iron present in the body (Flanagan and Valberg, 1988). In spite of these similarities, body homeostatic mechanisms of these elements differ and iron once absorbed is retained tenaciously, whereas zinc is both absorbed and excreted by the intestines (Flanagan and Valberg, 1988; Matseshe et al., 1980).

In rodents the capacity of the intestine to absorb iron is greatly increased by feeding a low-iron diet and in iron-deficient rats and mice, the oral absorption of zinc is also greatly increased (Forth and Rummel, 1973). It appears that in animals with a high capacity to absorb iron, a significant portion of zinc absorption occurs via the iron-absorbing mechanism and transport of zinc at least partly is inhibited by iron (Pollack et al., 1965; Hamilton et al., 1979).

The absorption of iron and zinc are not completely similar, however. Iron absorption is restricted to the duodenum, whereas zinc is absorbed throughout the upper small intestine (Forth and Rummel, 1973; Flanagan et al., 1983). Available data suggest that zinc is a less effective inhibitor of iron absorption than iron is of zinc absorption. Finally, in mice with sex-linked anemia (sla), the genetic lesion affects the absorption of iron but not zinc (Flanagan et al., 1984). These findings indicate that the absorption of zinc is not entirely restricted to that of the iron pathway.

In human studies, inorganic iron added to test solutions of zinc salts in Fe/Zn ratios of 2.25 significantly lowered zinc absorption (Solomons and Jacob, 1981; Solomons et al., 1983a; Valberg et al., 1984; Sandstrom et al.,

1985). Zinc and iron might not interact in humans to the same extent as in rodents and this could explain why higher Fe/Zn ratios were required to inhibit zinc absorption in the human studies (Flanagan and Valberg, 1988).

In contrast to the above results demonstrating an iron–zinc absorptive interaction in humans, in three studies where iron was given with food no effect on zinc absorption was observed. When oysters, providing about 54 mg "organic" zinc, were consumed with 100 mg ferrous iron, plasma uptake of zinc was unaltered (Solomons and Jacob, 1981). When turkey meat, containing 4 mg zinc, was consumed with either 17 or 34 mg ferric iron, zinc absorption was unchanged (Valberg et al., 1984). Finally, when ferrous iron at an Fe/Zn ratio of 25 was added to a composite meal containing 2.6 mg zinc, the absorption of zinc was not significantly changed (Sandstrom et al., 1985).

It appears, therefore, that under usual conditions, human zinc absorption is determined largely by the nature and extent of zinc complex formation with food in the intestinal tract and normally the influence of iron on zinc absorption may not be significant. Under unusual circumstances, however, if large iron supplements are ingested in the absence of food, it is likely that iron could detrimentally affect absorption of zinc.

According to Solomons (1988), however, an iron–zinc interaction in human subjects is of nutritional importance. He cites several findings supporting this conclusion. The first one is a study from Iran where Mahloudji et al. (1975) reported that growth of iron-deficient Iranian schoolboys was better when supplementation was with 20 mg iron daily, than when a combination of 20 mg iron and 20 mg zinc was used, suggesting that oral zinc may have decreased the absorption of iron in these children. The ^{59}Fe absorption studies of Aggett et al. (1983) lend support to this view.

The second evidence of interaction between iron and zinc in humans comes from the study of Prasad et al. (1978b) in four volunteers who were followed for several months on a semisynthetic soy protein-based zinc-deficient diet containing 3.5 mg zinc per day. The fall in plasma zinc was more precipitous in the first two subjects who received 130 mg iron per day (Fe/Zn ratio 37:1), relative to the other two who received 20.3 mg iron per day (Fe/Zn ratio 8:1), suggesting that iron excess may have contributed to zinc deficiency significantly in the first two subjects.

In another study, the growth rates of healthy, middle-class white infants in Denver, Colorado, were significantly different between those receiving 1.8 mg zinc per liter of infant formula versus those receiving 5.8 mg zinc per liter of infant formula (Solomons, 1988). The formula contained 12 mg iron/liter, suggesting that more zinc was required in the presence of iron for achieving optimal growth in the infants.

A final, population-level support for an antagonist iron–zinc interaction comes from observation of pregnant women receiving different levels of prenatal mineral supplements. Hambidge *et al.* (1983) reported an inverse relationship between the level of daily iron supplement and the plasma zinc level in the first and third trimesters of pregnancy. Campbell-Brown *et al.* (1985) observed that in their study the three women who supplemented 100 mg or more iron daily had the lowest plasma zinc levels relative to other pregnant women who supplemented less iron daily. Breskin *et al.* (1983) showed that prenatal supplements of 30 mg or more iron were associated with significantly lower plasma zinc levels in women compared with those who received no iron supplement or less than 30 mg per day. In contrast to the above observations, other investigators found no effect of iron supplementation (160 mg iron daily) on plasma zinc level in pregnant women (Solomons, 1988).

The absorption of trace elements involves intraluminal ligands, mucosal binding sites or channels, intracellular transport proteins, intracellular barrier proteins, basolateral energetic mechanisms, and circulating carrier proteins in the bloodstream (Solomons, 1988). Simultaneous presence of two ions with similar electronic configuration might lead to competition for movement at one or more of these specific levels.

A direct competition of iron and zinc in the intestine has been studied by several investigators using simple solutions of minerals and analyzing their effects on zinc uptake in the plasma (Abu-Hamdan *et al.*, 1984; Aggett *et al.*, 1983; Payton *et al.*, 1982; Sandstrom *et al.*, 1985; Solomons, 1983; Solomons and Jacob, 1981; Solomons *et al.*, 1983a,b; Valberg *et al.*, 1984). With only one exception (Payton *et al.*, 1982), all investigators observed an inhibitory effect of iron on zinc uptake. Consistent with the electronic configuration hypothesis, ferrous iron produced a greater inhibition of zinc uptake than did ferric iron (Hill and Matrone, 1970).

A high level of zinc supplement may affect iron storage directly. Toxic levels of zinc in the diet may shorten the life span of red blood cells and cause anemia because of the faster iron turnover (Settlemire and Matrone, 1967a). Zinc is known to interfere with iron uptake by the liver so that there is a decrease in the storage of iron as ferritin (Settlemire and Matrone, 1967b). Decreased iron content in the liver and kidneys in response to excessive zinc supplementation has been observed by several investigators (Cox and Harris, 1960; Magee and Matrone, 1960; Kang *et al.*, 1977; Hamilton *et al.*, 1979). Since transferrin in plasma is known to transport both iron and zinc, the interaction of iron and zinc may reflect competition at the transport level.

A marked increase in iron and a decrease in zinc concentration in various organs such as liver, bone, pancreas, and testis have been observed in zinc-deficient animals relative to pair-fed controls (Moses and Parker, 1964; Prasad *et al.*, 1967; Prasad *et al.*, 1969a,b). These changes are reversed following zinc supplementation. Alterations in iron absorption may be responsible for changes in iron levels of organs in zinc deficiency.

12.3 Zinc and Copper

The interaction between zinc and copper may be considered to be mutually antagonistic. Increased levels of copper in liver and bone have been observed in zinc-deficient experimental animals (Moses and Parker, 1964; Prasad et al., 1969a; Petering et al., 1971; Burch et al., 1975; Roth and Kirchgessner, 1977). The excretion of copper in milk is increased in zinc-deficient dairy cows (Kirchgessner et al., 1982). When the dietary zinc supply is excessively increased and the copper level is kept constant, copper deficiency is manifested by a reduction of copper concentrations in liver, heart, and serum, and the activities of copper metalloenzymes such as ceruloplasmin and cytochrome oxidase are decreased (Duncan et al., 1953; Van Reen, 1953; Cox and Harris, 1960). The effective dose level varies over a wide range from about 100 to more than 1000 mg Zn/kg dietary dry matter, because various dietary constituents may influence the availability of intestinal zinc and, thereby, the effects on copper metabolism. The changes imposed on the copper status may be reversed by raising the copper supply in order to narrow the zinc–copper ratio. Similarly, the zinc supply may be raised as a protective measure against copper toxicity (Suttle and Mills, 1966a,b; Bremner et al., 1976).

When the dietary copper supply is low, zinc will be present at a relatively high level compared with copper and in this situation the zinc status is not affected. Only a slight increase in zinc concentration in the liver and bone may be noted in copper-deficient animals (Schwarz and Kirchgessner, 1979). The effects of high copper intake when the zinc intake is normal are not well defined.

12.3.1 Mode of Copper–Zinc Interactions

Copper and zinc inhibit the intestinal absorption of each other. Depression of zinc absorption in the presence of excessive copper is observed only in rats offered an adequate supply of zinc, since in zinc-deficient rats zinc absorption could not be inhibited by copper (Evans et al., 1974; Schwarz and Kirchgessner, 1973; Schwarz and Kirchgessner, 1974a,b). Several studies have shown that in zinc deficiency, absorption of both zinc and copper administered separately is improved. Whereas absorption of copper is increased in copper deficiency, zinc absorption is not affected.

A completely satisfactory explanation for the copper–zinc interactions at the site of absorption is not available. Intestinal metallothionein level may play an important role. Metallothionein level in the intestine is directly related to the zinc status. After high doses of zinc supplementation, metallothionein level in the intestine increases (Richards and Cousins, 1975). Inasmuch as metallothionein has a higher affinity for copper, the increased intestinal metallothionein binds and traps copper which is ultimately excreted in the feces when the cells are sloughed off (Richards and Cousins, 1976). No such relationship between metallothionein and reduced zinc absorption after excess copper exposure has been reported (Hall *et al.*, 1979).

12.3.2 Zinc and Copper Interactions in Humans

Hypocupremia and hypoceruloplasminemia occurred in an adult with sickle-cell anemia (SCA) who received zinc (25 mg elemental zinc every 4 h) as an antisickling agent for 2 years (Prasad *et al.*, 1978a). The hypocupremia was associated with microcytosis and relative neutropenia (Figs. 12-1 and 12-2). Administration of copper resulted in an increase in RBC size and leukocyte counts. We have since observed hypoceruloplasminemia of varying degrees in several other SCA patients who were receiving oral zinc therapy. This complication was correctable with copper supplementation.

These observations were very interesting, in that they offered us a chance to decrease copper burden in patients with Wilson's disease by simply administering orally a relatively nontoxic element such as zinc. Wilson's disease is an autosomal recessive, inborn error involving low excretion of copper by the liver which leads to excessive accumulation of copper in the liver, brain, and other tissues and if untreated becomes fatal. Penicillamine, a copper

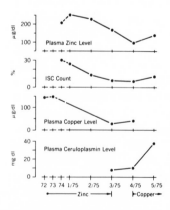

Figure 12-1. Plasma zinc, irreversible sickle cell (ISC) count, plasma copper, and plasma ceruloplasmin levels during the course of observation. Normal plasma zinc level (mean ± S.D.) is 113 ± 13.6 μg/dl, and normal plasma copper level is 116 ± 19 μg/dl. (Reprinted with permission from Prasad, A. S., Brewer, G. J., Schoomaker, E. B., and Rabbani, P., 1978. Hypocupremia induced by zinc therapy in adults, *J. Am. Med. Assoc.* 240:2167.)

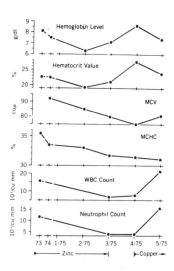

Figure 12-2. Hemoglobin level, hematocrit value, mean corpuscular cell volume (MCV), mean corpuscular hemoglobin concentration (MCHC), total leukocyte count, and total neutrophil count during the course of observation. Periods of zinc and copper administration are indicated. (Reprinted with permission from Prasad, A. S., Brewer, G. J., Schoomaker, E. B., and Rabbani, P., 1978. Hypocupremia induced by zinc therapy in adults, *J. Am. Med. Assoc.* 240:2167.)

chelator, is currently used for treatment of Wilson's disease. In about 30% of patients, acute sensitivity reactions including skin eruptions, fever, eosinophilia, leukopenia, thrombocytopenia, and lymphadenopathy occur within the first 2 weeks of therapy. If the drug is withdrawn for a few days, therapy frequently can be restarted, usually at a lower dose. However, in 10% of cases, intolerance to penicillamine is so great that the drug cannot be taken at all (Hirschman and Isselbacher, 1965). We therefore studied the efficacy of zinc treatment in Wilson's disease.

In our first studies we treated five patients with Wilson's disease with zinc (Brewer *et al.,* 1983). We administered 25 mg of elemental zinc as zinc acetate every 4 h during the day plus a 50-mg dose at bedtime. In studies on our first two patients with Wilson's disease, we assumed that direct competition between zinc and copper in the intestinal lumen would cause zinc inhibition of copper absorption and would immediately enhance the rate of fecal excretion of copper. We saw no such effect. Copper excretion increased after 21 days of zinc therapy in the first patient and after 7 days of zinc therapy in the second patient. We hypothesized that a buildup of zinc in body tissues must occur before copper absorption is inhibited.

We therefore modified the study protocol. The major modification was to pretreat the patients with zinc for 3 weeks before admission to the study. Penicillamine therapy was discontinued 1 week prior to admission.

At the time of admission, the patients were maintained on zinc therapy without penicillamine, and a copper balance study was done according to methods previously established. We induced a negative or neutral copper balance in all five subjects, who were receiving no therapy other than zinc.

A significant amount of copper is excreted into the gastrointestinal tract (endogenous excretion of copper). Whereas biliary copper is not reabsorbed to a large extent, the copper in saliva and gastric juices, totaling 1.5 mg/day, is normally reabsorbed. Reabsorption of nonbiliary copper as well as dietary copper are therefore targets for intestinal metallothionein which could bind copper and prevent its absorption. This hypothesis is consistent with our results inasmuch as pretreatment with zinc most likely resulted in increased synthesis of metallothionein in the intestinal cells (Brewer et al., 1983).

We further validated the copper balance data in Wilson's disease patients by measurement of ^{64}Cu uptake in blood following oral ingestion of 0.5 mCi ^{64}Cu acetate in 40 ml of cow's milk (Hill et al., 1986). The mean peak ^{64}Cu uptake into blood of nine Wilson's disease patients on D-penicillamine, Trien, or no medication was $6.04 \pm 2.74\%$, comparable with normal controls. Seven patients on zinc therapy had markedly and significantly reduced mean uptake of $0.79 \pm 1.05\%$ after treatment. These results demonstrated that the prevention of copper uptake into blood in Wilson's disease patients by zinc therapy can be evaluated by ^{64}Cu uptake and that peak uptake of less than 1% occurs in patients with neutral or negative copper balance.

Our further studies showed that 24-h urine copper and nonceruloplasmin plasma copper are the most useful parameters in monitoring zinc control of copper in Wilson's disease (Brewer et al., 1987a). In penicillamine-treated Wilson's disease patients, urinary copper is high because of the increased amount of mobilizable copper load in the body available for excretion and the effectiveness and dose of the penicillamine therapy. In the absence of penicillamine therapy (as is the case with zinc therapy), the quantity of copper in the urine becomes a reflection of the level of excess copper. With zinc therapy we see a fairly rapid decline of urinary copper in previously untreated patients, and a slow decline toward the normal range in patients who were previously treated with penicillamine, the urinary copper was 69 ± 24 μg/ day, which is considered to be in the normal range.

The range of nonceruloplasmin plasma copper (non-Cp Cu) of normal subjects is approximately 10–20 μg/100 ml. Non-Cp Cu is increased in untreated Wilson's disease patients and it is presumably the copper that causes toxicity in Wilson's disease (Brewer et al., 1987a). Thus, one objective of zinc therapy is to bring the level of non-Cp Cu to the normal range. In patients, as the treatment with zinc progresses, this variable tends to decrease to normal levels with continuation of the therapy.

In another study 12 patients with Wilson's disease, most of whom had received intensive treatment with penicillamine, were given zinc therapy as their sole treatment for copper control (Brewer et al., 1987b). Liver copper was assayed in serial biopsies during a 12- to 20-month follow-up period. Mean ($\pm$S.D.) baseline copper concentration was 255 ± 194 μg/g dry wt whereas after therapy it was 239 ± 185 μg/g dry wt. No subject showed hepatic reaccumulation of copper during zinc therapy (see Table 12-1 and Fig. 12-3). Copper balance, 24-h urinary copper, and non-Cp Cu indicated good copper control during zinc therapy. Hepatic zinc concentration increased two- to threefold over baseline values but no toxicity was seen. Hepatic zinc concentration appeared to reach a plateau after 12 to 18 months of zinc therapy. These results showed that oral zinc as the sole maintenance therapy in Wilson's disease prevents hepatic reaccumulation of copper. All patients on zinc remained clinically stable during the period of observation and no patient showed worsening or progression of neurologic disease or speech abnormality.

Our recent data (Table 12-2) indicate that a more simplified regimen such as 50 mg zinc three times a day is equally effective in controlling copper balance in patients with Wilson's disease and that administration of zinc six times a day at 4-h intervals was not necessary (Brewer *et al.*, 1987a; Brewer *et al.*, 1987b).

Table 12-1. Hepatic Copper Data in Patients with Wilson's Disease over 12 to 20 Months of Zinc Therapy

Patient number	Study interval (mo)	Liver copper (μg/g dry wt)		
		Start	End	Difference
Patients given intensive treatment with penicillamine before zinc therapy				
3	12	158	183	+25
5	18	678	630	−48
6	18	128	134	+6
7	15	210	253	+43
8	20	307	147	−160
9	15	162	40	−122
12	12	93	106	+13
16	12	35	70	+35
17	18	358	348	−10
22	12	58	87	+29
Mean (1 S.D.)	15.2 (3.1)	218 (191)	200 (177)	−18 (70)
Patients not given or not recently given treatment with penicillamine before zinc therapy				
20	12	395	428	+33
25	18	491	445	−46
Mean	15	438	436	−2
Grand mean (1 S.D.)	15.2 (3.1)	255 (194)	239 (185)	−16 (66)

[a] Adapted with permission from Brewer *et al.* 1987. Treatment of Wilson's disease with zinc. III. Prevention of reaccumulation of hepatic copper, *J. Lab. Clin. Med.* 109:526.

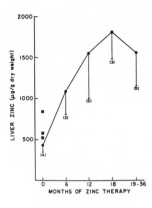

Figure 12-3. Means and S.D. of hepatic zinc values from liver biopsy specimens of patients with Wilson's disease (●) according to months of zinc therapy, and three control values (■). Number of values for each period are shown in parentheses. (Reprinted with permission from Brewer, G. J., Hill, G., Dick, R. D., Nostrant, T. T., Sams, J. S., Wells, J. J., and Prasad, A. S., 1987. Treatment of Wilson's disease with zinc. III. Prevention of reaccumulation of hepatic copper, *J. Lab. Clin. Med.* 109:529.)

A decreased plasma zinc level and increased plasma copper level have been reported in pregnancy, women on oral contraceptives, acute infection, malignancy, cardiovascular disease, renal disease, schizophrenia, and certain endocrine diseases such as acromegaly and Addison's disease.

12.4 Manganese and Zinc

Zinc levels in liver, kidney, spleen, and bone decline slightly as a result of manganese deficiency in experimental animals (Heiseke and Kirchgessner, 1978). Zinc retention may increase in response to diets supplemented with increased levels of manganese (Grace, 1973; Ivan and Grieve, 1975; Jarvinen and Ahlstrom, 1975). These findings are, however, controversial and more research is required in this area.

In some short-term studies, manganese levels in the liver, small intestine, heart, and bone decreased when zinc deficiency was induced in experimental animals (Schwarz and Kirchgessner, 1980). In other studies, however, no effect or even a positive effect of zinc depletion on manganese accumulation in certain organs was observed (Prasad *et al.,* 1967; Prasad *et al.,* 1969a; Roth and Kirchgessner, 1979). It is possible that zinc and manganese interact at the site of intermediary metabolism, leading to an increased requirement for manganese when zinc nutrition is suboptimal, and conversely, for zinc when manganese nutrition is inadequate. There is no evidence that manganese absorption changes with zinc deficiency or vice versa.

12.5 Cadmium, Lead, and Zinc

Zinc interacts negatively with cadmium. Zinc increases cadmium excretion and protects against its toxicity (Ahokas *et al.,* 1980; Lucis *et al.,* 1972;

Table 12-2. Zinc Therapy Regimen and Copper-Related Data in Patients with Wilson's Disease[a]

Patient number	Primary zinc therapy regimen	Study interval (mo)	Copper balance (μg/day)	24-h urine copper (μg/day)		Nonceruloplasmin plasma copper (μg/dl)	
				Start	End	Start	End
Patients given intensive treatment with penicillamine before zinc therapy							
3	50 mg × 2	12	−130	51	53	11.1	<1.0
5	50 mg × 3	18	−240	101	50	23.9	22.4
6	50 mg × 5	18	−490	51	51	37.8	16.4
7	50 mg × 3	15	−140	55	43	19.7	17.7
8	50 mg × 3	20	−120	51	63	10.2	3.7
9	50 mg × 3	15	−140	53	32	6.5	14.1
12	50 mg × 3	12	−350	53	60	27.4	11.7
16	50 mg × 3	12	−480	50	50	<1.0	3.0
17	50 mg × 3	18	−340	92	84	15.5	18.7
22	50 mg × 3	12	—	49	33	8.0	22.7
Mean (1 S.D.)		15.2 (3.1)	−270 (150)	60.6 (19.1)	51.9 (15.1)	16.1 (11.1)	13.1 (8.0)
Patients not given or not recently given treatment with penicillamine before zinc therapy							
20	50 mg × 5	12	−1752	287	227	27.7	26.7
25	50 mg × 3	18	—	172	133	39.7	17.7
Mean		15		230	180	33.7	22.2
Grand mean		15.2 (3.1)	−418 (490)	89 (72.3)	73 (55.5)	19.1 (12.4)	14.7 (8.3)

[a] Adapted with permission from Brewer et al. 1987. Treatment of Wilson's disease with zinc. III. Prevention of reaccumulation of hepatic copper, J. Lab. Clin. Med. 109: 526.

Schroeder *et al.*, 1970; Stowe, 1976). High ratios of cadmium to zinc in the kidneys have been reported in humans and rats with hypertension. In rats, zinc displaces cadmium from the kidney and decreases blood pressure (Schroeder *et al.*, 1970).

Marginal zinc deficiency in the rat leads to a greater accumulation of lead in rat pups whose mothers have been administered lead during lactation (Ashrafi and Fosmire, 1985). The clinical relevance of this observation to humans remains to be established.

12.6 Calcium and Zinc

At the intestinal level, calcium and zinc have an antagonistic relationship. In experimental animals, the absorption of zinc is decreased if the uptake of calcium is high and vice versa (Hanson *et al.*, 1958). In pigs, the relationship between parakeratosis and zinc deficiency was shown to be partly the result of high levels of dietary calcium in the presence of low zinc intake (Hanson *et al.*, 1958). The exacerbation of essential fatty acid deficiency by deficiency of zinc was shown to be enhanced in the presence of high calcium intake. In rats, fetal malformations induced by zinc deficiency were also dependent on calcium intake, in that when calcium intake was deficient, the malformation did not occur, whereas the malformations were increased when the intake of calcium was high (Hurley and Tao, 1972).

Zinc interacts with calcium at the red cell membrane level by suppressing calmodulin, a calcium-regulating protein (Baudier *et al.*, 1983). This property of zinc has been utilized to suppress formation of irreversibly sickled cells (ISCs), in which the intracellular concentration of calcium is known to rise and then calcium binds to hemoglobin and the membrane, forming ISCs. Oral administration of zinc to SCA patients is known to decrease the number of ISCs *in vivo* (Brewer, 1980; Brewer *et al.*, 1979).

12.7 Magnesium and Zinc

Mild deficiency of magnesium in the rat is associated with decreased levels of zinc in lower femur and whole carcass (Kubena *et al.*, 1985). Pyridoxal phosphokinase enzyme is catalyzed not only by magnesium but also by zinc (McCormick *et al.*, 1961).

12.8 Sodium and Zinc

Zinc (0.1 to 0.9 mM) infusion into the jejunum decreases sodium absorption in humans (Steinhardt and Adibi, 1984). Low dietary intake of so-

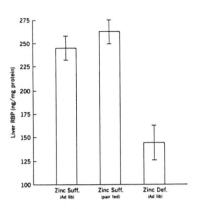

Figure 12-4. Effect of zinc deficiency on liver retinol-binding protein (RBP). Each bar represents the mean value for six or more samples. Vertical line at the top of each bar represents S.E. Liver RBP concentration was significantly lower in the zinc-deficient group ($p < 0.005$). [Reprinted with permission from Smith, J. C., Jr., 1982. Interrelationship of zinc and vitamin A metabolism in animal and human nutrition. A review, in *Clinical, Biochemical, and Nutritional Aspects of Trace Elements* (A. S. Prasad, ed.), Liss, New York, p. 243.]

dium is associated with increased urinary zinc excretion, suggesting that a sodium-dependent mechanism may control renal tubular reabsorption of zinc (Matustik *et al.,* 1982). The sodium/potassium content of muscle is increased in zinc-deficient rats and there is no increase in the aldosterone secretion in sodium-depleted rats as a result of zinc deficiency.

12.9 Vitamin A and Zinc

Zinc-deficient animals generally have low serum vitamin A and retinol-binding protein (RBP), low liver RBP, but a normal level of vitamin A in the

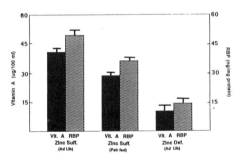

Figure 12-5. Effect of zinc deficiency on plasma vitamin A and retinol-binding protein (RBP). Each bar represents mean value for a group of seven rats. Vertical line at the top of each bar represents S.E. Each group of animals exhibited significantly different levels of plasma vitamin A or RBP ($p < 0.005$) from other groups. [Reprinted with permission from Smith, J. C., Jr., 1982. Interrelationship of zinc and vitamin A metabolism in animal and human nutrition. A review, in *Clinical, Biochemical, and Nutritional Aspects of Trace Elements* (A. S. Prasad, ed.), Liss, New York, p. 243.]

liver (Smith *et al.,* 1976) (see Figs. 12-4 and 12-5). It has been proposed that the release of vitamin A from the liver to the blood may be compromised in zinc deficiency. In human plasma, zinc has been reported to correlate positively with RBP (Smith, 1982).

The depression of plasma vitamin A does not appear to be a result of zinc deficiency *per se;* rather, it is a nonspecific response to food restriction and growth that are associated with zinc deficiency (Smith, 1982). Since liver vitamin A concentration is not altered by zinc deficiency, it appears that the absorption and transport of vitamin A to the liver are not affected.

The animal studies show that plasma and liver RBP concentration is reduced by zinc deficiency, suggesting that zinc plays a role in the synthesis of RBP.

In experimental animals, retinene reductase, presumably a zinc-metallo alcohol dehydrogenase of the retina, is sensitive to zinc status and its activity is decreased as a result of zinc deficiency. Thus, the conversion of vitamin A alcohol to vitamin A aldehyde may be impaired, resulting in an abnormality of dark adaptation in zinc-deficient subjects. Certain patients with cirrhosis of the liver and abnormal dark adaptation may not respond to vitamin A administration; on the other hand, zinc supplementation may correct this problem. Similarly, in zinc-deficient patients with sickle-cell disease, zinc supplementation is known to correct abnormal dark adaptation.

12.10 Vitamin B$_6$ and Zinc

Pyridoxal phosphate kinase, which has been shown to be activated by zinc (Neary and Divan, 1970), catalyzes the conversion of pyridoxine, pyridoxal, and pyridoxamine (collectively known as B$_6$). In zinc deficiency, the activity of pyridoxal phosphate kinase is decreased (Swenerton and Hurley, 1968). Pyridoxamine has also been reported to complex with zinc.

In vitamin B$_6$-deficient rats, tissue zinc levels have been observed to be either unchanged or generally decreased, except in the kidneys (Hsu, 1965; Ikeda *et al.,* 1979). Zinc absorption was increased by both vitamin B$_6$ deficiency and elevated vitamin B$_6$ intake (Cunnane, 1988). The mechanism of this curious phenomenon is not understood.

The gross manifestations of vitamin B$_6$ and zinc deficiency appear superficially similar. These include hypophagia, growth retardation, dermal lesions, alopecia, and immunodeficiency. Deficiencies of both may be partially corrected by essential fatty acid (EFA) supplementation (Cunnane, 1988). Decreased protein utilization, impaired metabolism of linoleic acid, and increased ratio of linoleic to arachidonic acid in tissue phospholipids have been reported to occur in both vitamin B$_6$ and zinc deficiencies. The essentiality of both zinc and vitamin B_6 from the adequate utilization of both protein and EFAs may partially account for similar manifestations as a result of these deficiencies.

12.11 Vitamin D and Zinc

Several observations indicate that interactions occur between vitamin D and zinc. It has been shown that the increase in urinary zinc excretion caused by calcium is blocked by vitamin D_3. Although in one study vitamin D_3 supplementation to uremic patients had no effect on zinc absorption, zinc status appeared to improve in uremic patients who received vitamin D supplementation (Hirschberg et al., 1985). It has also been reported that vitamin D_3 increases zinc concentration in the liver of rats following subtotal nephrectomy.

Zinc is known to stimulate vitamin D-induced DNA synthesis in femoral diaphyses (Cunnane, 1988). Recently it has been shown that zinc-finger proteins are involved in the genetic expression of vitamin D_3 receptors.

12.12 Vitamin E and Zinc

The interaction of vitamin E and zinc has not been investigated in depth. Both are antioxidants and possess membrane-stabilizing properties in the presence of free-radical excess, such as occurs following administration of carbon tetrachloride (Cunnane, 1988). In one study of zinc-deficient chicks, the dermal pathology and malonaldehyde release from the skin of the feet were significantly reduced by vitamin E.

12.13 Folic Acid and Zinc

In zinc-deficient rats, liver content of folic acid is decreased (Cunnane, 1988). Absorption of folic acid and plasma levels of the vitamin have been reported to be decreased in young men who were fed zinc-deficient diets. These findings suggest that zinc may be required for absorption of folic acid.

In another study, fecal zinc excretion was increased in men following folate supplementation (400 μg/2 days) and in male rats folic acid supplementation (500 mg/kg for 14 days in vivo) significantly decreased zinc absorption.

12.14 Biotin and Zinc

Symptoms of zinc deficiency in rats include growth retardation, alopecia, and dermatitis and are also seen in biotin deficiency, and biotin supplementation to zinc-deficient rats is known to partially alleviate these manifestations (Cunnane, 1988).

References

Abu-Hamdan, D. K., Mahajan, S. K., Migdal, S. D., Prasad, A. S., and McDonald, F. D., 1984. Zinc absorption in uremia: Effects of phosphate binders and iron supplements, *J. Am. Coll. Nutr.* 3:283.

Aggett, P. J., Crofton, R. W., Khin, C., Gvozdanovic, S., and Gvozdanovic, D., 1983. The mutual inhibitory effects on their bioavailability of inorganic zinc and iron, in *Zinc Deficiency in Human Subjects* (A. S. Prasad, A. O. Cavder, G. J. Brewer, and P. J. Aggett, eds.), Liss, New York, p. 117.

Ahokas, R. A., Dilts, P. V., and Lahaye, E. B., 1980. Cadmium-induced fetal growth retardation: Protective effect of excess dietary zinc, *Am. J. Obstet. Gynecol.* 136:216.

Ashrafi, M. H., and Fosmire, G. J., 1985. Effects of marginal zinc deficiency and subclinical lead toxicity in the rat neonate, *J. Nutr.* 115:334.

Baudier, J., Haglid, K., Haiech, J., and Gerard, D., 1983. Zinc ion binding to human calcium-binding proteins, calmodulin and S100b protein, *Biochem. Biophys. Res. Commun.* 114: 1138.

Bremner, I., Young, B. W., and Mills, C. F., 1976. Protective effect of zinc supplementation against copper toxicosis in sheep, *Br. J. Nutr.* 36:551.

Breskin, M. W., Worthington-Roberts, B. S., Knopp, R. H., Brown, Z., Plovie, B., Mottet, N. K., and Mills, J. L., 1983. First trimester serum zinc concentration in human pregnancy, *Am. J. Clin. Nutr.* 38:943.

Brewer, G. J., 1980. Calmodulin, zinc and calcium in cellular membrane regulation, *Am. J. Hematol.* 8:231.

Brewer, G. J., Aster, J. C., Knutsen, C. A., and Kruckberg, W. C., 1979. Zinc inhibition of calmodulin: A proposed molecular mechanism of zinc action on cellular function, *Am. J. Hematol.* 7:53.

Brewer, G. J., Hill, G. M., Prasad, A. S., Cossack, Z. T., and Rabbani, P., 1983. Oral zinc therapy for Wilson's disease, *Ann. Intern. Med.* 99:314.

Brewer, G. J., Hill, G. M., Dick, R. D., Nostrant, T. T., Sams, J. S., Wells, J. J., and Prasad, A. S., 1987a. Treatment of Wilson's disease with zinc. III. Prevention of reaccumulation of hepatic copper, *J. Lab. Clin. Med.* 109:526.

Brewer, G. J., Hill, G. M., Prasad, A. S., and Dick, R., 1987b. The treatment of Wilson's disease with zinc. IV. Efficacy monitoring using urine and plasma copper, *Proc. Soc. Exp. Biol. Med.* 184:446.

Burch, R. E., Williams, R. V., Hahn, H. K. J., Jetton, M. M., and Sullivan, J. F., 1975. Serum and tissue enzyme activity and trace element content in response to zinc deficiency in the pig, *Clin. Chem.* 21: Washington, D.C., 568.

Campbell-Brown, M., Ward, R. J., Haines, A. P., North, W. R. S., Abraham, R., and McFayden, I. R., 1985. Zinc and copper in Asian pregnancies: Is there evidence for a nutritional deficiency? *Br. J. Obstet. Gynecol.* 92:975.

Cox, D. H., and Harris, D. L., 1960. Effect of excess dietary zinc on iron and copper in the rat, *J. Nutr.* 70:514.

Cunnane, S. C., 1988. *Zinc: Clinical and Biochemical Significance*, CRC Press, Boca Raton, Fla., p. 113.

Duncan, G. D., Gray, L. F., and Daniel, L. J., 1953. Effect of zinc on cytochrome oxidase activity, *Proc. Soc. Exp. Biol. Med.* 83:625.

Evans, G. W., Grace, C. I., and Hahn, C., 1974. The effect of copper and cadmium on ^{65}Zn absorption in zinc-deficient and zinc-supplemented rats, *Bioinorg. Chem.* 3:115.

Flanagan, P. R., and Valberg, L. S., 1988. The intestinal interaction of zinc and iron in humans: Does it occur with food? in *Essential and Toxic Trace Elements in Human Health and Disease* (A. S. Prasad, ed.), Liss, New York, p. 501.

Flanagan, P. R., Haist, J., and Valberg, L. S., 1983. Zinc absorption, intraluminal zinc and intestinal metallothionein levels in zinc deficient and zinc repleted rodents, *J. Nutr.* 113:962.

Flanagan, P. R., Haist, J., MacKenzie, I., and Valberg, L. S., 1984. Intestinal absorption of zinc: Competitive interactions with iron, cobalt, and copper in mice with sex-linked anemia (sla), *Can. J. Physiol. Pharmacol.* 62:1124.

Forth, W., and Rummel, W., 1973. Iron absorption, *Physiol. Rev.* 63:724.

Grace, N. D., 1973. Effect of high dietary Mn levels on the growth rate and the level of mineral elements in the plasma and soft tissues of sheep, *N.Z. J. Agric. Res.* 16:177.

Hall, A. C., Young, B. W., and Bremner, I., 1979. Intestinal metallothionein and the mutual antagonism between copper and zinc in the rat, *J. Inorg. Biochem.* 11:57.

Hambidge, K. M., Krebs, N. F., Jacobs, M. A., Favier, A., Guyette, L., and Ickle, D. N., 1983. Zinc nutritional status during pregnancy: A longitudinal study, *Am. J. Clin. Nutr.* 37:429.

Hamilton, R. P., Fox, M. R. S., Fry, B. E., Jr., Jones, A. O. L., and Jacobs, R. M., 1979. Zinc interference with copper, iron and manganese in young Japanese quail, *J. Food Sci.* 44:738.

Hanson, L. J., Sorenson, D. K., and Kernkamp, H. C. H., 1958. Essential fatty acid deficiency— Its role in parakeratosis, *Am. J. Vet. Res.* 18:1921.

Heiseke, D., and Kirchgessner, M., 1978. Eisen- und Zinkgehalte in verschiedenen Organen der Ratte bei Mangan-Mangel, *Zentralbl. Veterinaermed. Reihe A* 25:307.

Hill, C. H., 1976. Mineral interrelationships, in *Trace Elements in Human Health and Disease*, Vol. II (A. S. Prasad, ed.), Academic Press, New York, p. 281.

Hill, C. H., 1988. Interactions among trace elements, in *Essential and Toxic Trace Elements in Human Health and Disease* (A. S. Prasad, ed.), Liss, New York, p. 491.

Hill, C. H., and Matrone, G., 1961. Studies on copper and iron deficiencies in growing chickens, *J. Nutr.* 73:425.

Hill, C. H., and Matrone, G., 1970. Chemical parameters in the study of in vivo and in vitro interactions of transition elements, *Fed. Proc.* 29:1474.

Hill, G. M., Brewer, G. J., Juni, J. E., Prasad, A. S., and Dick, R. D., 1986. Treatment of Wilson's disease with zinc. II. Validation of oral ^{64}copper with copper balance, *Am. J. Med. Sci.* 292:344.

Hirschberg, R., Von Herrath, D., Vob, K., Bosaller, W., Mauelshagen, U., Pauls, A., and Schaefer, K., 1985. Parathyroid hormone and 1,25-dihydroxyvitamin D_3 affect the tissue concentrations of zinc in uremic rats, *Nephron* 39:277.

Hirschman, S. Z., and Isselbacher, K. J., 1965. The nephrotic syndrome as a complication of penicillamine therapy of hepatolenticular degeneration (Wilson's disease), *Ann. Intern. Med.* 62:1297.

Hsu, J. M., 1965. Zinc content in pyridoxine deficient rats, *Proc. Soc. Exp. Biol. Med.* 119:177.

Hurley, L. S., and Tao, S. H., 1972. Alleviation of teratogenic effects of zinc deficiency by simultaneous lack of calcium, *Am. J. Physiol.* 222:322.

Ikeda, M., Hosotani, T., Ueda, T., Kotake, Y., and Sakeibara, B., 1979. Observations of the concentration of zinc and iron in tissues of vitamin B_6 deficient germ-free rats, *J. Nutr. Sci. Vitaminol.* 25:151.

Ivan, M., and Grieve, C. M., 1975. Effects of zinc, copper, and manganese supplementation of high concentrate ration on digestibility, growth, and tissue content of Holstein calves, *J. Dairy Sci.* 58:410.

Jarvinen, R., and Ahlstrom, A., 1975. Effect of the dietary manganese level on tissue manganese, iron, copper, and zinc concentrations in female rats and their fetuses, *Med. Biol.* 53:93.

Kang, H. K., Harvey, P. W., Valentine, J. L., and Swendseid, M. E., 1977. Zinc, iron, copper and magnesium concentrations in tissues of rats fed various amounts of zinc, *Clin. Chem.* 23: Washington, D.C., 1834.

Kirchgessner, M., Schwarz, F. J., and Schnegg, A., 1982. Interactions of essential metals in human physiology, in *Clinical, Biochemical, and Nutritional Aspects of Trace Elements* (A. S. Prasad, ed.), Liss, New York, p. 477.

Kubena, K. S., Landmann, W. A., Young, C. R., and Carpenter, Z. L., 1985. Influence of magnesium deficiency and soy protein on magnesium and zinc status in rats, *Nutr. Res.* 5:317.

Lucis, O. J., Lucis, R., and Shaikh, Z. A., 1972. Cadmium and zinc in pregnancy and lactation, *Arch. Environ. Health* 25:14.

McCormick, D. B., Gregory, M. E., and Snell, E. E., 1961. Pyridoxal phosphokinase I: Assay, distribution, purification and properties, *J. Biol. Chem.* 236:2076.

Magee, A. C., and Matrone, G., 1960. Studies on growth, copper metabolism and iron metabolism on rats fed high levels of zinc, *J. Nutr.* 72:233.

Mahloudji, M., Reinhold, J. G., Haghasenass, M., Ronaghy, H. A., Spivey-Fox, M. R. S., and Halsted, J. A., 1975. Combined zinc and iron supplementation of diets of 6- and 12-year-old school children in southern Iran, *Am. J. Clin. Nutr.* 28:721.

Matseshe, J. W., Phillips, S. F., Malagelada, J. R., and McCall, J. T., 1980. Recovery of dietary iron and zinc from the proximal intestine of healthy man: Studies of different meals and supplements, *Am. J. Clin. Nutr.* 33:1946.

Matustik, M. C., Chausner, A. B., and Meyer, W. J., 1982. The effect of sodium intake on zinc excretion in patients with sickle cell anemia, *J. Am. Coll. Nutr.* 1:331.

Moses, H. A., and Parker, H. E., 1964. Influence of dietary zinc and age on the mineral content of rat tissues, *Fed. Proc.* 23:132.

Neary, J. T., and Divan, W. F., 1970. Purification, properties and a possible mechanism for pyridoxal kinase for bovine brain, *J. Biol. Chem.* 245:5585.

Payton, K. B., Flanagan, P. R., Stinson, E. A., Chrodiker, D. R., Chamberlain, M. J., and Valberg, L. S., 1982. Technique for determination of human zinc absorption from measurement of radioactivity in a fecal sample of the body, *Gastroenterology* 83:1264.

Petering, H. G., Johnson, M. A., and Horwitz, J. P., 1971. Studies of zinc metabolism in the rat, *Arch. Environ. Health* 23:93.

Pollack, S., George, J. N., Reba, R. C., Kaufman, R. M., and Crosby, W. J., 1965. The absorption of nonferrous metals in iron deficiency, *J. Clin. Invest.* 44:1470.

Prasad, A. S., Oberleas, D., Wolf, P., and Horwitz, J. P., 1967. Studies on zinc deficiency: Changes in trace elements and enzyme activities in tissues of zinc deficient rats, *J. Clin. Invest.* 46: 549.

Prasad, A. S., Oberleas, D., Wolf, P., Horwitz, J. P., Miller, E. R., and Luecke, R. W., 1969a. Changes in trace elements and enzyme activities in tissues of zinc deficient pigs, *Am. J. Clin. Nutr.* 22:628.

Prasad, A. S., Oberleas, D., Wolf, P., and Horwitz, J. P., 1969b. Effect of growth hormone on nonhypophysectomized zinc deficient rats and zinc on hypophysectomized rats, *J. Lab. Clin. Med.* 73:486.

Prasad, A. S., Brewer, G. J., Schoomaker, E. B., and Rabbani, P., 1978a. Hypocupremia induced by zinc therapy in adults, *J. Am. Med. Assoc.* 240:2166.

Prasad, A. S., Rabbani, P., Abassi, A., Bowersox, E., and Spivey-Fox, M. R. S., 1978b. Experimental zinc deficiency in humans, *Ann. Intern. Med.* 89:483.

Richards, M. P., and Cousins, R. J., 1975. Mammalian zinc homeostasis: Requirement for RNA and metallothionein synthesis, *Biochem. Biophys. Res. Commun.* 64:1215.

Richards, M. P., and Cousins, R. J., 1976. Metallothionein and its relationship to the metabolism of dietary zinc in rats, *J. Nutr.* 106:1591.

Roth, H. P., and Kirchgessner, M., 1977. Zum Gehalt von Zink, Kupfer, Eisen, Mangan and Calcium in Knochen and Lebern von an Zink depletierter and repleiterter Ratten, *Zentralbl. Veterinaermed. Reihe A* 24:177.

Roth, H. P., and Kirchgessner, M., 1979. Zinc und Chromgehalte in Serum, Pankreas und Leber von Zn-Mangelratten nach Glucosetimulierung, *Z. Tierphysiol. Tierernaehr. Futtermittelkd.* 42:277.

Sandstrom, B., Davidson, L., Cederblad, A., and Lonnerdal, B., 1985. Oral iron, dietary ligands and zinc absorption, *J. Nutr.* 115:411.

Schroeder, H. A., Baker, J. T., Hansen, N. M., Size, J. G., and Wise, R. A., 1970. Vascular reactivity of rats altered by cadmium and a zinc chelate, *Arch. Environ. Health* 21:609.

Schwarz, F. J., and Kirchgessner, M., 1973. Intestinale Cu-Absorption in vitro nach. Fe-oder Zn-Depletion, *Z. Tierphysiol. Tierernaehr. Futtermittelkd.* 31:91.

Schwarz, F. J., and Kirchgessner, M., 1974a. Absorption von Zink-65 und Kupfer-64 im Zink-mangel, *Int. J. Vitam. Nutr. Res.* 44:258.

Schwarz, F. J., and Kirchgessner, M., 1974b. Wechselwirkungen bei der intestinalen Absorption von ^{64}Cu, ^{65}Zn and ^{59}Fe nach Cu-, Zn- oder Fe-Depletion, *Int. J. Vitam. Nutr. Res.* 44:116.

Schwarz, F. J., and Kirchgessner, M., 1979. Kupfer-, Zink-, Eisen- und Mangankonzentrationen im Serum in Knochen and der Leber nach Kupferdepletion, *Zentralbl. Veterinaermed. Reihe A* 26:493.

Schwarz, F. J., and Kirchgessner, M., 1980. Experimentelle Untersuchungen zur Interaktion Zwischen den Spurenelementen Zink und Mangan, *Z. Tierphysiol. Tierernaehr. Futtermittelkd.* 43:272.

Settlemire, C. T., and Matrone, G., 1967a. In vivo effect of zinc on iron turnover in rats and life span of the erythrocyte, *J. Nutr.* 92:159.

Settlemire, C. T., and Matrone, G., 1967b. In vivo interference of zinc with ferritin iron in the rat, *J. Nutr.* 92:153.

Smith, J. C., Jr., 1982. Interrelationship of zinc and vitamin A metabolism in animal and human nutrition: A review, in *Clinical, Biochemical, and Nutritional Aspects of Trace Elements* (A. S. Prasad, ed.), Liss, New York, p. 239.

Smith, J. C., Jr., Brown, E. D., McDaniel, E. G., and Chan, W., 1976. Alterations in vitamin A metabolism during zinc deficiency and food and growth restriction, *J. Nutr.* 106:569.

Solomons, N. W., 1983. Competitive mineral:mineral interactions in the intestine: Implications for zinc absorption in humans, in *Nutritional Bioavailability of Zinc*, ACS Symposium Series. American Chemical Society, Washington, D.C., p. 247.

Solomons, N. W., 1988. The iron:zinc interaction in the human intestine. Does it exist? An affirmative view, in *Essential and Toxic Trace Elements in Human Health and Disease* (A. S. Prasad, ed.), Liss, New York, p. 509.

Solomons, N. W., and Jacob, R. A., 1981. Studies on the bioavailability of zinc in humans: Effect of heme and nonheme iron on the absorption of zinc, *Am. J. Clin. Nutr.* 34:475.

Solomons, N. W., Pineda, O., Viteri, F., and Sandstead, H. H., 1983a. Studies on the bioavailability of zinc in humans: Mechanism of the intestinal interaction of nonheme iron and zinc, *J. Nutr.* 113:337.

Solomons, N. W., Marchini, J. S., Duarte-Favaro, R. M., Vannuchi, H., and Dutra de Oliveira, J. E., 1983b. Studies on the bioavailability of zinc in humans. VI. Intestinal interaction of tin and zinc, *Am. J. Clin. Nutr.* 37:566.

Steinhardt, H. J., and Adibi, S. A., 1984. Interaction between transport of zinc and other solute in human intestine, *Am. J. Physiol.* 247:G176.

Stowe, H. D., 1976. Biliary excretion of cadmium by rats: Effects of zinc, cadmium and selenium pre-treatments, *J. Toxicol. Environ. Health* 2:45.

Suttle, N. F., and Mills, C. F., 1966a. Studies on the toxicity of copper to pigs. 1: Effects of oral supplements of zinc and iron salts at the development of copper toxicosis, *Br. J. Nutr.* 20: 135.

Suttle, N. F., and Mills, C. F., 1966b. Studies on the toxicity of copper to pigs. 2: Effect of protein source and other dietary components on the response to high and moderate intakes of copper, *Br. J. Nutr.* 20:149.

Swenerton, H., and Hurley, L. S., 1968. Severe zinc deficiency in male and female rats, *J. Nutr.* 95:8.

Valberg, L. S., Flanagan, P. R., and Chamberlain, M. J., 1984. Effects of iron, tin, and copper on zinc absorption in humans, *Am. J. Clin. Nutr.* 40:536.

Van Reen, R., 1953. Effects of excessive dietary zinc in the rat and the interrelationship with copper, *Arch. Biochem. Biophys.* 46:337.

Techniques for Measurement of Zinc in Biological Samples 13

This chapter provides a brief description of techniques employed to assay zinc in plasma and blood cell in my laboratory. Many clinical laboratories use inaccurate and imprecise methods for assay of zinc. Since the levels of zinc in plasma and blood cells are very low, the methods must be very accurate in order to derive important biological information from this assay.

13.1 Determination of Zinc in Plasma

Zinc in biologic fluids can be measured by conventional chemical methods. These methods usually require large specimen volume and are time-consuming, technically demanding, and prone to contamination. The methods presented here use atomic absorption spectrophotometry, which is a simpler, faster, more precise, and more accurate technique (Prasad *et al.*, 1965; Whitehouse *et al.*, 1982; Wang *et al.*, 1989). The analytical conditions vary with different instruments or analytical systems. The methods described are suitable for use with different instruments and should produce comparable results.

13.1.1 Specimen

Fasting specimens are preferable inasmuch as diurnal variations of plasma zinc have been observed. Promptly separated plasma is more desirable than serum for zinc determination. Zinc-free heparin should be used as the anticoagulant. In order to remove zinc from heparin, a small resin column packed with Chelex 100 Dow-Rad, 100–200 mesh, sodium form, may be used. Hemolysis must be avoided; the zinc concentration in erythrocytes is about 10–15 times higher in plasma. All specimens must be collected and kept in chemically clean containers.

13.1.2 Reagents

1. Zinc metal, high purity (Fisher Scientific Co., Pittsburgh).
2. Trichloroacetic acid (TCA), AR, 10 g/dl (10% w/v). In 500-ml volumetric flask place 50.0 g TCA and bring to volume with distilled, deionized water.
3. Nitric acid, AR, zinc free (Fisher Scientific).
4. Hydrochloric acid, AR, zinc free (Fisher Scientific).
5. Glycerol, AR, 5 and 10% v/v. In 1000-ml volumetric flasks place 50 and 100 ml glycerol. Bring to volume with distilled, deionized water.

Standards (zinc stock standards are commercially available from Fisher Scientific).

1. Zinc, stock standard, 500 μg/ml. In 1000-ml volumetric flask dissolve 0.500 g zinc metal in minimum volume of aqueous 1:1 HCl. Bring to volume with 1% v/v HCl.
2. Zinc, intermediary standard, 500 μg/dl. In 1000-ml volumetric flask place 10.0 ml stock standard. Bring to volume with 1% v/v HCl.
3. Zinc, working standards. In 500-ml volumetric flasks place 400, 300, 200, and 100 ml intermediary standard. Bring to volume with 1% v/v HCl. These standards have concentrations of 400, 300, 200, and 100 μg/dl. Intermediary and working standards should be prepared before using.

13.1.3 Quality Assurance

Chemically clean plastic and glassware are essential for all analytical steps. All reagents must be prepared with distilled, deionized (Type I) water and checked for contamination. Controls of pooled plasma, sterilized by filtration (Millipore), and kept frozen ($-20°C$) in small (2–3 ml) vials are usually stable (± 1 S.D. from mean) for 90–120 days. Instrument calibration and air and gas filters and regulators should be checked every time a new series of standards or reagents is used. Traditionally, acid-washed glassware has been required for keeping all reagents and standards. Recently, plastic (i.e., high-density polypropylene, polycarbonate) containers and newly developed detergents (i.e., Contrad-70, Scientific Products, American Hospital Supply Corp., Evanston, Ill.) are gaining acceptance in trace element analysis.

13.1.4 Procedure

Two versions of the techniques for determination of zinc and copper are presented: one uses a protein-free filtrate, and the other uses a direct dilution of the specimen.

13.1.4.1 Protein-Free Filtrate Method

1. In 5-ml tubes place 1.0 ml working standards (in duplicate), controls, and plasma. Add 1.0 ml of 10% v/v TCA. Mix for 3 s on vortex mixer.
2. Place all tubes in 90°C water bath for 10 min.
3. Centrifuge (refrigerated centrifuge is desirable) at 2500–3000 rpm for 15 min.
4. Remove clear supernatant and transfer it to a 10-ml tube.
5. Add 1.0 ml of 10% v/v TCA to tube with protein precipitate. Mix and place in 90°C water bath for 10 min.
6. Repeat steps 4 and 5 twice. This gives a 1:4 dilution.
7. Set atomic absorption spectrophotometer at 213.9 mm for zinc.
8. Measure standards and controls. Prepare calibration graph to find concentration of plasma specimens.

Some instruments can be internally precalibrated, and the concentration of the specimen can be read directly. The linearity in the response of the instrument must be verified by the standard additions method or by measuring selected high or low concentrations of the element being measured.

A strip chart recorder is a desirable addition; it provides permanent records and shows deviations from analytical performance.

13.1.4.2 Direct Dilution Method

1. Standard preparation. To prepare working standards, dilute stock and intermediary standards with glycerol 5% v/v for zinc. Glycerol solution compensates for low viscosity of standards in comparison with diluted specimens.
2. Specimen preparation. Make plasma dilutions with distilled, deionized water, for zinc.
3. Measure standards and controls. Prepare calibration graph to find concentrations of plasma specimens. Make necessary conversions for dilution used for specimen. The direct dilution method must be checked against the TCA protein-free filtrate method. The result should agree within 1 S.D. from mean.

The protein-free filtrate method yields higher values, probably because matrix effects are minimized. On the other hand, the direct dilution technique is faster and simpler to operate.

13.2 Zinc in Platelets, Lymphocytes, and Granulocytes by Flameless Atomic Absorption Spectrophotometry

In this section a procedure is described for separating platelets (PLT), lymphocytes, and granulocytes (PMN) from a single sample of whole blood, and analysis for zinc in PLT, lymphocytes, and PMN by flameless atomic absorption spectrophotometry (Wang *et al.*, 1989). The method is based on two previously published papers (Prasad *et al.*, 1965; Whitehouse *et al.*, 1982).

13.2.1 Materials and Methods

13.2.1.1 Apparatus

For Zn measurements, an Instrumentation Laboratory (Thermo Jarrell Ash Corp., Franklin, Mass.) atomic absorption spectrophotometer Model 551 with autosampler Model 254 and furnace atomizer Model 655 were used. The autosampler was set to aspirate at 5.0 ml/min with a sample deposit time of 4 s. The furnace heating procedure was as follows: sample deposited and instant drying at 50°C; two-step pyrolysis with a 15-s ramp to 325°C and a 20-s ramp at 450°C and atomization at 1800°C using a pyrolytically coated delayed atomization graphite cuvette. The spectrophotometer was operated at 213.9 nm in the peak area (P/A) absorbance mode with a 1-nm bandpass.

A Baker Instruments Hematology Series System 8,000 cell counter with Series 108 dilutor (Baker Instruments Corp., Allentown, Pa.) was used for cell counts.

13.2.1.2 Reagents

All of the water used throughout the procedure was distilled (Barnstead) and passed through a Continental Water Systems (Hazel Park, Mich.) primary deionization unit and Millipore 0.2-μm postfilter cartridge with final filtration through an ultrapure mixed bed resin column (Barnstead Co., Division of Sybron Corp., Boston, Mass.).

13.2.1.3 Cell Separation and Sample Preparation

Venous blood (15 ml) was collected and placed in a 50-ml polypropylene tube (Sarstedt, No. 62.547.004, Princeton, N.J.) containing 300 μl of Zn-free heparin. The tube was capped, and gently inverted several times and centrifuged at 180g for 5 min. The PLT-rich plasma was removed and placed in a 14-ml polypropylene tube (Sarstedt, No. 55.538) and capped (Sarstedt, No. 65.793, push-in stoppers) for PLT isolation. Normal saline to the remaining blood was added very slowly (along the tube side) to a total volume of 30 ml, and centrifuged at 180g for 5 min. The supernate was discarded and normal saline added to a total volume of 24 ml. The tube was capped and gently inverted several times.

Four tubes of a discontinuous gradient were prepared using 14-ml polypropylene tubes. The lower gradient was 3 ml of Sigma Histopaque 1.119 and the upper gradient was 3 ml of Sigma Histopaque 1.077. These gradients were carefully layered using transfer pipettes (Saint-Amand Mfg. Co., Inc., San Fernando, Calif.) just prior to layering the diluted sample. Very gently 6 ml of the diluted blood using a transfer pipette was layered over the gradient of 1.077 in each of the four tubes. The tubes were capped and centrifuged at 180g for 20 to 30 min. The supernate was removed down to the normal saline–Histopaque 1.077 interface and the cloudy layer of lymphocytes was transferred from the four tubes to a 14-ml polypropylene tube using a transfer pipette. To harvest PMN, the remaining Histopaque 1.077 down to the 1.077/ 1.119 interface was discarded and the cloudy layer of PMN was transferred to two 14-ml polypropylene tubes using another transfer pipette. Lymphocyte and PMN tubes were filled to 13 ml with normal saline, capped and mixed gently.

The lymphocytes and PMN tubes were centrifuged at 600g for 15 min. The supernates were aspirated and pellets resuspended with 1 ml of a hypotonic phosphate-buffered saline (PBS) solution (27.89 mM NaCl, 1.83 mM Na$_2$HPO$_4$, 0.338 mM KH$_2$PO$_4$, 0.62 mM KCl). Water (4 ml) was added and mixed gently for 10 s using a transfer pipette. Five milliliters of a hypertonic PBS solution (237.4 mM NaCl, 15.83 mM Na$_2$HPO$_4$ 2.87 mM KH$_2$PO$_4$, 5.24 mM KCl) was added by inversion to restore the isotonicity. The tubes were centrifuged at 280g for 5 min. This procedure was repeated two or three times, as necessary, in order to remove any contaminant red blood cells (RBC). Once the pellet was free of RBC, the supernate was discarded, and the pellet resuspended with 1 ml of normal saline and the tube filled to 5 ml with normal saline. The tubes were centrifuged at 280g for 5 min. This procedure was done once or twice in order to wash the cells.

The PLT-rich plasma was centrifuged at 280g for 5 min in order to remove white blood cells (WBC) and RBC. The supernate was transferred to a 14-ml polypropylene tube, capped, and centrifuged at 600g for 20 min. The plasma was removed to a 5-ml polystyrene tube (Sarstedt, No. 55.476/013) for assay of Zn. Normal saline was used to wash PLT twice (the same as for lymphocytes and PMN, except that PLT were centrifuged at 600g for 20 min). In order to remove any contaminant RBC, the same procedure was used as that for lymphocytes and PMN to lyse RBC, but the samples were centrifuged at 600g for 20 min. Normal saline was used to wash the PLT twice.

Lymphocyte and PMN tubes were centrifuged at 280g for 5 min and PLT tubes were centrifuged at 600g for 20 min. The supernate was discarded and the pellets resuspended in 1–1.5 ml of normal saline. Portions (200 μl) of suspensions were placed in 1.5-ml polypropylene micro tubes (Sarstedt, No. 72.690) for cell counts. Other portions (500–1000 μl) of suspensions were placed in 1.5-ml micro tubes that were rinsed once with nitric acid solution (25 ml of 11.2 M nitric acid in 1000 ml of water) and three times with water. The tubes were capped and centrifuged at 600g for 20 min. An aspirator (water power) connected with a disposable microliter pipette tip (Rainin, Rt 20, Instrument Co., Inc., Woburn, Mass.) that had been rinsed with nitric acid solution and water (same as above) was used to carefully draw away the normal saline (not too close to the pellet in order to avoid disturbing the cells). Depending on cell number, 500–1000 μl of nitric acid (11.2 M) was added to the pellets so that the samples had 7000–10,000 lymphocytes/μl, 12,000–15,000 PMN/μl, or 400,000–500,000 PLT/μl. The digestion was carried overnight at room temperature, and then incubated for 4 h in a 60°C water bath. The samples were allowed to cool overnight at room temperature.

13.2.1.4 Analysis for Zn

Zn standards of 60, 120, 180, and 240 ng/ml were prepared by diluting stock Zn standard solution (Sigma, No. Z-2750, St Louis, MO.). Water (1950 μl) was placed into each of the 5-ml polystyrene tubes using pipette and tips (the tubes and tips were rinsed as described above). Portions (50 μl) of standards of 60, 120, 180, 240 ng/ml in duplicate were added to the tubes separately. The final concentrations of zinc standards were 1.5, 3.0, 4.5, 6.0 ng/ml. P/A absorbance of the zinc was measured by atomic absorption spectrophotometry. The samples of PLT, lymphocytes, and PMN were diluted in the same fashion as standards in duplicate. By using P/A absorbance, the spectrophotometer was programmed to read the samples in concentration unit. The instrument was autozeroed with water and autocalibrated with the midrange standard (3 ng/ml) before reading samples. After six samples were read, the instrument was re-autozeroed and re-autocalibrated. Zn values of PLT remaining in the lymphocyte and PMN fractions were subtracted from those of lymphocytes and PMNs separately. Zn content of each sample was expressed in terms of micrograms per 10^{10} cells.

13.2.1.5 Cell Type Homogeneity, Viability Test, and Matrix Effects

Wright's stain was used to determine relative percentages of lymphocytes and PMN in samples, and the trypan blue method (Hudson and Hay, 1980) was used to determine the viability of cells.

Five milliliters of nitric acid (11.2 M) was added to 500 mg of bovine liver standard (BLS; U.S. National Bureau of Standards, Washington, D.C., Standard Reference Material, No. 1577a) in a 50-ml polypropylene tube, mixed, and digested in the same manner as the samples. The sample was centrifuged at 900g for 20 min. The lipids floating in the solution were removed. The solution was transferred to a 14-ml polypropylene tube. The standards were prepared in the same manner as the Zn standards. BLS and Zn standards were measured during the same run.

At the last step of sample preparation, after normal saline was drawn from samples of PLT, lymphocytes, and PMN, 1.4 ml of water was added gently to each of the pellets. The water was then drawn away and nitric acid added. After digestion, absorbance of these samples was measured.

In order to determine the matrix effect of sodium chloride, 4 μl of normal saline was added to 500-μl aliquots of each sample of cells. The absorbance of these samples was measured.

13.2.2 Results

In this study, Zn values (μg/10^{10} cells, mean ± S.D.) in PLT, lymphocytes, and PMN were 3.3 ± 0.3, 50.9 ± 5.9, and 45.9 ± 6.4, respectively (Table 13-1), in normal healthy subjects between the ages of 23 and 40 years. Within- and between-run coefficients of variation (CVs) (%) were PLT 2.1 and 5.5, lymphocytes 3.9 and 6.6, and PMN 2.1 and 5.3, respectively. Reproducibility CV (%) of PLT, lymphocytes, and PMN were 6.6, 7.9, and 5.1, respectively (Table 13-2). The recovery of Zn added to the cell samples before digestion ranged from 94 to 106% (Table 13-3).

13.2.3 Matrix Effects

A small amount of sodium chloride was shown to affect absorbance of PLT, lymphocytes, and PMN when we used the P/H (peak-height) mode

Table 13-1. Comparison of Reported "Normal" Values of Zinc in Human PLT, Lymphocytes, and PMN[a]

Reference	Zinc (μg/10^{10} cells)		
	PLT	Lymphocytes	PMN
Nishi et al. (1981)		188 ± 30[b]	86 ± 23
Whitehouse et al. (1982)		115.5 ± 14.5 (24)	104.5 ± 12.5 (29)
Jepsen (1984)			1.1 ± 0.18 (13)[c]
Milne et al. (1985)	4.8 ± 1.8 (43)	74 ± 23 (45)	51 ± 11 (39)
Purcell et al. (1986)		123.0 ± 26.2 (29)	49.1 ± 11.3 (29)
Present study	3.3 ± 0.3 (16)	50.9 ± 5.9 (14)	45.9 ± 6.4 (15)

[a] Adapted with permission from Wang, H., Prasad, A. S., and DuMouchelle, E. A., 1989. Zinc in platelets, lymphocytes, and granulocytes by flameless atomic absorption spectrophotometry, J. Micronutr. Anal. 5:181.
[b] Values are means ± S.D. Number in parentheses in the number of subjects in the study.
[c] μmol per/10^{10} cells.

Table 13-2. Within- and between-Run Precision, and Reproducibility[a]

Sample		Within-run precision[b]			Between-run precision[c]		Reproducibility[d]	
	n	Mean ± S.D.	CV (%)	n	Mean ± S.D.	CV (%)	n	Mean ± S.D.	CV (%)
PLT	10	3.04 ± 0.063	2.1	5	3.66 ± 0.20	5.5	5	3.18 ± 0.21	6.6
Lymphocytes	10	48.61 ± 1.90	3.9	5	50.78 ± 3.55	6.6	5	44.86 ± 3.54	7.9
PMN	10	43.857 ± 1.18	2.1	5	38.12 ± 2.02	5.3		37.65 ± 1.90	5.1

[a] Adapted with permission from Wang, H., Prasad, A. S., and DuMouchelle, E. A., 1989. Zinc in platelets, lymphocytes, and granulocytes by flameless atomic absorption spectrophotometry, *J. Micronutr. Anal.* 5:181.
[b] Ten replicate analysis of one digested sample during the same run.
[c] Repeated analysis of one digested sample on five different days.
[d] Analysis of sample aliquots from a single blood specimen, showing overall precision. Each sample was isolated separately and simultaneously as if collected from separate individuals.

(Table 13-4). The P/A absorbance of Zn standard were similar to those of BLS (Table 13-5).

13.2.4 Comments

Besides preventing contamination with exogenous Zn during all phases of the analysis, two other problems can occur in the assay of Zn in cells. The first is how to obtain enough cells which are not clumped, not contaminated with RBC, and are homogeneous. The second problem is the matrix effect.

Several different centrifugation speeds and times have been used to separate lymphocytes from PMN when Ficoll–Hypaque was used (Whitehouse *et al.*, 1982; Cheek *et al.*, 1984; Purcell *et al.*, 1986). We found that the ho-

Table 13-3. Analytical Recovery Studies[a]

Sample	Zinc content (ng)			AV recovery (%)	CV (%)
	Original	Added	Found[b]		
PLT	22.00	60	76.93 (2.21)	93.67	3.0
Lymphocytes	12.80	30	43.87 (2.04)	102.5	4.6
PMN	21.80	30	55.00 (1.44)	106.2	2.6

[a] Adapted with permission from Wang, H., Prasad, A. S., and DuMouchelle, E. A., 1989. Zinc in platelets, lymphocytes, and granulocytes by flameless atomic absorption spectrophotometry, *J. Micronutr. Anal.* 5:181.
[b] Mean (S.D.) of three replicates of zinc added to the cells just before the digestion step.

Table 13-4. Matrix Effect of Sodium Chloride[a]

Sample	Absorbances[b]		
	P/A	P/H	p^c
PLT	0.378 ± 0.022	0.213 ± 0.012	<0.01
PLT + NaCl	0.366 ± 0.028	0.180 ± 0.008	<0.01
Lymphocytes	0.159 ± 0.029	0.098 ± 0.004	<0.05
Lymphocytes + NaCl	0.148 ± 0.022	0.073 ± 0.004	<0.01
PMN	0.275 ± 0.026	0.151 ± 0.006	<0.01
PMN + NaCl	0.271 ± 0.020	0.139 ± 0.012	<0.01

[a] Adapted with permission from Wang, H., Prasad, A. S., and DuMouchelle, E. A., 1989. Zinc in platelets, lymphocytes, and granulocytes by flameless atomic absorption spectrophotometry, *J. Micronutr. Anal.* 5:181.
[b] Measured in triplicate, mean ± S.D.
[c] Statistical analysis: two-tailed independent *t*-test.

mogeneity of cells was better when we used a lower centrifugation speed (180*g*, 20–30 min). This also gave better yields.

The lymphocyte fraction is easily contaminated with PLT, so it is very important to remove PLT from this fraction for accurate measurements of Zn in lymphocytes. The number of PLT remaining in the lymphocyte fraction can increase the apparent Zn content of lymphocytes (Milne *et al.*, 1985). Presence of PLT as contaminants can also cause cell clumping, so that the lymphocyte cell count is underestimated and Zn levels of the lymphocytes are overestimated.

In order to overcome the problem with PLT and RBC contamination, we took several precautions. The PLT-rich plasma was removed from the lymphocyte pool and RBC were lysed and cells were washed twice. Additionally, we subtracted the Zn values of platelets that remained as contaminants in the lymphocyte pool. In most cases, PLT contamination may contribute as much as 8–14% of Zn to the lymphocytes. There are fewer PLT in the PMN fraction, but it is better to subtract PLT Zn from the apparent PMN Zn also. In most cases, PLT contamination may contribute 2–6% of the levels in the PMN fraction. Usually, we did not have problems with cell clumping as long as the blood sample was processed immediately after drawing.

For lymphocytes and PMN, lysing and washing should not be performed more than four times; otherwise, the viability of cells decreased and Zn values also decreased. PLT may be lysed and washed two more times without affecting the Zn values (personal observation). It is necessary to wash PLT before lysing, to loosen PLT pellets so that the contaminating RBC are lysed efficiently.

Nitric acid does not pose a matrix problem inasmuch as both the standards and samples have the same amount of nitric acid.

Table 13-5. Comparison of Absorbances of Zinc Standard
and Bovine Liver Standard[a]

	Absorbances[b]			
	P/A		P/H	
Concentration of zinc (ng/ml)	A Zinc standard	B BLS	C Zinc standard	D BLS
1.5	0.243 ± 0.008	0.250 ± 0.10	0.159 ± 0.007	0.199 ± 0.008
3.0	0.424 ± 0.009	0.412 ± 0.007	0.275 ± 0.008	0.323 ± 0.008
4.5	0.593 ± 0.007	0.580 ± 0.010	0.361 ± 0.007	0.457 ± 0.010
6.0	0.749 ± 0.010	0.740 ± 0.008	0.449 ± 0.009	0.555 ± 0.009
	Statistical analysis (two-tailed independent t-test)			
	A vs. C	A vs. B	B vs. D	C vs. D
1.5	$p < 0.01$	NS	$p < 0.01$	$p < 0.01$
3.0	$p < 0.01$	NS	$p < 0.01$	$p < 0.01$
4.5	$p < 0.01$	NS	$p < 0.01$	$p < 0.01$
6.0	$p < 0.01$	NS	$p < 0.01$	$p < 0.01$

[a] Adapted with permission from Wang, H., Prasad, A. S., and DuMouchelle, E. A., 1989. Zinc in platelets, lymphocytes, and granulocytes by flameless atomic absorption spectrophotometry, *J. Micronutr. Anal.* 5:181.
[b] Measured in triplicate, mean ± S.D.

BLS is a biological tissue. There are differences between the P/H readouts of BLS and those of Zn standards. However, the P/A readouts of BLS are very close to those of Zn standards (see Table 13-5). This means that the results will be low if we use Zn standard to measure Zn in BLS by P/H, and the results will be much more accurate if we use P/A. We measured Zn in BLS by P/A on 10 different days (in 2 months) and the results (119.20 ± 4.74 μg/g Zn per g dry wt, mean ± S.D.) were comparable to the certified Zn value of 123 ± 8 μg/g. Zn standard is a pure solution, and is different from BLS, which is an organic tissue. Therefore, for analysis of Zn in organic tissues (such as PLT, lymphocytes, and PMN), one should use P/A absorbance of Zn standard for accurate results.

Sodium chloride can also cause significant matrix effect. Table 13-4 shows that if the samples have a small amount of normal saline (4 μl of normal saline added to 500 μl of samples), P/A absorbance are not affected, but P/H absorbance seem to be affected. We cannot remove all of the normal saline in the sample, but we can remove the majority. We have used a very thin tip to draw away normal saline. After drawing, we cannot see any drops of normal saline. The pellet volumes of PLT, lymphocytes, and PMN are only about

5–15 μl, and we estimate that the maximum amount of normal saline in the pellet would approximate 4 μl.

Our results of Zn in PMN agree with those reported by Milne *et al.* (1985) and Purcell *et al.* (1986). Zn levels of lymphocytes and PLT in this study are, however, much lower than those reported earlier (see Table 13-1). This is because we have taken greater precaution to remove PLT from the lymphocyte pool and we corrected the apparent Zn values of lymphocytes by subtracting the Zn content of PLT which remained as contaminants. We also lysed RBC and washed PLT carefully in order to eliminate contamination with RBC in the PLT fraction which ultimately gave lower values for Zn in the PLT in this study.

In summary, the method described in this chapter yields clump-free and contamination-free lymphocytes, PMN, and PLT. Our results are reliable and repeatable.

References

Cheek, D. B., Hay, H. J., Lattanzio, L., Ness, D., Ludwigsen, N., and Spargo, R., 1984. Zinc content of red and white blood cells in aboriginal children, *Aust. N.Z. J. Med.* 14:638.

Hudson, L., and Hay, F. C., 1980. *Practical Immunology,* 2nd ed., Blackwell, Oxford, p. 29.

Jepsen, L. V., 1984. Determination of zinc in erythrocytes, granulocytes and serum by flame atomic absorption spectrophotometry, *Scand. J. Clin. Lab. Invest.* 44:299.

Milne, D. B., Ralston, N. V. C., and Wallwork, J. C., 1985. Zinc content of cellular components of blood: Methods for cell separation and analysis evaluated, *Clin. Chem.* 31: Washington, D.C., 65.

Nishi, Y., Hatano, S., Horina, N., Sakano, T., and Usui, T., 1981. Zinc concentration in leukocytes: Mononuclear cell, granulocytes, T-lymphocytes, non-T lymphocytes and monocytes, *Hiroshima J. Med. Sci.* 30:65.

Prasad, A. S., Oberleas, D., and Halsted, J. A., 1965. Determination of zinc in biological fluids by atomic absorption spectrophotometry in normal and cirrhotic subjects, *J. Lab. Clin. Med.* 66:508.

Purcell, S. K., Hambidge, K. M., and Jacobs, M. A., 1986. Zinc concentration in mononuclear and polymorphonuclear leukocytes, *Clin. Chim. Acta* 155:179.

Wang, H., Prasad, A. S., and DuMouchelle, E. A., 1989. Zinc in platelets, lymphocytes, and granulocytes by flameless atomic absorption spectrophotometry, *J. Micronutr. Anal.* 5:181.

Whitehouse, R. C., Prasad, A. S., Rabbani, P. I., and Cossack, Z. T., 1982. Zinc in plasma, neutrophils, lymphocytes and erythrocytes as determined by flameless atomic absorption spectrophotometry, *Clin. Chem.* 28: Washington, D.C., 475.

Index

Absorption, of zinc, 193–201
  in acrodermatitis enteropathica, 12
  copper and, 263–264, 265–266
  dietary ligands and, 193, 194–199
  endogenous secretion in, 194, 200–201
  glucose and, 108
  intracellular transport in, 194, 199
  iron and, 194, 260–263
  metallothionein and, 88–89
  in pregnancy, 199
  protein and, 201–202
  in renal failure, 235–236
  in zinc excess, 199
Acquired immune deficiency syndrome
    (AIDS), opportunistic infections of,
    170–171
Acrodermatitis enteropathica
  alkaline phosphatase deficiency of, 46
  chemotaxis abnormalities of, 182
  cottonseed oil therapy for, 196
  essential fatty acids in, 119–121, 123
  following total parenteral nutrition, 221,
    222, 223, 224
  immune defects of, 168
  psychologiccal disturbances of, 158
  RNAse activity in, 33
  symptoms, 168, 219, 220
  zinc absorption in, 12
  zinc therapy for, 43
    effect on myelination, 154
Actinomycin, as metallothionein inhibitor,
    84, 85
Adenosine-5′-tetraphosphate 5′-adenosine,
    185–186, 188
Adenosine deaminase, 42–43, 46, 183

Adenosine diphosphate, in zinc
    supplementation, 185, 187
Adenosine monophosphate deaminase, in
    zinc deficiency, 43–45, 232
Adenosine triphosphatase
  in sickle cell anemia, 141, 142
  zinc-related inhibition of, 133
Adenosine triphosphate
  as aspartate transcarbamylase enhancer,
    21–22
  in zinc erythrocyte uptake, 137
  in zinc supplementation, 185, 187
Adrenals, in zinc deficiency, 109–110
Adrenocorticotropic hormone, interleukin-
    1β-stimulated release of, 99
Aging, zinc's role in, 251; *see also* Elderly
    persons
Albumin
  zinc binding by, 137, 211–212
  zinc transport and, 199
Alcohol
  effect on zinc absorption, 198–199
  teratogenicity of, 221, 222
  See also Alcoholism
Alcohol abstinence syndrome, 159
Alcohol dehydrogenase
  retinal, 231
  in zinc deficiency, 29, 30, 34–35
  zinc's function in, 20, 22–23
Alcoholism, as zinc deficiency cause,
    158–159, 220–223
  alcoholic seizures and, 159
  liver cirrhosis and, 229
  mental deterioration and, 158
  in pregnancy, 252